D0945417

Groundwater Contamination

Optimal Capture and Containment

Steven M. Gorelick
R. Allan Freeze
David Donohue
Joseph F. Keely

Library of Congress Cataloging-in-Publication Data

Groundwater contamination : optimal capture and containment / Steven M. Gorelick . . . [et al.].
p. cm.
Includes bibliographical references and index.
ISBN 0-87371-872-0
1. Groundwater—Pollution. 2. Aquifers. 3. Water quality management.
I. Gorelick, Steven M.
TD426.G737 1993
628.1′68—dc20 93-23950
CIP

Direct all inquiries to CRC Press, Inc., 2000 Corporate Blvd., N.W., Boca Raton, Florida 33431.

PRINTED IN THE UNITED STATES OF AMERICA
1 2 3 4 5 6 7 8 9 0
Printed on acid-free paper

Acknowledgments

This book is a modified and edited version of a report prepared by the authors for the Air Force Engineering Services Center through Chem-Nuclear Geotech, Inc. We gratefully acknowledge the comments of Bruce Nielsen, Mike Elliot, and Doug Downey as project officers. We offer tremendous thanks to Claire Tiedeman who worked as a Stanford University research assistant to conduct simulations which served as the example contaminant capture schemes which appear in Chapter V.

The Authors

Steven M. Gorelick is an Associate Professor in the School of Earth Sciences at Stanford University where he runs a diverse hydrogeology program. After receiving his M.S. and Ph.D. degrees in Hydrology from Stanford, he worked in hydrogeology at the U.S. Geological Survey where he has over ten years of service. Dr. Gorelick is a fellow of both AGU and GSA, as well as recipient of a Presidential Young Investigator Award and the AGU Macelwane Medal. He is active as an environmental consultant for both government and industry in the areas of contaminant hydrogeology and water resources management. He is a popular lecturer and short-course instructor. At Stanford he conducts research and teaches in the areas of hydrogeology, groundwater management, transport simulation, and aquifer remediation. He has published extensively in these areas. Dr. Gorelick's research balances science and engineering. On the engineering side his recent work has focused on a newly patented method for in-situ removal of Volatile Organic Compounds from groundwater. On the scientific side he has worked collaboratively to apply simulation methods to problems ranging from regional paleoclimatic and hydrogeologic reconstructions to understanding pore-scale flow phenomena.

R. Allan Freeze is President of R. Allan Freeze Engineering Inc. based in White Rock, BC, Canada. Recent private-sector consulting clients have included Eastman Kodak, Ciba-Geigy, Atochem North America, and Texas Eastern Gas Pipelines. Recent public-sector clients have included Environment Canada, U.S. Environmental Protection Agency, U.S. Department of Energy, Colorado State Engineer's Office, and B.C. Hydro. Involvement with leading consulting engineering firms includes Golder Associates, CH2M Hill, Intera Technologies, S.S. Papadopulos and Associates, and Geraghty and Miller. Recent projects have included site assessment and remedial design at contaminated sites in Rochester, NY, Toms River, NJ, Tacoma, WA, San Fernando Valley, CA, and Laramie, WY.

Prior to establishing his consulting practice, Allan Freeze worked for Environment Canada in Calgary, AB, the IBM Thomas J. Watson Research Center in Yorktown Heights, NY, and the University of British Columbia in Vancouver, BC. During his 18-year career at U.B.C., he was Director of the Geological Engineering Program for six years and Associate Dean of Graduate Studies for three years.

During his academic career Allan Freeze published over 100 research papers. He has received the Horton and Macelwane Awards from the American Geophysical Union, the Meinzer Award from the Geological Society of America, and the Theis Award from the American Institute of Hydrology. He is a Fellow of the Royal Society of Canada. He is a former Editor of the journal *Water Resources Research*, and a former President of the Hydrology Section of the American Geophysical Union. He is coauthor, with John Cherry, of the widely used textbook "Groundwater."

David Donohue is a Senior Hydrogeologist with Chem-Nuclear Geotech, Inc. at the Department of Energy Grand Junction Projects Office. Mr. Donohue has a BS degree in Civil and Environmental Engineering from the University of Wisconsin-Madison, and a MS degree in Hydrogeology from Utah State University. His past efforts have involved modeling groundwater flow and discharge processes in closed desert basins and frequency domain analysis of hydrogeologic time series data. He has recently presented lectures on linear optimization of aquifer management systems at the U.S. Air Force Engineering Services Center and the Department of Energy's Technology Information Exchange Conference. Mr. Donohue is currently conducting several projects for the Department of Energy and the U.S. Air Force including: aquifer management optimization under uncertainty; coupled application of groundwater flow models, geostatistical models, and geographic information systems; and modeling and optimization of remedial actions for a large stratified aquifer system containing inorganic and organic contaminants in both non-aqueous and dissolved phases.

Joseph F. Keely, Ph.D., P.Hg., FAIC/CPC, is Vice President and Director of Computer Modeling at the hydrogeological engineering firm of Boateng & Assoc., Inc. (Mercer Island, Washington), where his client service work involves mathematical modeling of subsurface fluid flow and contaminant transport, statistical and graphical analyses of field data, and design, optimization, and evaluation of site characterizations and remediations. In addition to client service, Dr. Keely continues to pursue a wide variety of research interests, including drilling and sampling improvements, pump-and-treat design strategies, remediation compliance criteria and measures of effectiveness, and socioeconomic influences on technical developments, business practices, professional ethics, and regulatory policies. Dr. Keely also routinely travels and lectures throughout the nation and abroad. His private time is spent reading philosophy, playing musical instruments, hiking, piloting small aircraft, and writing rock and blues lyrics.

Table of Contents

Appendix

List of Figures

List of Tables

Glossary of Notations

Symbol	Explanation	Introduced on Page
α	Angle between regional flow direction and positive x axis	125
i	Annual discount rate	108
η_a	Aquifer compressibility	23
$\rho_y(x)$	Autocorrelation function of Y	331
V	Average linear velocity (seepage velocity)	28
$\bar{A}_v$	Average void area	28
T_0	Basic time lag	74
$B_j(t)$	Benefits of alternative j in year t	116
ρ_b	Bulk density of porous medium sample	23
D*	Coefficient of molecular diffusion	46
$\mathbf{Q}$	Column vector of pumping rates	147
P(x,y)	Complex head potential	124
$\Phi(x,y)$	Complex velocity potential	124
η_w	Compressibility of water	24
$f(\mathbf{q})$	Concentration distribution function from simulation	342
J_i	Contaminant flux due to hydrodynamic dispersion in the x_i direction	46
Q_c	Contaminant mass flux	83
$Cf_j(t)$	Cost associated with failure of alternative j in year t	116
$C_j(t)$	Cost of alternative j in year t	116
A	Cross-sectional area	28
a	Cross-sectional area of sample in a permeameter test	73
ρ	Density	23
ρ_s	Density of solid phase of a porous media sample	23

L	Distance between hydraulic head (H) measurement points used in calculating hydraulic gradients	73
s	Drawdown	50
S_i	Drawdown at a location i	147
r_{ij}	Drawdown response coefficient at location i due to unit well at location j	147
μ	Dynamic viscosity	23
n_e	Effective porosity	21
z	Elevation head	29
β	Factor for well-screen and gravel-pack geometry	77
n_f	Fracture porosity	22
d_{10}	Grain-size diameter	72
K	Hydraulic conductivity	24
H	Hydraulic head value at a particular point in space and time	73
h	Hydraulic head variable	29
D_{ij}	Hydrodynamic dispersion coefficient	46
Y_i	Integer variable for well i	196
D_L	Longitudinal component of coefficient of mechanical dispersion	48
α_L	Longitudinal dispersivity	48
F_i	Mass flux of contaminants in x_i direction	46
D_{ij}^m	Mechanical dispersion coefficient	46
d	Monthly discount rate	174
γ	Normalized utility function	320
n	Number of wells	125
Z	Objective function	141
Z_j	Objective function for alternative j	107
K_{ow}	Octanol-water partition coefficient	44
f_{oc}	Organic carbon content	44
ω	Packed-off borehole interval volume	77
k	Permeability	21
R	Piezometer radius	73
T	Planning horizon (Section 4 and Appendix A only)	108
μ_y	Population mean of a random variable	330
σ_y^2	Population variance of a random variable Y	330
n	Porosity	21
ϕ	Potential function	124
p	Pressure	29
Ψ	Pressure head	29

$Pf_j(t)$	Probability of failure of alternative j in year t	116
Q	Pumping rate for discharge from or injection to a well	51
q_i	Pumping rate decision variable at well i (Appendix B only)	342
r	Radial distance from a well	51
I	Regional hydraulic gradient	125
R	Response matrix of drawdown-response coefficients	149
R_f	Retardation factor	44
RHS	Right-hand side	189
$R_j(t)$	Risks of alternative j in year t	116
$\mathbf{r}_{\mathbf{diff}}{}^{t}$	Row vector of differences in drawdown-response coefficients	159
$\mathbf{r}_{\mathbf{i}}^{t}$	Row vector of drawdown-response coefficients	147
$\overline{Y}$	Sample mean of y_i's	331
S_y^2	Sample variance of y_i's	331
b	Saturated thickness of an aquifer	25
c	Solute concentration in a pore fluid	46
c′	Solute concentration in a source or sink	46
W	Source or sink strength in units of volumetric flow rate per unit volume of porous media	49
W′	Source or sink strength in units of volumetric flow rate per unit area of porous media	49
q	Specific discharge (Darcy velocity)	24
S_r	Specific retention	77
S_s	Specific storage	25
S_y	Specific yield	21
i	$\sqrt{-1}$ (page 124 only)	124
S	Storativity	26
$\Psi(x,y)$	Stream function	124
w(x,y)	Modified stream function	124
$L_{i,n}$	Total lift at well i during period n	173
N	Total number of management periods	172
I	Total number of wells	172
n_t	Total porosity	21
Z′	Total pumping cost	173
V_t	Total volume of porous medium sample	21
T	Transmissivity	25
D_T	Transverse component of coefficient of mechanical dispersion	48
α_T	Transverse dispersivity	48

C^*	Undiscounted unit cost	174
$C_{i,n}$	Unit cost of pumping per unit lift at well i during period n	172
$L^*_{i,n}$	Unmanaged lift at well i during period n	173
S	Vector of drawdowns	149
V_v	Volume of voids in a porous medium sample	21
Q	Volumetric flow rate	27
c^*_j	Water-quality standard at point j	342
U(u)	Well function	51
x_1,x_2,x_3	x, y, and z coordinate directions, respectively	46

CHAPTER I

Introduction

This book summarizes a set of techniques that can be used to design efficient and cost-effective capture and containment systems for groundwater remediation. The impetus for the book is the increasing nationwide concern regarding groundwater contamination.

The contaminants found in groundwaters at many sites are representative of virtually all major industrial by-products. Although the characteristics of groundwater contamination and the hydrogeologic conditions may be unique to each site, the design techniques presented in this book are quite general and should be applicable to a large number of these sites.

A. Objective

The objective of this book is to present a description of a set of tools and techniques that can be used to design the most efficient and cost-effective withdrawal and injection systems for groundwater remedial action operations.

A large number of remedial action alternatives are available for application to groundwater-contamination problems. Systems based on a pump-and-treat approach are an essential alternative of choice for many groundwater-contamination problems. For many sites the pump-and-treat strategy must be exercised for decades as a means to capture the contaminated groundwater and will not result in complete aquifer remediation. Under such circumstances capture and containment must be efficiently designed because it will be operated for a long period.

Unfortunately, a widely applicable and rigorous approach to the hydrologic design of efficient capture systems has not been adopted by groundwater hydrologists. In many instances, professional engineering judgment is the basis for the design of a remedial action well field. The locations and estimates of pumping rates for a well or set of wells are determined on the basis of the available data and the experience of the design engineer. Subsequently, a numerical or analytical model is used to predict the impacts of proposed remedial action on the contaminant plume. The hydrologist modifies the well system and repeats the process until an acceptable plume response is achieved. This approach lacks mathematically formalized checks and balances that might be used to ensure that the hydraulic design is optimal, both with respect to cost and with respect to physical control of the contaminant plume. Because pump-and-treat systems often evolve into *long-term* operations, the potentially high costs dictate the need to attempt to maximize their efficiency.

The recommended approach is based on a combination of simulation and optimization. Simulation is carried out with the usual types of groundwater models for flow and transport. Optimization is based on standard linear-programming techniques. The mechanics of the suggested approach are presented in Chapter V. Chapter IV sets the framework for the problem and places the simulation-optimization technique into the context of other available solution technologies. There are limitations on the application of the simulation-optimization methodology and these are recognized throughout the book. Chapter VI provides more qualitative guidelines for cases that are not easily amenable to the simulation-optimization approach.

B. Background

1. Groundwater Contamination: U.S. Air Force Installations as an Example

Although every contaminant hydrogeology problem is unique, many of them share common features. The methods described here apply to groundwater contamination problems at many publicly or privately owned sites. The features common to many contamination problems that lead to wide applicability include (1) type of contaminant (dissolved fuels, solvents, and metals), (2) type of porous media (unconsolidated sedimentary deposits), and (3) type of aquifer (shallow, unconfined).

Typical of many contamination problems are those at U.S. Air Force installations. As part of its overall mission, the United States Air Force operates 137 installations in the United States and in foreign countries. The Air Force also owns 13 contractor-operated plants, which support these installations. Many of these bases and plants are nearly self-sufficient complexes that provide on-site industrial capabilities. Typical industrial operations at these sites include metal casting, metal fabrication, metal plating, solvent handling, fuel storage and handling, fire fighting, electronics manufacturing, food processing, and waste disposal.

During the last decade, both the Air Force and the public have expressed increasing concern about environmental contamination resulting from past and present activities at Air Force-owned sites. Because of its potential to affect public health and the environment, groundwater contamination is a major subject of this concern.

Although the characteristics of the groundwater contamination and the nature of the hydrogeologic conditions are unique to each Air Force site where contamination exists, several generalizations apply. The typical contaminated hydrogeologic system occurs in a relatively shallow water-table aquifer in unconsolidated sedimentary deposits. Other types of affected hydrogeologic flow systems include near-surface fractured-rock aquifers contaminated by infiltration from the surface and confined or semiconfined aquifers threatened by contaminant leakage through confining beds. Because water-table aquifers are usually close to the surface and have no protective confining layer, they represent the most frequently affected groundwater flow systems at Air Force installations.

The contaminants found in groundwater beneath U.S. Air Force installations are representative of virtually all major industrial byproducts, including metals, volatile and semivolatile organics, petroleum hydrocarbons, pesticides, polychlorinated biphenyls (PCBs), asbestos, radionuclides, and other select inorganics. However, in terms of the total volume of groundwater affected by a particular contaminant or group of contaminants and the magnitude of the contaminant concentration, the bulk of Air Force contamination problems are limited to a relatively small number of compounds.

The most common contaminants at Air Force sites comprise three major contaminant groups. Petroleum hydrocarbon contaminants are the most common and usually include benzene, ethyl benzene, toluene, and xylene. Trichloroethylene (TCE) and related solvents are the second most common contaminants. These solvent compounds originate from degreasing, paint handling, metal plating, and other similar operations at Air Force installations. Toxic metals and other inorganic contaminants represent the third most common contaminant type. Specific constituents from this group found in contaminated groundwater include lead, cadmium, chromium, and cyanide. There are, of course, other contaminants found at Air Force sites, but these three groups are by far the most common.

In terms of demographics, typical Air Force installations do not vary significantly from average urban communities. In fact, most Air Force installations are located near urban centers, and can be considered as extensions of them. Land uses are both industrial and residential, but agricultural land use is essentially nonexistent. Potable and industrial water sources at Air Force installations consist of both surface water and groundwater, while some base-operated water-supply systems receive water from off-site municipal sources. To the extent that these conditions are similar to those of an average urban community, unique demographics are not usually a concern when evaluating groundwater contamination problems at Air Force sites.

2. Groundwater Remedial Action Under the Air Force Installation Restoration Program

The U.S. Air Force initiated the Installation Restoration Program (IRP) in response to growing public concern about the quality of the environment and health risks associated with past Air Force waste disposal and handling practices. The Defense Environmental Quality Program Policy Memorandum 81–5, issued on 11 December 1981, set forth the policy for the program. The IRP (U.S. Air Force, 1985) is a process that provides a general framework for characterizing and remediating groundwater contamination and other types of environmental contamination such as that associated with soils, surface waters, or the atmosphere.

With respect to the groundwater component of its environmental remediation activities, the Air Force has adopted the protocols of the U.S. Environmental Protection Agency's (EPA's) Comprehensive Environmental Response, Compensation, and Liability Act (CERCLA) in conducting its IRP (U.S. EPA, 1987a, 1987b, 1987c, 1988a, 1988b). Figure 1 shows the sequential framework for a standard CERCLA investigation conducted at a potential National Priorities List (NPL) site according to EPA guidelines (U.S. EPA, 1986). The Air Force follows similar preliminary-assessment, site-inspection, and hazard-ranking steps prior to developing a Remedial Investigation/Feasibility Study (RI/FS) plan. Figure 2 outlines the components of the RI/FS process used at NPL sites (U.S. EPA, 1988a); a similar process is carried out at Air Force sites. In fact, some Air Force sites are NPL sites, and at these sites the CERCLA process and the IRP process formally converge.

All NPL sites (non-Air Force and Air Force) and all Air Force sites that are not candidates for No Further Action (Figure 1) require an RI/FS. Under the new EPA guidance document for the conduct of an RI/FS (U.S. EPA, 1988a), this stage comprises the bulk of the site investigation work. During an RI/FS, investigators conduct detailed

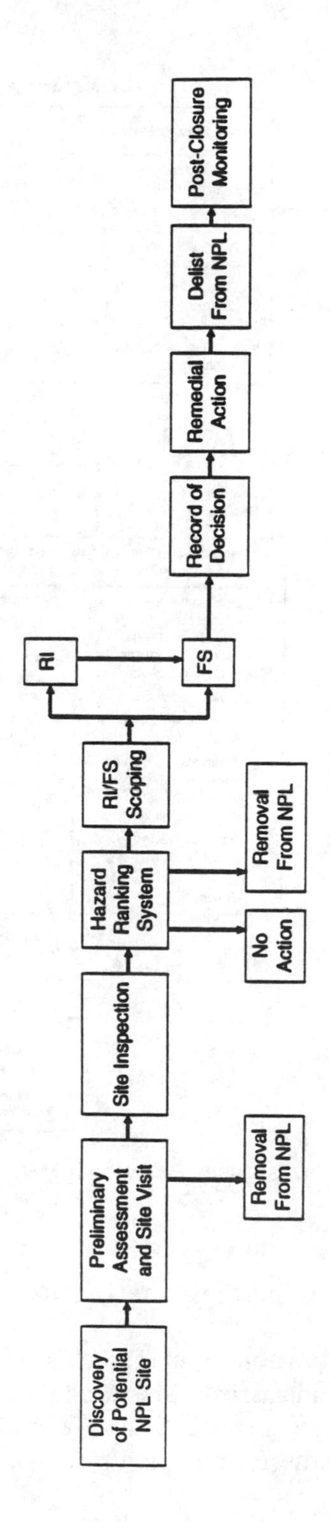

Figure 1. Framework for EPA site investigation and remediation program at NPL sites under CERCLA (after U.S. EPA. 1986).

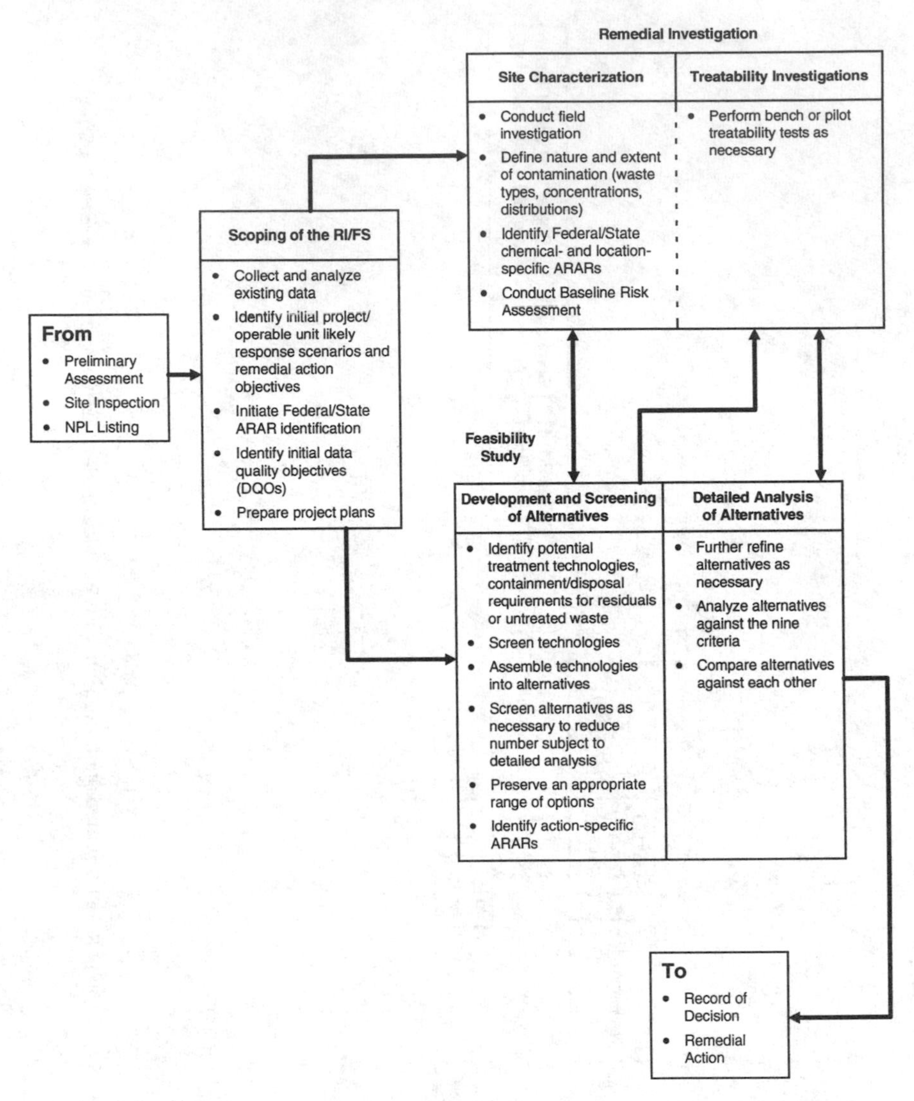

Figure 2. Components of the RI/FS Investigation Process for NPL Sites (after U.S. EPA, 1988a).

studies of the contaminant transport aspects of the hydrogeologic system.

During the Feasibility Study portion of an RI/FS, a set of candidate remedial technologies must be identified. The candidate technologies identified in the IRP guidance document (U.S. Air Force, 1985) include (1) impermeable caps, (2) groundwater pumping systems, (3) impermeable barriers, (4) subsurface collection drains, (5) surface-water diversion and collection systems, (6) permeable treatment beds, (7) grading activities, (8) revegetation activities, and (9) bioreclamation systems. The design team is expected to develop remedial action alternatives using combinations of those remedial technologies that show promise during the screening process. A detailed evaluation of the alternatives is then carried out on the basis of (1) engineering feasibility, (2) cost analysis, (3) environmental impact, (4) public health risk analysis, and (5) an assessment of regulatory compliance against all Applicable or Relevant and Appropriate Requirements (ARARs). The No Further Action alternative is usually included in the evaluation step in order to provide a baseline for comparison.

Once an alternative remedial action is selected, the design and construction phase ensues. This includes the development of (1) engineering plans, (2) a health and safety plan, (3) a site security plan, (4) quality-assurance/quality-control (QA/QC) plans, (5) operational monitoring plans, and (6) post-closure monitoring plans.

The discussion of engineering design in thc following chapters presumes a generic approach to the hydrogeologically controlled aspects of the remediation process. We do not further discuss CERCLA or IRP protocols, public health risk analysis, or QA/QC issues.

3. Engineering Design Approaches for Groundwater Remedial Action Systems

Six general classes of groundwater remedial action, apart from physical containment, are available for sites involving contamination problems that must be remediated. They include

- Plume stabilization through hydraulic control to minimize spreading.
- Diversion of flow or redirection of contaminant plume to protect a well or other resource.
- Contaminant removal to clean up an aquifer.
- In situ biological treatment to reduce contaminant levels in the groundwater.
- Wellhead treatment to clean up groundwater withdrawn for a specific use.
- Monitoring of contaminant levels until need for one of the above remedial actions is confirmed, or until natural attenuation reduces contaminant levels.

Plume stabilization through hydraulic control is a mitigation measure only. The intent of this remedial action is to minimize the spread of contaminants and protect the surrounding uncontaminated groundwater resources. Because the goals of this action are relatively modest in comparison to some of the others, it has been one of the most successful methods used in practical applications. This technique uses withdrawal and injection wells, drains, interceptor trenches, and/or cutoff walls to manipulate the hydraulic gradient and create a no-flow boundary surrounding the downgradient edge of the contaminant plume. To meet regulatory constraints, the groundwater withdrawn from the aquifer is usually treated prior to reinjection or discharge.

Flow diversion and plume redirection are similar in practice to plume stabilization. The objective of this remedial action is to protect specific groundwater resources that could be affected by the influx of contaminated groundwater. Such resources might take the form of a well field for municipal water supply, or a surface-water supply reservoir that is recharged by groundwater flow. As with plume stabilization, flow diversion/plume redirection is accomplished through hydraulic-gradient control.

The third category of remedial action, contaminant removal, is the one most typically sought by members of the public who are affected by a groundwater contamination problem. The objective of this remedial action is to remove as much contamination as possible and return the groundwater to its original quality. Contaminant-removal systems consist of withdrawal wells, trenches and drains, and above-ground treatment or disposal facilities. These systems have been relatively unsuccessful due to the difficulty associated with fully describing the contaminant transport process in "real world" aquifers and the persistence of sorbed and pure-phase chemicals on and in the solid matrix of the geologic media. In most cases, complete contaminant removal, even if feasible, is prohibitively expensive. However, in many cases it may be desirable (and/or required) to reduce contaminant concentrations to specific target levels that are protective of human health and environmental quality.

In situ biological treatment is a remedial approach that is currently under investigation. In situ biological treatment is receiving increasing attention from regulatory agencies because it involves the actual destruction of contaminants rather than a transfer of contaminants from one medium or phase to another. Some encouraging research results are emerging, especially with respect to remediation of fuel-based hydrocarbons. For the remediation of contaminated groundwater, biological restoration can use natural or injected microorganisms in the groundwater to degrade contaminants to non-toxic by-products. In most instances, oxygen and methane are added to the groundwater flow system to enhance the growth and reproduction of microbes.

In some instances, aquifer conditions or economics may dictate that no action be taken to address a groundwater contamination problem. Instead, wellhead treatment may be used to treat the problem at the point of use. The advantage of this approach is that it minimizes the cost by treating only the water that is used. Such an approach may be short-sighted, however, as future population growth may affect treatment costs, and natural discharge of contaminants to surface waters is not addressed. In addition, the public rarely condones this method if the contaminated well supplies municipal or private drinking water.

Natural attenuation relies on the natural processes within an aquifer to restore groundwater quality. The processes that provide this restoration are dilution, adsorption, and biological degradation.

Dilution is the reduction in solute concentration with time and distance traveled as a result of mixing, dispersion, and diffusion.

Adsorption is the process by which solute particles become sorbed to the solid matrix of the aquifer. However, the sorption process is usually reversible and, therefore, does not provide a reliable means of permanent groundwater restoration.

Biological degradation is the natural transformation of chemical compounds via microbial activity. This process may involve the production of many intermediate compounds along a decay chain, with the possibility that some daughter compounds may be more toxic and persistent than the parent compound. The best-known example of a toxic daughter is vinyl chloride, which is an intermediate product in the decay chains of perchloroethylene (PCE) and TCE. Its biodegradation has been shown to occur but is slow under anerobic conditions. The natural process of TCE degradation to vinyl chloride is known to occur, and this may pose a potential health risk as well as a treatment problem for more conventional remediation methods (Roberts et al., 1989; McCarty et al., 1991).

Diffusion, dispersion, adsorption, and biological degradation are briefly described in Chapter II.F.

To ensure that contamination does not migrate beyond predetermined boundaries, a stringent monitoring program must accompany any no-action alternative that relies on natural attenuation.

4. Remedial Action Construction, Operational Management, and Closure

Because of the variability and uncertainty associated with each hydrogeologic environment, the groundwater hydrologist's role does not end with completion of a remedial action design. After the remedial action is designed, the system is constructed and operation begins. At this stage, a groundwater monitoring program begins to assess the performance of the remedial action system.

Data collected from the groundwater monitoring program may provide the best means to address uncertainty in design. Groundwater samples collected from specified "points of compliance" and analytical data for key indicator parameters are used to determine if the system is meeting regulatory requirements and design objectives. If the system fails to meet regulatory requirements or to perform as specified in the design plans, the performance-assessment data can be used to modify the design to remedy these deficiencies. Chapter VII discusses operational management and performance-assessment monitoring.

For remediation systems that try to achieve a specific cleanup objective within a stated time frame (as contrasted to perpetual-care systems such as plume stabilization or plume redirection), a closure plan is necessary once the objectives are satisfied. The closure plan must address the actions needed to ensure that water-quality standards are met after the remedial action system is shut down.

For aquifers undergoing remedial action designed to reduce contaminant concentrations to target levels, the long-term nature of the exercise warrants a rather comprehensive post-closure monitoring plan. At facilities with ongoing activities such as most Air Force installations, the post-closure monitoring plan should be designed to detect any new release of contamination and to address the potential for any post-closure contaminant migration resulting from the cessation of remedial action.

C. Scope

1. Application and Limitations

To allow a detailed treatment of the topics most applicable to remedial action system design, our scope is restricted to quantitative design strategies that have been fully developed and tested by the research community. Specifically, we focus on design issues associated with hydrogeologic aspects of capture and containment systems, with an emphasis on optimization of well-field designs (including location and pumping rates for withdrawal and injection wells). This presentation will show bias towards remedial actions intended to achieve hydraulic control of contaminant plumes; but the techniques are also applicable (albeit with considerable additional complexity) to remedial actions predicated on contaminant removal and/or contaminant concentration reduction.

The design methods presented are confined to those that address only uniform-density and miscible-contamination problems. There are two reasons for this. First, there is a limitation on the analytical tools available to address problems of variable density and immiscible non-aqueous-phase liquids. Both analytical and numerical modeling capabilities for these problems are still in the development stage and have not reached a point where optimization techniques can be routinely applied by practicing hydrologists. Second, uniform-density miscible-contaminant problems represent a common group of problems for which remedial actions are currently being designed. The pump-and-treat technology is widely used for dissolved contaminants, whereas recovery systems for some non-aqueous-phase liquids are still in the developmental stages.

While the simulation component of the simulation-optimization techniques described in this book can use both analytical or numerical models of groundwater flow systems, the emphasis is on numerical models. Numerical models have greater capability to address the hydrogeologic variability within a given site and between individual sites. Three-dimensional models are available in analytical or numerical form, but design methodologies presented here include only two-dimensional horizontal flow-system applications. The methodologies are presented in sufficient detail to allow most hydrologists to make modifications for problems involving three-dimensional contaminant-transport problems.

This book is limited to deterministic groundwater flow and contaminant-transport models. Optimization techniques have been applied using stochastic models of the contaminant-transport process, but our scope is limited to technology that has already been fully developed in order to allow widespread application of the design techniques.

This book is also limited to advection-dominated contaminant transport. Excluding an explicit treatment of dispersion and diffusion in designing optimized pumping schemes is not a severe limitation in terms of practical applications. Among the most troublesome and common contamination problems are those involving relatively high hydraulic-conductivity water-table aquifers in unconsolidated deposits. Because advection is commonly the dominant transport process in such hydrogeologic environments, an advection-based design approach is appropriate.

Lastly, the discussions of optimization of pumping-scheme design will use only linear and quadratic optimization techniques. The complexities that will be described in Chapter V associated with nonlinear optimization techniques (excluding quadratic optimization) preclude their immediate use in many practical design problems.

2. Strengths of This Work and Relation to Current Practices

The previous discussions of limitations may lead to a misunderstanding on the part of the reader that the design techniques presented here are not powerful enough to address difficult contamination problems. This is not the case. Standard design practice has previously been based on a combination of engineering judgment and a trial-and-error iteration process using analytical and numerical models to evaluate the adequacy of successive design alternatives. The techniques presented here improve upon this approach by using optimization theory to bring a rigorous and objective measure of *efficiency* to the design process. We provide a complete methodology for designing efficient and cost-effective pumping schemes for practical groundwater contamination problems in commonly occurring hydrogeologic environments. Discussion of the optimization techniques includes a description of the required data and preliminary design tasks, methods of performance assessment and design modification, and design concerns for complex settings. As long as flow and solute-transport models can be employed to accurately simulate contaminant migration, the simulation-optimization methodology should provide a sound basis for designing efficient and cost-effective pumping schemes for a wide variety of problems.

Table 1 summarizes the conditions for which the methodology presented in this book is especially well-suited, as outlined in the above discussions.

Table 1. Conditions Well-Suited to Application of Simulation-Optimization Methodology Presented in this Book.

Item	Conditions
Aquifer	• Porous media • Shallow, unconfined, or confined aquifer • Saturated groundwater zone • Two-dimensional representation in horizontal plane
Contaminants	• Benzene, toluene, ethyl benzene, xylene (BTEX); constituents of petroleum hydrocarbons; organic solvents; and metals • Miscible dissolved phase
Contaminant transport	• Advection-dominated system, with or without retardation • Diffusion, dispersion, and biodegradation of lesser importance
Remedial alternative	• Capture and containment approach • Extraction wells and/or injection wells • Designed to achieve plume stabilization through gradient control
Method of analysis	• Deterministic framework • Numerical simulation model of advective transport • Linear programming optimization scheme

REFERENCES FOR CHAPTER I

McCarty, P. L., Semprini, M. E. Dolan, T. C. Harmon, C. Tiedeman, and S. M. Gorelick, *In-Situ Methantropic Bioremediation for Contaminated Groundwater at St. Joseph, Michigan,* Presented at the International Symposium on In-Situ and On-Site Bioreclamation, San Diego, CA, March 1991.

Roberts, P. V., L. Semprini, G. D. Hopkins, D. Grbic-Galic, P. L. McCarty, and M. Reinhard, *In-Situ Aquifer Restoration of Chlorinated Aliphatics by Methantropic* Bacteria, EPA/600/2-89/033; Center for Information Research Information, Cincinnati, OH, 1989.

U.S. Air Force, *Air Force Installation Restoration Program Management Guidance,* Air Force Engineering and Services Center, Tyndall Air Force Base, FL, 1985.

U.S. Environmental Protection Agency, *Superfund Federal-Lead Remedial Project Management Handbook*, Office of Emergency and Remedial Response, OSWER Directive 9355.1–1, 1986.

______, *Data Quality Objectives for Remedial Response Activities, Development Process,* Office of Emergency and Remedial Response, OSWER Directive 9355.0–7B, 1987a.

———, *Data Quality Objectives for Remedial Response Activities, Example Scenario: RI/FS Activities at a Site With Contaminated Soils and Ground Water,* Office of Emergency and Remedial Response, OSWER Directive 9355.0–7B, 1987b.

———, Expanded Site Inspection, Transitional Guidance for Fiscal Year 1988, Office of Emergency and Remedial Response, OSWER Directive 9345.1–02, 1987c.

———, Guidance for Conducting Remedial Investigations and Feasibility Studies Under CERCLA, Office of Emergency and Remedial Response, OSWER Directive 9355.3–01, 1988a.

———, Preliminary Assessment Guidance, Fiscal Year 1988, Office of Emergency and Remedial Response, OSWER Directive 9345.0–01, 1988b.

CHAPTER II

Review of Basic Concepts and Principles in Contaminant Hydrogeology

This chapter reviews the basic principles of groundwater hydrology and contaminant transport. Some readers may wish to skip this chapter and move directly to the presentations of Chapters III or IV.

Chapters II.A, II.B, and II.C treat the fundamental concepts of groundwater flow in aquifers. Readers who are familiar with hydrogeological environments, aquifer properties, and the physics of steady-state and transient groundwater flow may wish to skip these subchapters and move directly to Chapter II.D.

Chapters II.D, II.E, and II.F treat the fundamental concepts of contaminant transport. The transport equations are presented in Chapter II.F.5.

Chapters II.G and II.H review steady-state and transient well hydraulics in the context of pump-and-treat remedial well networks.

Chapter II.I provides a brief discussion of the types of computer models required in the simulation-optimization procedures.

A. Hydrogeological Environments

1. *Aquifers and Aquitards*

Aquifers are the fundamental hydrogeologic unit in which most processes of interest in contaminant hydrogeology occur. Freeze and Cherry (1979) define an aquifer as a "saturated permeable unit that can transmit significant quantities of water under ordinary hydraulic gradients." This is, by necessity, a subjective and qualitative definition that was originally tailored to suit the needs of water-supply hydrogeology. In contaminant hydrogeology, this definition may well extend to geologic units not typically considered to be aquifers in a water-supply sense. From a remedial perspective, a glacial till unit of high clay content and little groundwater flow might be considered as an aquifer if it were located next to a surface-water body used for a drinking water supply. The definition of an aquifer has a clear meaning only when accompanied by a description of the hydrogeologic setting, the nature of the contamination problem, and the attendant regulatory and public concerns.

With such a broad definition, almost any saturated geologic material can be regarded as an aquifer. The only exceptions might be unfractured crystalline rock and massive clay formations. Materials commonly regarded as aquifers include unconsolidated sedimentary deposits, fractured or porous sandstone, volcanic rocks such as basalt and tuff, and dolomite or limestone strata. Many other rock types, in certain locations, may also meet the definition of an aquifer.

The term aquitard is used to describe low-permeability stratigraphic units that transmit significantly less water than adjacent aquifers. Aquitards usually define the upper or lower boundary of an aquifer. When an aquifer is defined as a thick formation composed of several stratigraphic units of varying permeability, the term aquitard sometimes refers to low-permeability strata within the aquifer. Like the term aquifer, the use of the term aquitard is also subjective and has a clear meaning only in the context of a specific hydrogeologic system. The two terms must be viewed as relative; a silt layer might be considered an aquifer in a silt-clay system and an aquitard in a silt-sand system.

2. *Porous-Media and Fractured-Rock Aquifers*

Geological materials may be classified as either consolidated or unconsolidated. Consolidated formations are those in which the solid matrix is either porous or fractured rock. Unconsolidated formations are composed of uncemented sedimentary deposits of gravel, sand, silt, and clay of alluvial, marine, lacustrine, or glacial origin. All unconsolidated formations and many porous and/or highly fractured rock formations can be treated from a hydrogeological perspective as a porous-media continuum. Sparsely fractured formations may require treatment as a network of discrete fracture planes. In most locations, the uppermost saturated groundwater flow system consists of unconsolidated aquifers or highly fractured consolidated aquifers that can be treated as porous media. These aquifers are also the most frequently affected by contamination. These aquifers are also the best candidates for successful remedial actions. Therefore, we will emphasize remedial technologies and strategies for addressing groundwater contamination problems in porous-media aquifers. We will not address remedial strategies for geologic formations that must be treated as discrete fracture networks.

3. Confined and Unconfined Aquifers

Aquifers are further classified on the basis of conditions at their upper surface. An aquifer bounded on its upper surface by an aquitard is a confined aquifer; an aquifer with a water table as its upper boundary is an unconfined aquifer. In an unconfined aquifer, the water table separates the underlying saturated zone from the overlying unsaturated zone, or vadose zone. The water table is defined as the surface at which the absolute pressure of the fluid in the pores is equal to the local atmospheric pressure.

Several additional distinctions between a confined aquifer and an unconfined aquifer are of interest in the design of remedial well fields. By definition, the upper surface of a confined aquifer remains saturated and the zone of saturation does not fluctuate significantly over time in response to natural or induced changes in the flow system; in an unconfined aquifer, the zone of saturation changes through time under the influence of water-table fluctuations.

This fluctuating surface has important implications for an unconfined aquifer underlying contaminated soil. As the water table fluctuates through time, the saturated part of the flow system repeatedly comes in contact with the contaminated soil, which acts as a renewable source of contamination for the groundwater. Each rise of the water table serves to recharge the contaminant load in the groundwater that might otherwise be depleted by flow-through or, at worst, be supplemented only by contaminants transported via infiltration from above.

The two aquifer types also differ in the mechanisms that release water from storage. In a water-table aquifer, draining of some pore spaces accompanies a withdrawal of water from the aquifer. In a confined aquifer, two simultaneous processes accompany the withdrawal of water. First, the pore space is reduced due to compaction of the bulk matrix (as water pressure decreases, stresses on the solid grains increase and cause a decrease in the bulk volume). Second, the pore water expands due to the reduction in the pressure of the fluid.

The drainage of pore fluids in unconfined aquifers under groundwater pumping creates a cone of depression in the water table. The cone of depression changes the saturated thickness of an unconfined aquifer. This is a process described by a nonlinear equation. If this change is significant in comparison to the undisturbed saturated thickness, difficulties arise in the combined application of standard simulation models and linear optimization procedures. This feature of water-table aquifers has the potential to complicate the remedial action design process in unconfined aquifers relative to that for confined aquifers where the processes are described by a linear equation, but can be handled as discussed in V.F.6.

For situations in which a light non-aqueous-phase liquid (LNAPL) forms a contaminant plume floating on the water table, the development of a cone of depression adds another problem by leaving residual contamination adsorbed onto the soil or otherwise trapped by capillary forces. If this residual contamination is not removed, it will serve as a new source of contamination if and when the pump is shut off and the water level returns to its original position.

The most important and obvious distinction between a confined aquifer and an unconfined aquifer is the lack of a low-permeability confining layer overlying unconfined aquifers. This leaves the aquifers susceptible to direct contamination from liquid chemical spills, lagoons, landfill leachate, and other similar contaminant sources. For this reason, most groundwater contamination problems occur initially in water-table aquifers, as opposed to confined aquifers. Subsequent migration across underlying confining layers often leads to later contamination of deeper confined aquifers.

4. Saturated and Unsaturated Zones

In an unconfined aquifer, the region above the water table is referred to as the unsaturated zone and the region below the water table is referred to as the saturated zone. The distinction is based on the difference in the fluid pressure of the two zones. In the unsaturated zone, the pressure in the fluid is less than atmospheric pressure (i.e., gage pressure is less than zero) due to the surface-tension forces that hold the water in the pore spaces. In the saturated zone, the fluid pressure is greater than atmospheric pressure due to the weight of the overlying water. Groundwater contamination problems that can be dealt with via the type of pump-and-treat systems described in this book occur in the saturated zone. It is in this zone, then, that we must be able to accurately describe and successfully control groundwater transport. Vapor pumping can be used to remove contaminants from the unsaturated zone, but this technology is outside the scope of this book.

5. Hydrogeologic Controls: Lithology, Stratigraphy, and Structure

The distribution of contaminant plumes (and the feasibility of removing or stabilizing them with pump-and-treat systems) are controlled in part by the presence and geometry of the aquifer-aquitard systems at a site, and these, in turn, are controlled by the lithology, stratigraphy, and structure of the regional and local geologic environment. Davis and De Wiest (1966) provide an excellent discussion of hydrogeological environments.

The lithology of a geologic deposit is defined by its mineral composition, grain sizes, and grain packing. The lithologic description of an aquifer or aquitard material leads to a definition of the rock or porous media type. Lithology is a function of basic rock or media properties on a small scale, including consideration of grain-size distribution and whether the material is consolidated or unconsolidated, fractured or unfractured, weathered or unweathered, and sorted or unsorted.

The stratigraphy of an aquifer-aquitard system describes the geometric grouping of the various lithologic units. It is controlled by the depositional environments that created the deposits. In sedimentary systems of marine or lacustrine origin, the stratigraphy is often horizontally layered. If the layering is on a very fine scale, it may be necessary to group layers into hydrostratigraphic units that possess distinct and definable physical and hydrogeologic characteristics. In alluvial or glaciofluvial sediments, the depositional environment leads to stratigraphic configurations that feature irregular and discontinuous lenses of high- and low-permeability materials.

The stratigraphic sequences at a site are often disturbed by the presence of small-scale structural features such as fractures, and/or by large-scale structural features such as folds and faults.

The lithology, stratigraphy, and structure at a site are determined from geologic mapping, borehole logging, and geophysical studies at the site scale; and from an interpretation of the site location in the regional stratigraphic and structural setting of the area. We assume that this process of geologic interpretation has been completed as a part of the Remedial Investigation (RI) process at the site. Emphasis is on the design of remedial action at sites that have been fully characterized, so that the geometry, thickness, and lateral extent of the aquifers of interest are relatively well known.

Directions of contaminant migration in a contaminated aquifer prior to remediation are controlled by the regional groundwater flow system that develops in the hydrogeologic environment. The flow system configuration depends on the geometry and hydraulic-conductivity distribution of the component aquifers and aquitards, and these in turn depend on the lithologic, stratigraphic, and structural controls. Of particular importance is the distribution over the ground surface of recharge areas and discharge areas for the uppermost unconfined aquifers. The nature of the recharge-discharge regime in an area and its relation to contaminant sources is an important aspect of the hydrogeological interpretation carried out during the RI process.

It is also possible for the time-dependent processes of recharge and discharge at a site to dramatically alter groundwater flow directions in the vicinity of a contaminant plume. For example, consider a plume in an aquifer adjacent to a stream or reservoir that exhibits large seasonal fluctuations in water level. During low-water periods, the water body may act as a groundwater discharge zone, causing the plume to migrate toward it. During high-water periods, recharge from the river or reservoir could cause contaminant migration away from the water body. Failure to consider such a flow reversal when designing a remedial system could result in contaminants escaping the containment zone and reaching the surface-water body *or* migrating in the opposite direction. Changes in the recharge-discharge regime can also be induced by human activities; examples include seasonal pumpage for irrigation, or municipal pumpage that is operated only to satisfy peak demands. Because of these potential impacts these processes must be understood at the conceptual level to ensure adequacy of a remedial design.

B. Physical Properties of Aquifers

Groundwater flow and contaminant transport processes at a local or site scale are controlled by several fundamental aquifer properties that must be quantified if a remedial design is to perform as intended. Many texts on groundwater hydrology provide comprehensive discussions of the influence of these properties and techniques for measuring them, including de Marsily (1986), Bear (1979), Freeze and Cherry (1979), Todd (1980), and Domenico and Schwartz (1990). The following chapters present a brief review of these properties.

1. Properties of Porous Media

Four of the basic properties of porous media of interest in contaminant studies are permeability, porosity, compressibility, and density. Each are briefly discussed below.

Permeability is commonly used in a *qualitative* sense to describe the ability of a porous medium to transmit fluid. The term is sometimes incorrectly used interchangeably with the term hydraulic conductivity. This can lead to confusion. Permeability, as correctly used, is briefly discussed in this subchapter; hydraulic conductivity is fully defined in Chapter II.B.3.

Permeability (k) describes the capacity for flow of *any* fluid through a specific porous medium. It is a function of the porous media properties only and has the dimensions of [L^2]. Permeability is measured using a permeameter in the laboratory. It may also be calculated from empirical equations based on porous media characteristics, such as grain size and pore size (Bear, 1979), but these relationships are very imprecise. Several groundwater hydrology texts present tables of values of k for various geologic materials (Freeze and Cherry, 1979; Bear, 1979).

Porosity is the second basic porous-media property of interest. The concept is deceptively simple; yet when considered in detail, porosity can be surprisingly complex. At the basic level, total porosity (n_t) is defined as the ratio of the volume of voids (V_v) in a soil or rock sample to the total volume (V_t) of the sample:

$$n_t = V_v/V_t \, . \tag{1}$$

Because some pore spaces in a soil or rock matrix are isolated and others are essentially dead-end channels, total porosity is not the correct parameter for describing the fraction of a porous medium that is open to groundwater flow. Effective porosity (n_e) is defined as the volume fraction of pore spaces that are open to groundwater flow in a continuum sense. Effective porosity includes only those pore spaces that are continuously interconnected. In analyzing saturated groundwater flow velocities (Chapter II.C.2) in either confined or unconfined aquifers, effective porosity is the correct parameter for quantifying the relative volume of porous media through which groundwater flows. It is always less than or equal to the total porosity.

A parameter used in water-table and unsaturated-flow problems that is related to porosity is specific yield (S_y). Specific yield is defined as the ratio of *drainable* void volume to the total volume of a representative sample of soil or rock. This term is most often used to characterize the storage properties of unsaturated soils or water-table aquifers. Because pore sizes for most geologic materials vary about some average, and water-retaining capillary forces are a function of pore size, drainage of water from these spaces is not instantaneous. Specific yield, which is measured as a fraction of the total volume, must then also vary in time. It is customary to use specific yield to define the maximum drainable porosity of a saturated medium.

Some authors use the terms effective porosity and specific yield (or drainable porosity) interchangeably. This is not a wise practice as the two terms can have different values, depending on how and at what time during a drainage event the specific yield is measured. Only if the specific yield value is calculated for complete drainage of a water-table aquifer and only if this aquifer has a minimal number of dead-end pore spaces can effective porosity and specific yield be considered the same.

Fractured rock settings manifest yet another type of porosity, fracture porosity (n_f). In highly fractured rocks that exhibit many continuous intersecting fracture planes, n_f is often used to define the porosity of an "equivalent porous media." Flow in the *fractured* rock media may then be analyzed by applying traditional porous-media flow theory to the equivalent *porous* media with porosity n_f (Freeze and Cherry, 1979). The volume of fractured rock media used to determine n_f must be large enough to ensure that flow within the volume is hydraulically equivalent to flow within a similar volume of porous media.

The fracture flow problem also provides the needed setting to introduce the concepts of primary and secondary porosity. Primary porosity applies to the void space within the solid phase of a fractured rock body, such as the pore spaces within a fractured sandstone strata. Secondary porosity applies to the void space associated with the fractures themselves. Primary and secondary porosity are also used to characterize the dual-porosity structure of some unconsolidated deposits where the individual solid grains exhibit an internal porosity that is distinct from the bulk porosity.

The third basic porous-media property is compressibility. Three porous-media compressibility terms can be defined: one for the solid phase of a porous medium, one for the porous media itself (including the void spaces), and one for the aquifer as a whole.

Compressibility, in general, is defined as strain divided by stress. It is the inverse of the modulus of elasticity. It reflects the amount of deformation that will be experienced due to a given stress.

Compressibility of the solid phase of a porous media is defined as the percent change in volume of solid grains *divided* by the change in stress that is producing the deformation. This component of compressibility is negligible compared to the other components and is therefore commonly ignored.

Compressibility of the porous media, including the void spaces, is defined as the percent change in the total volume of porous media divided by the change in effective stress causing the deformation. This aspect of compressibility is due to the compression of individual solid-phase particles into new packing arrangements (i.e., a rearrangement of void-space sizes and shapes). Effective stress is that portion of the total stress at any plane in a porous medium that is borne by the solid phase. Effective stress equals the total stress minus the pressure of the pore fluids.

Applying the concept of porous media compressibility to the entire thickness of an aquifer, we can define aquifer compressibility (η_a) as the percent change in aquifer thickness, divided by the associated change in effective stress. Bear (1972) and Freeze and Cherry (1979) provide thorough discussions of porous media and aquifer compressibility.

Because the fluid-pressure fluctuations in unconfined aquifers are relatively small in comparison to the total stress, compressibility is of minor importance for these systems. Compressibility cannot be ignored in confined aquifers where fluid pressures can represent a significant portion of the magnitude of the total stress. Because most contamination problems are found in shallow unconfined aquifers, compressibility does not commonly play a role in remedial action design.

Density is the fourth intrinsic porous-media property of interest and can be defined for either the bulk porous media or the solid phase. On a mass basis, bulk density (ρ_b) is the dry mass per unit volume (including void spaces) of undisturbed porous media. For unconsolidated deposits, bulk density is influenced by the mineralogy and by the percentage of void spaces in a sample. Density of the solid phase (ρ_s) is defined as the dry mass of solids per unit volume of solids. Bulk density and solid-phase density are of interest in calculating porosities (Chapter III.C.2).

2. *Intrinsic Properties of Groundwater*

Several intrinsic properties of groundwater are of interest in aquifer studies in general and groundwater remedial actions in particular. These properties are density, compressibility, and viscosity.

The density (ρ) is the mass per unit volume of water. It is a function of temperature, solute concentration, and pressure. The density of water is 1.0 gram per cubic centimeter (g/cm^3) for relatively pure water at temperatures and pressures in the range of most flow systems. Variations in temperature, pressure, and solute concentration commonly found in groundwater contamination problems are typically not sufficient to change density enough to impact the flow system. However, exceptions exist. High-solute concentrations (e.g., landfill leachates with total dissolved solids greater than 10,000 parts per million [ppm] or seawater intrusion problems) and high temperatures (e.g., water infiltrating from a lagoon containing warm water from an industrial cooling process) can lead to density variations that impact flow. We assume constant fluid density.

Dynamic viscosity (μ) is defined as the coefficient of proportionality between the shear stress and the velocity gradient of a flowing fluid. Viscosity is a function of temperature and pressure, although for groundwater flow problems this functional dependence is negligible. For groundwater at temperatures between 13°C and 17°C, dynamic viscosity ranges from 1.207×10^{-3} to 1.086×10^{-3} newton-second per square meter ($N \cdot s/m^2$).

From a practical perspective, the importance of viscosity is limited to its influence on hydraulic conductivity, particularly for fluids other than groundwater. Examples where such an interest might arise are in the flow of LNAPL over the surface of an unconfined aquifer, or in the flow of DNAPL as it percolates through a water-tablc aquifer toward an impermeable stratum at the aquifer base. Chapter II.B.3 discusses the relationship between viscosity and hydraulic conductivity.

The compressibility of water (η_w) plays a role in the study of flow in confined aquifers. It is one component in the storativity term (Chapter II.B.3). Compressibility is a function of temperature and pressure, although this dependence is commonly ignored. For standard groundwater temperatures and pressures, $\eta_w = 4.4 \times 10^{-10} m^2/N$.

3. *Definitions of Common Hydraulic Parameters*

The parameters discussed in the two previous chapters represent the fundamental physical characteristics of porous media and groundwater. In the practical application of the principles of physics and mathematics to groundwater flow analysis, a set of parameters derived from these fundamental properties is commonly used. These parameters include hydraulic conductivity, transmissivity, specific storage, storativity, and specific yield. Only basic definitions are presented here. More detailed discussions of these parameters are available in comprehensive groundwater texts by Bear (1972, 1979), Freeze and Cherry (1979), Todd (1980), de Marsily (1986), and Domenico and Schwartz (1990).

Hydraulic conductivity (K), is the coefficient of proportionality between specific discharge (q) and the hydraulic gradient (dh/dx) in Darcy's law. Hydraulic conductivity is a measure of the capacity of a porous medium to transmit flow of a specific fluid. Experimental studies have shown K to be a function of the intrinsic permeability (k) of the porous medium and the specific weight (ρg) and viscosity (μ) of the fluid. K is given by

$$K = k\rho g/\mu, \tag{2}$$

and has dimensions of [L/T]. In this book, interest centers solely on the hydraulic conductivity with respect to the flow of water. Most groundwater-hydrology texts present ranges of conductivity values for various types of porous media.

If K-values may be considered constant throughout an aquifer domain, the domain is said to be *homogeneous*. If K varies through space, the aquifer is *heterogeneous*. Heterogeneity may be caused by stratigraphic layering; by trends in grain-size distribution, fracture density, or other geologic trends; or by more-or-less random variations in K-values due to changing conditions during the deposition, erosion, weathering, diagenetic, or structural development of the strata.

If K-values are constant in all directions at a given measurement point, the aquifer is said to be *isotropic* at that point. If K varies with direction, the aquifer is *anisotropic*. Small-scale anisotropy is caused by grain orientations or preferential fracture orientations within individual stratigraphic layers. Larger scale anisotropy can be caused by microstratigraphy within aquifers. Horizontal K-values usually exceed vertical K-values in horizontally layered sedimentary sequences.

Transmissivity (T) is the counterpart of hydraulic conductivity used to quantify the flow capacity of an aquifer, as opposed to an isolated volume of porous media. Transmissivity is defined as the average hydraulic conductivity of a vertical cross section of an aquifer multiplied by the saturated vertical thickness of the aquifer, (b). Transmissivity is given by

$$T = Kb. \tag{3}$$

The dimensions of T are $[L^2/T]$.

Hydraulic conductivity and transmissivity are the parameters commonly used in groundwater flow equations to quantify the capacity of an aquifer to transmit flow. For systems in which the hydraulic conditions governing the flow are changing with time (transient flow problems), the mechanism of water storage and release in the saturated material must also be known.

Specific storage (S_s) is the basic storage parameter for confined aquifers. It is defined as the volume of water produced by a unit volume of porous media during a unit decline of hydraulic head. Recalling that the release of water from storage in a confined aquifer is due to compression of the solid matrix and decompression of the water, S_s which has dimensions of $[L^{-1}]$, is given by

$$S_s = \rho g \, (\eta_a + n\eta_w), \tag{4}$$

where η_a is the compressibility of the aquifer, η_w is the compressibility of water, and n is the porosity of the aquifer.

In the same way that transmissivity is an extension of hydraulic conductivity to a specific aquifer cross section, storativity (S) is an extension of specific storage. Storativity is obtained by multiplying the specific storage by the saturated thickness of the aquifer:

$$S = S_s b. \tag{5}$$

Storativity is defined as the volume of water released from a column of aquifer material of unit horizontal cross-sectional area per unit decline in hydraulic head. This quantity is sometimes called the storage coefficient.

Because the release of water from storage in an unconfined aquifer is due to the drainage of pore spaces, neither specific storage nor storativity are appropriate storage terms for transient water-table aquifer problems. Specific yield (S_y), which was introduced in the discussion of porosity, is the appropriate storage parameter for this case. Ignoring the complexities associated with this quantity, specific yield is defined simply as the volume of water released from storage per unit of horizontal cross-sectional area per unit decline in water-table elevation. S_y is much larger than S because the drainable porosity of an aquifer represents a much greater volume than that associated with compaction of the solid matrix and expansion of compressed water.

There are several difficulties associated with the concept of specific yield. First, S_y is a function of the previous drainage history of the aquifer. Second, S_y may be a function of the water-table position, relative to the ground surface (Gillham, 1984). Lastly, S_y may exhibit hysteresis where the value of S_y for a recharge event is different than that for a drainage event. These functional relationships are difficult to quantify. Standard groundwater texts provide limited discussion of these difficulties.

C. Fundamental Concepts of Groundwater Flow

1. Saturated Flow and Unsaturated Flow

Groundwater movement can occur as either saturated or unsaturated flow. To date, the great majority of contaminant-related hydrogeologic studies have focused on saturated flow. Nonetheless, unsaturated flow processes can play an important role in groundwater contamination problems and remedial action designs.

The unsaturated zone above the water table is the region in which unsaturated flow occurs. This zone is characterized by water pressures that are less than atmospheric pressure and pore spaces that are only partially filled with water.

Water movement in the unsaturated zone follows the same laws of physics that govern flow in the saturated zone: namely, that water moves from locations of higher to lower hydraulic potential (see Chapter II.C.2). Bear (1972, 1979) and Freeze and Cherry (1979) provide detailed discussions of the characteristics of fluid flow in the unsaturated zone. Except for a few special cases, remedial actions for contamination problems in the unsaturated zone will not be discussed in this book. However, the capacity for this region to act as a source for contaminants reaching saturated aquifers via recharge demands that attention be given to the unsaturated zone as part of a complete characterization study prior to remedial action design.

The saturated zone refers to the region below the water table. It may include both unconfined and confined aquifers. It is characterized by water pressures that are greater than atmospheric pressure and fully saturated pore spaces. The mechanics of fluid flow in the saturated zone are commonly governed by Darcy's law (see Chapter II.C.2). The design of remedial actions for contaminants transported in the saturated zone is our focus.

2. *Fundamental Principles of Groundwater Flow*

Groundwater flow generally obeys Darcy's law. The important principles of Darcy's law are shown in Figure 3, which presents an idealized drawing of an experimental setup that could be used to confirm Darcy's law. Water introduced at the cylinder's upper end at a volumetric flow rate Q (dimensions of L^3/T) will eventually discharge from the opposite end at the same rate. Darcy observed that the volumetric flow per unit area perpendicular to the flow ($q = Q/A$) was proportional to the change in water level in the two manometers ($h_1 - h_2$) and inversely proportional to the flow distance between these two points (Δl):

$$q = -K \frac{\Delta h}{\Delta l}, \tag{6}$$

where the difference in hydraulic head ($\Delta h = h_2 - h_1$) is the source of the negative sign (flow is in the direction from higher to lower hydraulic head) and K is the proportionality coefficient. The term on the left is defined as the specific discharge (q). It has dimensions of [L/T], but

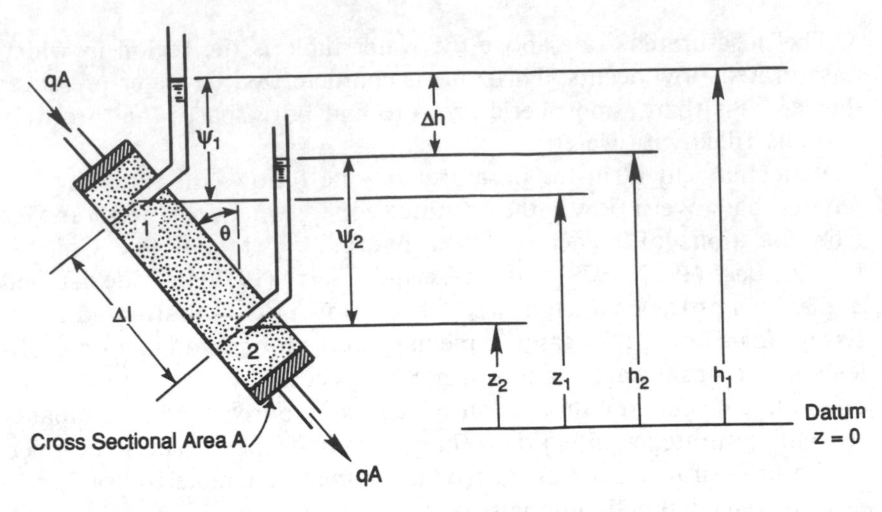

Figure 3. Idealized illustration of experimental setup used to confirm Darcy's Law (after Freeze and Cherry, 1979).

despite these dimensions, it is *not* the velocity of the groundwater. Expressing Darcy's law using the more customary differential notation gives

$$q = -K \frac{dh}{dl}. \tag{7}$$

The coefficient of proportionality (K) is hydraulic conductivity, introduced in Chapter II.B.3. The quotient dh/dl is the hydraulic gradient.

It is important to reiterate that q in Equation (7) is not the velocity of the groundwater. This is apparent by observing that q is defined as the ratio of the total discharge through the column (Q) to the total cross-sectional area (A). If the column is an open pipe, q is the fluid velocity in the pipe. If the column is filled with a porous solid, less area is open to flow. This open area is defined as the void area (A_v). The average void area of numerous cross sections is equal to the effective porosity. The average void area is therefore given by $\overline{A}_v = n_e A$. Dividing the volumetric discharge by the average area open to flow gives an estimate of the average groundwater velocity (V) parallel to the average flow direction:

$$V = \frac{Q}{n_e A} = \frac{q}{n_e}. \tag{8}$$

The term V is called the average linear velocity and is the closest approximation to the true groundwater velocity that is commonly used in practice. The average linear velocity gives an indication of the speed at which contaminants are transported by the bulk flow of groundwater, i.e., advection.

The hydraulic head (h) in Equations (6) and (7) provides a direct indication of the potential energy at points 1 and 2 in Figure 3. In general, hydraulic head is the sum of the elevation head (z) and pressure head (p/ρg):

$$h = z + (p/\rho g). \tag{9}$$

Pressure head is denoted as $\psi = p/\rho g$ and is equal to the height above the measurement point to which water will rise in a manometer (Figure 3).

In any given horizontal or vertical cross section, lines can be drawn through points of equal hydraulic head to form equipotential lines. In an isotropic medium containing a fluid of constant density, these lines are perpendicular to the velocity vectors at all points. For a plan view of a water-table aquifer, lines of equal hydraulic head are contour lines of the water-table surface. If the aquifer is isotropic, flow is perpendicular to these contours in the direction of decreasing hydraulic head. Streamlines (or flowlines) depict the directions of groundwater flow at a given instant in time.

3. Steady and Transient Flow; Uniform and Nonuniform Flow

The conditions associated with a groundwater flow system can either change with time or remain constant. A system in which the hydraulic head remains constant in time is at steady state. In such a system, the velocity of the groundwater also remains constant with time because the velocity is directly proportional to the gradient. If the hydraulic head changes with time, the system is transient. In such a system, the velocity vectors change with time.

If a flow system is at steady state, the pattern of the streamlines will not change with time, and each streamline will map the entire path followed by a particle of groundwater. This line is referred to as a pathline and, for steady state, the pathlines will coincide with the streamlines.

If a flow system is transient, the pathlines will change with time. Each path can then best be visualized as a connected series of short vectors defining the direction of groundwater movement at individual points. No single streamline can describe the entire path followed by an individual water particle over a finite period of time. If the paths are integrated over time, they will yield a pathline that describes the route of a single water particle. This pathline is unique to a single water particle and will not coincide with any of the instantaneous streamlines.

If a well is pumped in an aquifer, it will induce a period of transient flow, but will tend toward a final steady state. During transient flow, the water pumped by the well is partially obtained from storage. At steady state, the pumpage comes into equilibrium with inflows from the aquifer boundaries, recharge from injection wells and infiltration, and/or leakage from adjacent aquifers. If a system is to be treated as transient and studied with the aid of a mathematical model, then either the storativity, S, or the specific yield, S_y, or both must be included to account for changes in aquifer storage that accompany changes in hydraulic head.

An important distinction between steady and transient flow systems relates to the transport of contaminants in the groundwater. Contaminant migration follows the pathlines. For a steady system, the streamlines obtained by constructing a flow net will be coincident with the pathlines, and they will therefore map the movement of contaminants dissolved in the groundwater. For a homogeneous and isotropic system, construction of flow nets (graphically or with the aid of a computer) is relatively simple (Freeze and Cherry, 1979; Bear, 1979) and provides a useful framework for understanding the contaminant transport process.

For a transient system, pathlines and streamlines are not coincident. The pathline of a contaminant particle cannot be determined from a flow net, but must be obtained by tracking the movement of groundwater particles through the system. This tracking involves obtaining multiple solutions of a model of the system at successive times over the period of interest. The difficulty associated with determining contaminant migration pathways in transient flow systems often leads hydrogeologists to idealize the system as a steady-state flow problem. This is usually done by using a temporal average of the hydraulic heads over the period of interest, thus averaging out the fluctuations in time.

A steady-state groundwater flow system can be classified as uniform or nonuniform. A uniform flow field in three dimensions exhibits plane-parallel equipotential surfaces; in two dimensions, a uniform flow field exhibits straight, parallel equipotential lines. Under these conditions, velocity vectors in the uniform flow region are identical in magnitude and direction. Nonuniform flow systems are those in which the velocity vectors are not identical in magnitude and direction. Equipotential surfaces in such systems are curved surfaces. The lack of uniformity in flow results from hydrogeologic boundaries, differential recharge or discharge, or porous media heterogeneities.

This distinction between uniform flow and nonuniform flow is significant from the perspective of contaminant-transport and remedial action design for several reasons. Foremost is that the type of flow system influences the degree of difficulty associated with analyzing the contaminant-transport process. For a uniform flow problem, a flow net can be constructed manually and this flow net maps the advective migration routes for contaminant particles. Conversely, mapping a nonuniform flow system commonly requires solving an analytical or numerical model of the system. Once pumping wells are present in an aquifer, the radial components of flow to the well preclude the continued existence of a uniform flow field.

There are also significant differences in the way contaminated groundwater moves in uniform and nonuniform flow systems. A volume of contaminated groundwater will remain undistorted while advecting through a uniform flow field. The same block of contaminated water beneath a water-table mound caused by landfill leachate recharge (a nonuniform flow system) would spread radially outward from the mound. The contamination in the nonuniform flow system would be spread over a larger area than in the uniform flow system due to the diverging flowlines, but concentrations would be lower away from the source. This example illustrates the obvious importance of the uniform flow concept in choosing initial alternatives for a remedial action design. The uniformity distinction also impacts the choice of mathematical tools for remedial action design, as presented in Chapter IV.

D. Nature of Contaminant Sources

The focus of this book lies in the design of pumping schemes for efficient and cost-effective removal and control of contaminated groundwater. As such, the emphasis in the technical material is on characterizing and manipulating the transport of dissolved solutes that have already entered the groundwater system. However, the perspective of the hydrogeologist who is designing a remedial system must also include a conceptual understanding of the manner in which contaminants reach the groundwater. This perspective is critical for an effective remedial system.

1. Source Types

A small number of different, although not mutually exclusive, types of sources account for the majority of groundwater contamination sites. The most common point-source types include:

- Leaking underground storage tanks.*
- Leaking pipelines (including sewers).*
- Spills or other major releases of liquids at the ground surface.*
- Industrial operations.*
- Landfills.
- Holding ponds and lagoons.
- Injection wells.

The four asterisks denote the source types that cause most groundwater contamination problems at U.S. Air Force installations. Fuels, fuel-related hydrocarbons, and chlorinated solvents represent the major contaminants originating from these types of sources at Air Force sites. Other source types may also contribute contaminants ranging from coal-tar residues (e.g., pentachlorophenol) to toxic inorganics (e.g., arsenic and chromium) to radionuclides (e.g., thorium–232 and its daughter products). Although fuel and solvent spills constitute a large percentage of the sources found at Air Force sites, other contaminants found at Air Force source areas are generally quite similar to those found in source areas at non-Air Force sites.

The most significant question is whether or not a source can be cost-effectively remediated as part of an entire remedial strategy. Answering this question is not an objective of this book, but it must be answered during the process of developing a groundwater remedial action if that action is to be effective. Designing a pump-and-treat system for a plume of xylene that originated from a source that had been removed by excavation would be approached differently than designing a system for a plume originating at a landfill expected to generate leachate for another 20 years.

2. Point Sources Versus Distributed Sources

When a contaminant source area is localized in a large groundwater flow system, it is classified as a point source. Generally, the source types listed in the previous chapter are classified as point sources. In all cases, the spatial extent of the source area is relatively small and fairly well defined.

Some contamination problems originate from spatially distributed sources. These distributed sources encompass large areas relative to the flow system of interest. Examples of distributed sources include pesticides and fertilizers in both agricultural and urban settings, road chemicals such as paving by-products and road salts, leakage from septic systems and aging sewer systems, storm-water runoff from urban areas, and atmospheric fallout transported to aquifers via recharge (e.g., tritium from aboveground testing of nuclear weapons).

There are significant differences between point and distributed sources in terms of groundwater remedial action design. Point-source problems are more favorable candidates for pump-and-treat systems. Distributed sources are typically candidates for passive remedial action that include preventative source control (such as reduced fertilizer use) and natural attenuation. The most common approach to groundwater remediation for distributed source contamination has been to treat the groundwater with well head treatment as it is removed for use.

3. Long-Term Sources Versus Temporary Sources

The design life of an engineered component of a system is defined as the length of time the component is intended to function according to its design standards. The design life of a groundwater remedial action system is a function of several variables, including the permanence of the contaminant source.

The duration of contaminant releases from a source is a function of (1) the magnitude of concentration that constitutes "contamination" from a regulatory perspective, (2) the volume of contaminant at the source, (3) the dissipation mechanism, and (4) the source-removal technology available.

An example of a source that might be considered short-term is a spill of acetone over a shallow water-table aquifer. Virtually all product reaching the water table would go into solution in the groundwater because acetone is extremely soluble in water, and the product remaining in the unsaturated zone would quickly volatilize and discharge to the atmosphere due to acetone's high vapor pressure. Subsequently, the contamination in the groundwater would be the sole source of concern. A remedial action for such a problem might be designed to remove the dissolved and adsorbed acetone; once done, the groundwater restoration would be complete.

Alternatively, a mass of pentachlorophenol-contaminated clayey soil beneath a large building, an enormous landfill leaking chromium leachate, or a layer of TCE DNAPL spread over the bottom of an aquifer might pose long-term sources of contaminant release. These examples illustrate cases where the contaminant sources could not likely be removed or quickly "remediated" by conventional pump-and-treat remediation.

Designing a remedial action for problems similar to these last three examples might realistically require the assumption that the remedial system be permanent. That is, there is no mechanism or approach to predict the point in time at which all contaminants will be removed from the system. Remedial actions for which no reliable shutoff date can be reliably predicted may be designated as "perpetual care" problems. Perpetual care is the rather unpleasant but realistic prospect for many groundwater problems involving contaminants that are persistent in the environment or that originate from sources that cannot be economically remediated. For situations in which long-term operational costs can be very high, the optimization techniques presented in this book may be especially valuable for cost-effectively containing the problem.

4. Hydraulic Characteristics of Sources

Many of the point sources listed in Chapter II.D.1 can create or exhibit unique hydraulic behavior in the saturated zone. This behavior often influences the manner in which contaminants enter and are subsequently transported by the flow system. The most common hydraulic impacts associated with contaminant sources are:

(a) Floating of buoyant immiscible plumes of lighter-than-water non-aqueous-phase liquids (LNAPLs) such as fuels.

- LNAPL plume forms on surface of water-table aquifer.
- LNAPL plume is capable of migrating in downgradient direction on the water-table surface.
- LNAPL plume cannot be effectively recovered by groundwater withdrawal from depth; it requires skimming of pure product.
- LNAPL plume acts as a source for dissolved constituents entering the saturated flow system.
- Thick LNAPL plume is capable of altering flow and transport within the groundwater system.
- LNAPL plume requires the use of complicated multiphase flow models, if detailed flow analysis is needed.
- Residual contamination adsorbed onto soils in the region of the plume after LNAPL removal acts as source for continued groundwater contamination.

(b) Sinking and spreading of immiscible plumes of denser-than-water non-aqueous-phase liquids (DNAPLs) such as chlorinated solvents.

- DNAPL plume forms on impermeable aquifer bottom after product sinks through aquifer.
- DNAPL plume will flow in down-dip direction of impermeable unit; may not coincide with local or regional groundwater flow directions.
- Recovery of DNAPL plume is difficult, if not impossible.
- DNAPL plume acts as a mobile and "permanent" source for dissolved constituents entering the groundwater.
- Residual contamination adsorbed onto soils or trapped in between grains after DNAPL removal or migration from an area acts as a source for dissolved constituents entering the groundwater.

(c) Groundwater mounding in water-table aquifers beneath landfills, surface impoundments, and disposal pits.

- Mounding of water table occurs in response to increased recharge.
- Recharge water forming the groundwater mound is the vehicle for contaminant migration to the aquifer.
- Elevated hydraulic heads serve to perturb the regional flow field and produce a radial spread of contamination.
- Mounding of the water table may lead to saturation of wastes or contaminated porous media, and increase the magnitude of the contamination problem.

Because the hydraulic effect and behavior of source types (a) and (b) above are complicated multiphase flow problems, rigorous mathematical analysis is not commonly applied to these problems. Instead, remedial action designs are often based on experience and engineering judgment supported by simplified mathematical models of the fluid flow systems. Performance of a remedial action for LNAPL or DNAPL plume recovery is then closely monitored and used to make ongoing system modifications. This "operational management" ameliorates the shortcomings of the system design resulting from incomplete knowledge of the behavior of the two-phase flow system.

5. Known Sources Versus Unknown Sources

The ideal remedial strategy for a groundwater contamination problem should be based on a large volume of data that characterize all aspects of the problem, including the source characteristics discussed in this chapter. In heavily developed industrial areas, however, it is entirely possible that the nature of the source (or, more commonly, sources) of contamination found in the groundwater is unknown. Factors that contribute to such a deficiency include combinations of (1) a lack of knowledge of historical industrial activities, (2) changing land uses, (3) a highly transient flow system resulting from human stresses, and (4) closely spaced candidate industries that are unwilling to accept responsibility and divulge information.

A lack of information on the nature of contaminant sources may influence the selection and design of a remedial technology. Even more significantly, inability to pinpoint a source may result in lengthy and costly litigation aimed at assigning financial responsibility for the treatment of the problem. Design and implementation of a remedial strategy are then usually delayed until a final decision assigning responsibility is reached by a court order. In the interim, the groundwater contamination problem increases in areal extent, making the final remedial action more difficult and expensive. For complicated groundwater contamination problems at heavily industrialized sites where an unknown source of contamination continues to feed contaminants to an aquifer, perpetual care is the likely outcome for groundwater remedial action.

E. Contaminant Properties

Once a contaminant comes into contact with groundwater, several contaminant-specific characteristics influence its capacity to be transported by flowing groundwater. These characteristics can be defined for a pure contaminant product, such as liquid TCE or fuel, or they can be defined for the dissolved constituents in groundwater, such as dissolved chromium ions.

1. Chemical Properties of Contaminants

From the perspective of groundwater pollution, the most significant contaminant characteristic is solubility. The solubility of a solute is defined as the mass of the solute that will dissolve in a unit volume of solution under specified conditions. The solubility defines the maximum possible concentration that commonly occurs in groundwater for any given contaminant. It is typically expressed in units of $[M/L^3]$, where the mass is that of the solute and the volume is that of the solution. Table 2 lists the solubilities for some of the common organic contaminants at fuel-spill sites.

Solubility is largely governed by the similarities or differences in the type and strength of intermolecular bonding in the solute compound and in the solvent. As a general rule, solutes that exhibit bonding characteristics similar to a given solvent will be highly soluble in that liquid. Many other factors are involved, however, and can complicate this matter

Table 2. Solubilities of Organic Compounds Commonly Found at Fuel-Spill Sites (After Verschueren, 1983).[a]

Compound	Solubility in Milligrams per Liter
Benzene	1,780
Ethyl Benzene	152
Methylene Chloride	20,000
M-Xylene	NA[b]
O-Xylene	175
P-Xylene	198
Toluene	515
Trichloroethylene	1,100
Vinyl Chloride	1,100

[a]Solubility is a function of temperature. For gas-phase compounds such as vinyl chloride, solubility is also a function of partial and total pressure.
[b]NA = not available.

significantly. The reader is referred to an introductory text on aqueous chemistry (Stumm and Morgan, 1970) for more complete details.

For groundwater problems, nonpolar liquids commonly exhibit low solubilities due to the polar bonding of water. Many organic liquids fall in the category of nonpolar liquids.

Some organic liquids, such as petroleum and the fuels derived from it, are composed of a variety of different molecules or compounds. Other organic liquids, such as TCE, contain only a single type of molecule. Jet fuel is an example of a multicompound liquid and is composed of a mixture of aliphatic, cyclic, and aromatic hydrocarbons. Each group includes many distinct species of hydrocarbon molecules. Four aromatic compounds commonly found at jet-fuel contamination sites are benzene, toluene, ethyl benzene, and xylene (BTEX). Solubilities for the BTEX compounds are listed in Table 2.

Regardless of whether a liquid is composed of a single type of molecule, such as TCE, or a mixture, such as jet fuel, it is the nature of the intermolecular bonding in the liquid that contributes to its generally low solubility in water.

Liquids composed of molecules with intermolecular forces similar to those in water exhibit high or infinite solubility in water. Examples include alcohols, acetone, and ethyleneglycol. Infinite solubility in water indicates that an unlimited amount of the solute, for example methyl alcohol, will dissolve or go into solution in a finite amount of water. In these instances, the designation of "solvent" and "solute" becomes somewhat arbitrary. Liquids with infinite solubilities in water are referred to as being miscible with water. Liquids with finite solubilities in water are referred to as being immiscible with water.

The term miscible is sometimes applied in a semiquantitative manner to describe liquids with finite solubilities as highly miscible or slightly miscible. To avoid confusion, we will use solubility to provide a quantitative definition of the amount of contaminant that will dissolve in water (Table 2). The term "miscible" will be used to describe only those liquids with infinite solubility in water. The term "immiscible" will be used to describe liquids with finite solubilities in water—even if the solubility is high.

Liquids that are immiscible with water are of special interest in groundwater flow problems. The failure of such liquids to completely dissolve in groundwater produces a stratified flow problem, with each liquid behaving as a separate phase. Low-density immiscible liquids (referred to as light non-aqueous-phase liquids or LNAPLs) will float on the surface of the higher density groundwater. High-density liquids (referred to as dense non-aqueous-phase liquids or DNAPLs) sink through the groundwater until they reach the aquifer bottom. Jet fuel is an example of an LNAPL, and TCE is an example of a DNAPL. While these liquids do not go completely into solution in groundwater, they do contain compounds with limited solubilities in water.

For a contamination problem involving jet fuel, it is often the low concentrations of BTEX resulting from fuel reaching the water table that cause concern and warrant efforts to recover the spilled fuel. Similarly, a spill of TCE will percolate to the bottom of the aquifer and form a distinct mass of pure contaminant liquid. This liquid will release small concentrations of TCE into the groundwater. This contaminated groundwater may migrate large distances from the spill site and degrade water quality in a distant water-supply well, while the source of the contamination remains at or near the original spill site. Again, the low concentrations of dissolved contaminants within the groundwater would provide the motivation for attempting to recover the pure contaminant product. Furthermore, daughter products from degradation may be found, which would complicate the matter of source identification.

When a remediation problem such as either of the two previous examples does involve a mass of pure liquid contaminant that is in contact with groundwater, the flow system must be treated as a multiphase problem. This commonly requires the use of complicated numerical models for detailed analysis.

Solubilities of solids in groundwater range from high (for minerals such as the halide salts) to low (for metals such as lead, cadmium, and chromium). As with a liquid, the solubility of a solid in water determines the amount of a solid that can go into solution in groundwater. However, solubilities of solids are considerably less interesting than solubilities of liquids because the low solubilities common to most solid-phase contaminants of interest do not produce the multiphase flow problems associated with insoluble liquids. For extremely high concentrations of soluble compounds (for example, common salts in seawater), concentrated solutions can produce density-driven flow. For many contamination problems involving low-concentration organic plumes, however, dissolved solids in groundwater are not found at high enough concentrations to measurably affect the groundwater density.

Solubility data for various liquids and solids are available from standard chemistry texts, chemical data literature (Verschueren, 1986; NIOSH, 1985; and Weast and Astle, 1979), and product information files kept by the manufacturer and user. A Material Safety Data Sheet (MSDS) is one type of reference that is commonly available from the files of a manufacturer or user of a given compound. An MSDS may provide basic chemistry information, including solubility.

The solubility of a material in water can be predicted using the law of mass action. If a solid is introduced into water, there will be a dissolution reaction that will proceed until the reaction is at equilibrium. The law of mass action can be used to express the relationship among the solute concentrations and the solvent concentration when the reaction is at equilibrium. The equilibrium point for the dissolution reaction for any contaminant is identified by the thermodynamic equilibrium constant for the reaction.

Once they have entered the groundwater flow system, dissolved contaminants are available for further chemical interactions with native waters, with other contaminants, or with the mineral components of the host aquifer. The law of mass action also predicts these potential chemical reactions, and each has an equilibrium constant associated with it. Among the possible reactions are precipitation-dissolution reactions and oxidation-reduction reactions. The equilibrium constants for these reactions are a function of temperature, pressure, pH, and Eh. These environmental conditions in the aquifer, together with the equilibrium constant for a given reaction, predict whether such a reaction will take place, and, if so, what its implications are for the contaminant concentrations in the aquifer. Equilibrium constants do not provide information on the rate of reactions. Equilibrium constants are available in table form for standard temperatures and pressures for many reactions (cf. Freeze and Cherry, 1979; Stumm and Morgan, 1970; Garrels and Christ, 1965).

A more detailed presentation of aqueous and contaminant geochemistry is beyond the scope of this book. Many of the reactions of interest

Table 3. Specific Gravities of Organic Compounds Commonly Found at U.S. Air Force Sites (After Verschueren, 1983).

Compound	Specific Gravity
Benzene	0.88
Ethyl Benzene	0.87
Methylene Chloride	1.33
M-Xylene	0.86
O-Xylene	0.88
P-Xylene	0.86
Toluene	0.87
Trichloroethylene	1.46
Vinyl Chloride	0.91

lead to retardation of the migration velocities of contaminant plumes. These issues are further discussed in Chapter II.F.3.

2. Physical Properties of Contaminants

The density of an immiscible liquid contaminant will determine its behavior upon reaching the water table. LNAPLs, the immiscible liquids with densities less than water ($\rho_{water} = 1.0\ g/cm^3$), will float. Petroleum hydrocarbons are the most common LNAPLs.

DNAPLs, the immiscible liquids with densities greater than water, will sink toward the bottom of the aquifer. Chlorinated hydrocarbons (solvents) are the most common DNAPLs.

The specific gravity of a liquid is the ratio of its density to that of water. Table 3 lists the specific gravities for the same set of organic compounds listed in Table 2.

Chapter II.B.2 introduced viscosity as the liquid property that acts to resist the effects of shear stress within a mass of liquid. In contaminant hydrogeology, viscosity is used mainly to assess the *relative* mobility of pure liquid contaminants. Recall from Equation (4) that the hydraulic conductivity for a specific-liquid/porous-media combination is inversely proportional to the viscosity. The greater the viscosity, the less mobile a liquid will be in the subsurface environment. Viscosity gives the hydrogeologist a qualitative "feel" for how fast an LNAPL or DNAPL spill will migrate through an aquifer and how fast it will "pancake out" once it reaches a barrier to flow (the water table for LNAPLs and the aquifer bottom or perching layer for DNAPLs).

When the flow of LNAPLs or DNAPLs is studied quantitatively (by applying numerical multiphase flow models), it is necessary to know the viscosity and the density of the non-aqueous-phase liquid (NAPL) in order to determine the conductivity of the porous medium to NAPL flow, using Equation (2).

To determine the NAPL conductivity using Equation (2), the intrinsic permeability (k) is replaced by relative permeability. For a particular fluid in a multiphase flow problem, relative permeability is defined as the ratio of effective permeability to intrinsic permeability (of the wetting fluid) at saturation. Effective permeability is a function of the degree of saturation and is defined as the intrinsic permeability of the porous medium at a particular degree of saturation. For a given fluid in a given porous medium, there will be a characteristic curve that defines the effective permeability as a function of the degree of saturation. Effective and relative permeabilities are generally determined by experiment or from published experimental results. Corey (1986) provides a thorough discussion of effective and relative permeabilities.

For some liquid contaminants, the viscosity values required for a multiphase flow analysis can be found in the chemical data literature (Weast and Astle, 1979), although the number of compounds for which data are available is limited. Viscosity can also be determined by measurement in the laboratory.

Volatility, ignitability, explosivity, and toxicity are additional contaminant properties of interest when conducting remedial site investigations and developing remedial strategies. All four of these properties are applicable to organic contaminants in general. For inorganic contaminants, only toxicity is of general interest.

Volatility, expressed in terms of the vapor pressure of a liquid, is of concern in the design of vacuum extraction systems for removing organic contaminant residues from the unsaturated zone. Organic compounds with high vapor pressures (such as benzene and TCE) are good candidates for remediation via vapor extraction systems.

Volatility is also a concern when determining health and safety protocols for personnel working near contaminated materials. The propensity for high vapor concentrations in the ambient air requires consideration for respiratory protection when working in contaminated environments where volatile compounds are present.

Ignitability and explosivity of contaminants (or contaminated materials) are two additional health and safety concerns for personnel working in a contaminated environment. Ignitability is the ease with which a substance ignites and burns continuously in the presence of a flame or spark. It is defined by the temperature at which this occurs. Ignitability is also of interest in developing remedial strategies that involve incineration (of contaminated soils, for example) as a remedial technology.

Explosivity is a rather nebulous property that describes the tendency for a material (solid, liquid, or gas phase) in a confined space to combust at a sufficient rate to create an explosion. Explosivity for volatile organic compounds (VOCs) is usually measured in terms of a Lower Explosive Limit (LEL), which is the minimum gas-phase concentration in a confined space that will produce an explosion in the presence of a source (flame or spark).

Toxicity is a measure of the potential of a contaminant to cause adverse health affects in living organisms. Regulatory agencies use toxicity data to establish maximum acceptable concentration limits for compounds in groundwater. Toxicity data are also used during public health studies (part of the RI/FS process) to determine the environmental health impact of a particular contaminant. Contaminant toxicity can play an indirect role in defining the scale and performance criteria (cleanup levels) of individual pump-and-treat remedial actions.

F. Principles of Contaminant Transport

Contaminants dissolved in groundwater are transported by three processes: advection, mechanical dispersion, and molecular diffusion. Mechanical dispersion and molecular diffusion are referred to collectively as hydrodynamic dispersion. All three processes operate simultaneously in flowing groundwater. In the process of being transported, many contaminants react with other compounds or ions in solution. Contaminants may also adsorb onto or desorb off the solid matrix. Chemical reactions, exchange, and adsorption/desorption processes can significantly slow the rate of contaminant transport. A brief review of these transport processes and reaction mechanism follows.

1. Advection of Dissolved Solutes

Advection is the movement of dissolved solute with flowing groundwater. It is the component of the total transport process that can be mathematically accounted for by groundwater flow parallel to the hydraulic gradient in an isotropic medium with a velocity equal to the average linear velocity (V).

In coarse-grained homogeneous aquifers, advective transport dominates the transport process. In the design of capture and containment systems, it is common to treat advection as the sole mechanism for contaminant transport.

2. Dispersion and Diffusion of Dissolved Solutes

The two processes that constitute hydrodynamic dispersion are diffusion and mechanical dispersion. They are usually secondary in importance to advection in the context of capture and containment systems. Molecular diffusion is the process in which dissolved contaminants move from areas of high concentration to areas of low concentration in response to the presence of a concentration gradient. Diffusion can operate in the presence or absence of groundwater motion. Because the molecular forces driving diffusive transport are small in comparison to those driving rapid groundwater advection, diffusion is significant primarily in low-velocity hydrogeologic environments. Diffusion is also significant during contaminant movement by matrix diffusion into or out of a low hydraulic-conductivity clay lens in a sand aquifer. Such lenses could become contaminated from a long exposure to contaminated groundwater in the surrounding sand. A remedial action that could rapidly restore water quality in the permeable sand aquifer would be hampered as the direction of diffusion reverses and contaminants migrate back to the sand unit. Scenarios such as this require that diffusive transport be considered when designing a remedial action.

Mechanical dispersion is the component of transport that results from the convoluted paths that water and contaminant particles follow while flowing through porous and fractured media. Whereas advection is the component of transport resulting from the "average" flow of groundwater, mechanical dispersion is the component that is due to the spatial variation of flow paths and the variations in velocity that characterize groundwater movement at all scales. Any property of the porous media that creates velocity variations increases mechanical dispersion. Freeze and Cherry (1979) and Bear (1979) provide instructive illustrations of the nature of dispersion.

Mechanical dispersion serves to spread a contaminant plume over a greater area (both parallel and orthogonal to the hydraulic gradient) than would be occupied if only advection was occurring. This spreading dilutes contaminant concentrations as a plume moves through an aquifer.

As the heterogeneity of an aquifer increases, the irregularity of groundwater flow paths increases. The increase in flow path irregularity causes a subsequent increase in mechanical dispersion. In heterogeneous aquifers, hydrodynamic dispersion may approach advection in importance as a transport mechanism.

The degree of tortuosity of a porous medium administers some control on the magnitude of dispersion. As groundwater flows through a coarse uniform gravel, it will follow moderately irregular paths that are roughly parallel to the hydraulic gradient. Water flowing through a sparsely fractured rock mass, where the fracture interconnections are widely spaced, will follow some paths that are far from parallel to the average hydraulic gradient. The greater tortuosity of media such as fractured rock contributes to greater dispersion.

3. Retardation of Dissolved Solutes

Changes in concentration of dissolved solutes can occur because of chemical reactions that take place entirely in the aqueous phase, or because of the transfer of solute to or from the solid matrix of the porous medium. Among the reactions that can alter contaminant concentrations in groundwater are: (1) adsorption-desorption reactions, (2) acid-base reactions, (3) solution-precipitation reactions, (4) oxidation-reduction reactions, and (5) microbiological transformations. Of these, we will discuss only the first and the last.

The adsorption of dissolved solute onto aquifer matrix materials results in a reduction of concentration in the aqueous phase and a retardation of the velocity of contaminant migration relative to that of the groundwater flow. The higher the fraction of contaminant sorbed, the more retarded is its transport. If the sorptive reaction behaves linearly and is at equilibrium, the solute will move at an average velocity equal to the average linear velocity of the groundwater flow divided by a retardation factor, R_f.

The degree of retardation experienced by a particular organic contaminant in a particular aquifer will generally depend on the organic carbon content, f_{oc}, of the aquifer materials. The higher the organic carbon content, the more sites there are for adsorption. A greater number of adsorption sites leads to a higher retardation factor. Contaminants may also adsorb onto clay surfaces.

For an aquifer with a given organic carbon content, retardation is contaminant-dependent. More-hydrophobic compounds are more highly retarded than less-hydrophobic compounds. The degree of hydrophobicity is indicated by the octanol-water partition coefficient, K_{ow}. Tables of K_{ow} values are available for a wide range of potential groundwater contaminants (Verschueren, 1983), and several relationships that relate K_{ow}, f_{oc}, and R_f in a context applicable to contaminant hydrogeological studies, are also available (Freeze and Cherry, 1979; Mackay et al., 1985).

Retardation factors for the most commonly occurring organic compounds at contaminated groundwater sites generally fall in the range between 1 and 10. These contaminants would be expected to migrate at rates between 100 percent and 10 percent of the average linear velocity of the groundwater flow.

It is generally assumed that sorptive reactions are reversible; however, they may be slow. As contaminant concentrations in aquifer waters are decreased under remedial actions, the desorption process, whereby sorbed solutes reenter the aqueous phase, can create a lingering source of contaminants that delays complete remediation.

Sorption is not the only process that can lead to altered contaminant concentrations in aquifers. Organic contaminants can be biologically transformed into other compounds by microorganisms attached to solid surfaces within the aquifer. Some of these microorganisms thrive under aerobic conditions; others prefer anaerobic conditions. The efficiency of a particular biotransformation process is usually reported in terms of a half-life. Rates are believed to range widely with half-lives varying from a few days to many years.

Reduction of the concentration of a particular organic contaminant due to biological transformation does not ensure a reduction in toxicity. Many of the daughter products of transformation processes are as toxic, or more toxic, than their parents. Biotransformation of trichloroethylene, for example, can result in the formation of vinyl chloride, which is more toxic (and more stable) than its parent.

4. Facilitated Transport of Contaminants

Facilitated transport occurs when the mobility of a contaminant increases as a result of physical, chemical, or biological changes in the aquifer. Examples of facilitated transport include cosolvation, particle transport, and dissolution.

Cosolvation is the process by which the mobility of one contaminant is enhanced by the presence of a solvent. Hydrophobic organic contaminants (HOCs) such as PCBs have low solubility in water and tend to be almost immobile in cases where they are the only contaminant present. However, these HOCs are much more soluble when chlorinated solvents are present. Cosolvation is typically most significant near the source where solvent concentrations are likely to be high.

Particle transport occurs when flowing groundwater moves colloid-sized particles onto which contaminants are adsorbed or chemically bonded. Examples of colloids include natural and anthropogenic organic particles and macromolecules, and clays and metal oxides. Transport of these particles is limited by pores with smaller dimensions than the colloids. In aquifers with large pore sizes, particle transport may result in extensive movement of contaminants that would otherwise exhibit severe retardation.

Dissolution is a type of facilitated transport that occurs when changes in groundwater chemistry transfer contaminants from the solid phase to a dissolved phase. Changes in the pH or Eh of groundwater are common causes of dissolution. Another example of a chemical change leading to dissolution is the depletion of oxygen as a result of biological activity. This process can contribute to the desorption of redox-sensitive contaminants.

5. Contaminant Transport Equation

Consider a three-dimensional flow field in which groundwater is moving in the x_1, x_2, and x_3 directions with average linear velocity components V_1, V_2, and V_3 as defined in Equation (8). The subscripts 1, 2, and 3 represent the x, y, and z coordinate directions, respectively. The continuity equation for the mass flux of contaminants, F_i ($i = 1, 2, 3$), through an elemental control volume of porous media in a three-dimensional flow field is given by Freeze and Cherry (1979, Appendix 10) as

$$\frac{\partial F_1}{\partial x_1} + \frac{\partial F_2}{\partial x_2} + \frac{\partial F_3}{\partial x_3} + c'W = -n_e \frac{\partial c}{\partial t} \qquad (10)$$

where c' is the solute concentration in a source or sink fluid, W is the source or sink strength in units of volumetric flow rate per unit volume of porous media (positive for outflow), n_e is the effective porosity, and c is the solute concentration in the pore fluid. The mass flux of contaminants, F_i, is the mass of contaminants passing through a unit cross-sectional area per unit time (M/L^2T) in the x_i direction. F_i is given by

$$F_i = q_i c + J_i \qquad i = 1, 2, 3 \qquad (11)$$

where $q_i c$ is the contaminant flux due to advection in the x_i direction, and J_i is the contaminant flux due to hydrodynamic dispersion in the x_i direction. The dispersive flux is given by

$$J_i = -n_e D_{ij} \frac{\partial c}{\partial x_j} \qquad i, j = 1, 2, 3 \qquad (12)$$

where D_{ij} is the hydrodynamic dispersion coefficient and is equal to the sum of the coefficient of mechanical dispersion, D^m_{ij}, and the coefficient of molecular diffusion, D^*. It is important to note that this development of the transport equation has not accounted for the effects of retardation, facilitated transport, or any chemical reactions.

Substituting (12) into (11) and expressing q_i in terms of the average linear velocity, V_i [Equation (8)], the mass flux is expressed as

$$F_i = V_i n_e c - n_e D_{ij} \frac{\partial c}{\partial x_j}. \tag{13}$$

Writing Equation (10) in tensor form with (13) substituted for F_i gives the general form of the three-dimensional advection-dispersion equation:

$$\frac{\partial}{\partial x_i}\left(D_{ij}\frac{\partial c}{\partial x_j}\right) - \frac{\partial}{\partial x_i}(cV_i) - \frac{c'W}{n_e} = \frac{\partial c}{\partial t}. \tag{14}$$

For two-dimensional transport problems, Equation (14) can be integrated with respect to x_3 to produce the two-dimensional transport equation. Written in expanded form, this equation appears as

$$\frac{\partial}{\partial x_1}\left(bD_{11}\frac{\partial c}{\partial x_1} + bD_{12}\frac{\partial c}{\partial x_2}\right) + \frac{\partial}{\partial x_2}\left(bD_{21}\frac{\partial c}{\partial x_1} + bD_{22}\frac{\partial c}{\partial x_2}\right)$$

$$- \frac{\partial}{\partial x_1}(bcV_1) - \frac{\partial}{\partial x_2}(bcV_2) - \frac{c'Wb}{n_e} = \frac{\partial}{\partial t}(cb). \tag{15}$$

In general, the hydrodynamic dispersion tensor for three-dimensional transport problems, D_{ij} in (14) is a second-order tensor with 9 components. For two-dimensional problems represented by Equation (15), which assumes an isotropic porous media, the dispersion tensor consists of four coefficients. These coefficients are defined as

$$D_{11}\ (=D_{xx}) = D_L\frac{(V_1)^2}{|\mathbf{V}|^2} + D_T\frac{(V_2)^2}{|\mathbf{V}|^2} + D^* \tag{16}$$

$$D_{22}\ (=D_{yy}) = D_T\frac{(V_1)^2}{|\mathbf{V}|^2} + D_L\frac{(V_2)^2}{|\mathbf{V}|^2} + D^* \tag{17}$$

$$D_{12} = D_{21}\ (= D_{xy} = D_{yx}) = (D_L - D_T)\frac{V_1V_2}{|\mathbf{V}|^2} + D^* \tag{18}$$

where

$$D_L = \alpha_L\ |\mathbf{V}| \tag{19}$$

$$D_T = \alpha_T\ |\mathbf{V}| \tag{20}$$

and $\mathbf{V}$ is the average linear velocity vector. D_L and D_T are the longitudinal and transverse components of the coefficient of mechanical dispersion $[L^2/T]$. The factors α_L and α_T are referred to as the longitudinal and transverse dispersivities, respectively, and have units of length [L].

For a uniform flow problem in which the x_1 or x-axis of the coordinate system is aligned with the direction of flow and in which diffusion is negligible compared to mechanical dispersion, the dispersion coefficients D_{ij} for $i \neq j$ are zero in (15) and (18). In this situation $V_1 = V_x = |\mathbf{V}|$ and $V_2 = 0$. The dispersion coefficients remaining in (15) are then obtained by solving (16) and (17) with $V_1 = |\mathbf{V}|$ and $V_2 = 0$.

$$D_{11} = D_L \frac{(V_1)^2}{|\mathbf{V}|^2} = D_L. \tag{21}$$

$$D_{22} = D_T \frac{(V_1)^2}{|\mathbf{V}|^2} = D_T. \tag{22}$$

Solving either (14) or (15) requires the solution of an appropriate flow equation in order to first define the velocity field. The existence of either uniform or steady flow conditions will simplify this task considerably.

A solution of (14) provides a prediction of the three-dimensional concentration pattern, c(x, y, z, t). Solution of (15) provides the predicted concentration pattern in two dimensions, c(x, y, t). Either (14) or (15) can be adapted to account for the retardation of dissolved contaminants by the addition of a retardation factor. Readers are directed to Freeze and Cherry (1979) or de Marsily (1986) for details.

In most "real-world" transport problems, the flow field is not uniform and steady. Heterogeneous geological environments create nonuniform flow, and time-dependent changes in boundary conditions and system stresses create transient flow. In such cases, the average linear velocity, V, is a function of position and time. If we can calculate the hydraulic-head pattern, h(x,y,t), in an aquifer, then we can calculate $V_x(x,y,t)$ and $V_y(x,y,t)$ through Equations (7) and (8). Prediction of the hydraulic-head pattern requires solution of the flow equation, which is now briefly developed.

The continuity equation for the mass flux of water through an elemental control volume in a two-dimensional representation of an aquifer in a transient flow field is given by Freeze and Cherry (1979) as

$$\frac{\partial}{\partial x}(\rho q_x) + \frac{\partial}{\partial y}(\rho q_y) = \rho S_s \frac{\partial h}{\partial t} \tag{23}$$

where ρ is the density of water, S_s is the specific storage of the porous media [as defined by Equation (4)], and q_x and q_y are components of the specific discharge vector. Expanding the terms on the left-hand side by the chain rule and recognizing that terms of the form $\rho\partial q_x/\partial x$ are much greater than terms of the form $q_x\partial\rho/\partial x$ allows us to eliminate ρ from both sides of (23). Inserting Darcy's law, Equation (7), we obtain:

$$\frac{\partial}{\partial x}\left(K_{xx}\frac{\partial h}{\partial x}\right) + \frac{\partial}{\partial x}\left(K_{xy}\frac{\partial h}{\partial y}\right) + \frac{\partial}{\partial y}\left(K_{yx}\frac{\partial h}{\partial x}\right) + \frac{\partial}{\partial y}\left(K_{yy}\frac{\partial h}{\partial y}\right) = S_s\frac{\partial h}{\partial t}. \qquad (24)$$

Recalling the definitions of transmissivity, T [Equation (3)], and storativity S [Equation (5)], and allowing for recharge to the aquifer at a rate W′, (24) becomes:

$$\frac{\partial}{\partial x}\left(T_{xx}\frac{\partial h}{\partial x}\right) + \frac{\partial}{\partial x}\left(T_{xy}\frac{\partial h}{\partial y}\right) + \frac{\partial}{\partial y}\left(T_{yx}\frac{\partial h}{\partial x}\right) + \frac{\partial}{\partial y}\left(T_{yy}\frac{\partial h}{\partial y}\right) = S\frac{\partial h}{\partial t} + W'. \qquad (25)$$

Given the principal components of the transmissivity tensor, T_{xx} and T_{yy}, the storativity, S, and the recharge rate, W′, one can solve (25) for h(x,y,t). W′ is the source/sink rate in units of length per time.

It is possible to write (25) in indicial notation. In this case, the coordinate directions become (x_1, x_2) instead of (x, y), and the transmissivity terms T_{xx} and T_{yy} are recognized as the diagonal term in a tensor representation of T. In indicial notation, (25) becomes:

$$\frac{\partial}{\partial x_i}\left(T_{ij}\frac{\partial h}{\partial x_j}\right) = S\frac{\partial h}{\partial t} + W' \qquad i, j = 1,2 \qquad (26)$$

It is this form of the flow equation that is used in the simulation-optimization presentation in Chapter V.

A full solution for advective-dispersive transport in a two-dimensional, horizontal aquifer under nonuniform, transient flow requires the coupled solution of Equations (14) and (26).

If the groundwater flow field is steady rather than transient, and if recharge W′ is negligible, then the nonuniform flow field in heterogeneous media is described by Equation (25) with the right-hand side set to zero:

$$\frac{\partial}{\partial x}\left(T_{xx}\frac{\partial h}{\partial x}\right) + \frac{\partial}{\partial x}\left(T_{xy}\frac{\partial h}{\partial y}\right) + \frac{\partial}{\partial y}\left(T_{yx}\frac{\partial h}{\partial x}\right) + \frac{\partial}{\partial y}\left(T_{yy}\frac{\partial h}{\partial y}\right) = 0. \qquad (27)$$

A full solution for advective-dispersive transport in this case requires the coupled solution of Equations (14) and (26).

For heterogeneous anisotropic media in which $T_{xx} = T_{xx}(x,y)$, $T_{yy} =$

$T_{yy}(x,y)$, $T_{xy} = T_{xy}(x,y) = T_{yx}(x,y)$, and $S = S(x,y)$, these equations cannot be solved analytically, and it has become standard practice to use numerical models of the kind described in Chapter II.H.

In most moderately uniform, unconsolidated, sand-and-gravel aquifers, advective transport overwhelms the effects of dispersion. This is especially true under the high gradients that are developed around pumping wells and injection wells in pump-and-treat remedial schemes. Under these circumstances, it may not be necessary to solve the transport equation (15) at all. It may be sufficient to solve only the flow Equation (25) or (27). In advection-controlled systems, the solution, h(x,y,t) or h(x,y), provides all the information needed to define hydraulic gradients, flow directions, flow paths, and flow velocities; and under the assumption that the solutes are carried by the flow without dilution or retardation, the flow information defines the concentration patterns. With the flow solution one can approximately define contaminant pathlines, plume-boundary locations, plume-front migration velocities, travel times, capture zones, and contaminant-mass flux rates into remedial extraction wells. Chapters IV and V will show how flow models alone (and then how flow models coupled to optimization models) can be used to design capture and containment systems.

G. Aquifer Hydraulics

1. Transient Drawdown Cones Around Withdrawal Wells

There is probably no better known hydrogeologic theory than that associated with the cone of drawdown that develops in aquifer heads around a pumped well. The theory is fully explored in most groundwater texts (cf. Freeze and Cherry, 1979; Todd, 1980). We present only a brief review here.

Consider a homogeneous, isotropic confined aquifer at depth with transmissivity, T, and storativity, S. Furthermore, consider that this idealized aquifer is of constant thickness and infinite areal extent. Assume static initial conditions, with hydraulic head, $h(x,y) = h_0$, throughout the aquifer. A fully penetrating, constant-discharge well located at position (x_0,y_0) begins withdrawing water from the aquifer at a rate Q at time $t = 0$. We are interested in predicting the drawdown s(x,y,t) at any given time, t, where the drawdown is defined as

$$s(x,y,t) = h_0 - h(x, y, t). \tag{28}$$

For the idealized case we have introduced, there is a radial symmetry to the flow system that develops around the withdrawal well. If we set the well at $x_0 = 0$, $y_0 = 0$, then we can define a radial distance, r, where

$$r = \sqrt{x^2 + y^2} \tag{29}$$

and we can try to predict h(r,t). The h(r,t) values can easily be converted to values of h(x,y,t) through (29), and to drawdown values, s(x,y,t), through (28).

The equation that describes the transient flow of water in a heterogeneous, anisotropic aquifer is given in Cartesian coordinates as Equation (25). For a homogeneous, isotropic aquifer in radial coordinates, this equation becomes (Freeze and Cherry, 1979)

$$\frac{\partial^2 h}{\partial r^2} + \frac{1}{r}\frac{\partial h}{\partial r} = \frac{S}{T}\frac{\partial h}{\partial t} + W'. \tag{30}$$

The analytical solution to this partial differential equation, subject to static initial conditions, and boundary conditions representative of a well pumping at a constant rate Q at $r = 0$, was developed by Theis in 1935 and is known as the Theis solution:

$$h\,(r,t) = h_0 - \frac{Q}{4\pi T}\,U(u), \tag{31}$$

where

$$u = \frac{r^2 S}{4tT}. \tag{32}$$

Tables of the well function, U(u), are included in all the groundwater texts referenced above.

For the idealized system described thus far, Equations (31) and (32) and a table of U(u) versus u are all that is required to predict h(r,t), h(x,y,t), or s(x,y,t), given values for T, S, and Q.

However, the assumptions underlying the Theis solution are quite severe. Let us review them:

1. The aquifer must be homogeneous and isotropic with constant T and S.
2. The aquifer must be infinite in horizontal extent.
3. The aquifer must be fully confined.
4. There is only one well, and it must pump at a constant rate Q.

5. The well must fully penetrate the confined aquifer.
6. The initial conditions must be static.

It is possible to remove many of these assumptions and still develop closed-form analytical solutions that can be used to predict drawdown cones around pumping wells. Freeze and Cherry (1979) review the available solutions for: (1) bounded aquifers, (2) leaky confined aquifers, (3) unconfined aquifers, (4) multiple pumping wells, (5) nonconstant pumping rates, and (6) partially penetrating pumping wells. These more-complex analytical solutions will not be presented here.

While analytical solutions are invaluable in carrying out scoping calculations on idealized representations of actual aquifers, defensible predictions of drawdown for real aquifers usually require consideration of the stratigraphic layering of aquifer-aquitard systems, the areal heterogeneity within aquifers, and the possible presence of anisotropic conditions. Under these complexities, analytical solutions cannot be obtained, and hydrogeologists must turn to finite-element or finite-difference numerical models. Such models have been in existence since the early 1960s and have now become just as much a part of standard hydrogeological practice as is the Theis equation. The U.S. Geological Survey model, MODFLOW (McDonald and Harbaugh, 1984), has become in many ways the industry standard. Its two-dimensional predecessor by Trescott, Pinder, and Larson (1976) has also been used extensively. Suffice it to say here that MODFLOW or any one of dozens of available competing numerical flow models, can provide predictions of drawdown cones in complex two-dimensional or three-dimensional hydrogeological systems. They can do so for heterogeneous, anisotropic conditions; in confined, leaky, or unconfined aquifers; and with any number of pumping wells operating on any pumping-rate schedule.

Most remedial pump-and-treat schemes involve more than one pumping well. Because Equation (25) is linear, the computed drawdowns, s(x,y,t), due to each of the wells in a multiple-well system may be superimposed on one another. The total drawdown at any location (x,y) in an aquifer is the sum of all the drawdowns that would be created at that point by each of the wells if they were each pumping alone.

Injection wells create buildup cones in aquifer heads. A buildup cone is the exact mirror image of a drawdown cone. The buildup s(x,y,t) due to injection at rate Q is identical to the "negative drawdown" that would be calculated for pumping at rate –Q. In systems with both pumping wells and injection wells, the computed drawdowns and buildups are also superimposed upon one another.

2. Steady-State Drawdown Cones Around Withdrawal Wells

Given sufficient time, the drawdown cone that develops around a withdrawal well in an aquifer will approach a steady-state configuration. At steady state, the well will no longer be drawing water from storage in the aquifer; rather, the pumping rate will be in equilibrium with all inflows and outflows, including those from the aquifer boundaries, from recharge through the unsaturated zone, and/or from leakage through overlying or underlying aquitards. In large low-permeability aquifers, steady-state conditions may take years to attain; in smaller high-permeability aquifers, steady-state conditions may be achieved in a matter of days, especially if the pumping centers are located near constant-head boundaries in the form of rivers or lakes.

The equation that describes the steady-state flow of water in a homogeneous, isotropic aquifer in radial coordinates is obtained by setting the right-hand side of Equation (30) equal to zero:

$$\frac{\partial^2 h}{\partial r^2} + \frac{1}{r}\frac{\partial h}{\partial r} = 0. \qquad (33)$$

The analytical solution to this equation (Todd, 1980) was developed by Thiem in 1906 and is known as the Thiem equation. If there is a well pumping at $r = 0$ at a rate Q, and if there is a constant-head boundary at $r = r_0$ with $h = h_0$, then the head at any radial distance from the pumping well is given by

$$h(r) = h_0 - \frac{Q}{2\pi T}\ln\left(\frac{r_0}{r}\right).$$

Given values of r_0, h_0, Q, and T, one can calculate h(r) with (34). Head values can be converted to drawdown values, s(r), by

$$s(r) = h_0 - h(r). \qquad (35)$$

As with transient drawdown calculations, analytical expressions for steady-state drawdown patterns are limited to simple cases that meet the six assumptions listed in the previous subsection. For complex heterogeneous, anisotropic hydrogeological environments, nonsymmetric flow regions and boundary conditions, and/or multiple extraction and injection wells, hydrogeologists turn to numerical models.

Any finite-difference or finite-element numerical model requires a discretization of the aquifer into a grid of n blocks or elements, each of which may have its own hydrogeological properties. The partial differential Equation (25) or (27) is then replaced by a set of n algebraic finite-difference equations, one equation for each block. The hydraulic head values at the center of each block constitute the unknowns. Solution of the n equations in n unknowns provides a map of the hydraulic head field (or the drawdown) in the aquifer. The set of equations that must be solved are usually laid out in matrix form:

$$\mathbf{A}\,\mathbf{H} = \mathbf{B} \tag{36}$$

where **A** is an n × n matrix of finite-difference coefficients that includes consideration of the system geometry and the hydrogeological parameters, **H** is a 1 × n vector of unknown heads, and **B** is a 1 × n vector of boundary conditions and pumping rates.

Readers not already familiar with finite-difference methods will probably not be much enlightened by this brief development. More complete explanations are available in Freeze and Cherry (1979), Wang and Anderson (1982), and Bear and Verruijt (1987). Equation (36) is introduced here because it makes a brief reappearance (in slightly different form) in Chapter V.G.2 in the discussion of the embedding method of steady-state simulation optimization.

3. *Implications for Capture and Containment Systems*

If flow modeling is to be used for the design of well networks in capture and containment scenarios, then it is necessary to review the concepts of streamlines and pathlines (Chapter II.C.3) as they pertain to transient aquifer hydraulics.

Recall that in steady-state flow systems, hydraulic gradients do not change as a function of time, and therefore neither do streamlines. Under these circumstances, the streamlines of groundwater flow and the pathlines of contaminant migration are identical.

In transient flow systems, hydraulic gradients change as a function of time as do streamlines. Under these circumstances, streamlines have significance only in the sense that they provide a snapshot of the changing flow system at a particular moment in time. Pathlines are tangent to the streamlines at any moment in time, but they reflect the movement of the water and solute particles as they change directions from the current streamline pattern to the emerging pattern for the next moment in time. Under these circumstance, streamlines and pathlines do not coincide.

In the design of pumping networks for contaminant capture, our primary interest lies in identifying the pathlines that are diverted into the extraction wells. It is our desire to design an extraction-well network that will capture all contaminated pathlines. If we try to treat the transient period of aquifer response, this can prove difficult. However, in high-permeability aquifers, for certain sets of boundary and initial conditions, it is likely that the transient flow conditions created by the initiation of pumping will quickly converge to a steady state. Once steady-state flow is achieved, pathlines and streamlines become identical, and the analysis of capture zones becomes more straightforward.

One of the basic cases for analysis is that of an extraction well operating in a steady, uniform flow field. In other words, instead of static initial conditions with $h = h_0$ everywhere in the aquifer, the initial conditions in this case reflect a uniform gradient across the site. Under these circumstances, once a new steady state is achieved, some of the streamlines will be diverted into the extraction well and some will not. Those that are diverted define the steady-state capture zone. This concept has considerable significance in the design of pumping-well networks and is discussed in detail in Chapter IV.

H. Computer Models

Designing an efficient system for the control of contaminated groundwater requires an understanding of the contaminant distribution and transport mechanisms within an aquifer. It is also necessary to carry out a quantitative evaluation of the impact that alternative capture system designs will have on contaminant transport. In practice, both tasks are almost always attempted through the use of models of the groundwater system.

A mathematical model consists of an equation or system of equations that mathematically describe the processes of interest, in this case groundwater flow and contaminant transport. Obtaining output from the model involves solving the equations subject to the boundary and initial conditions imposed by the physical system being modelled. The variety of models and solution techniques available for groundwater studies is wide, as is the range of complexity and the degree of accuracy that can be anticipated.

1. Flow, Transport, and Geochemical Speciation Models

There are three types of models that might be applied to groundwater contamination problems: flow models, transport models, and geochemical speciation models.

Flow models are used to study groundwater movement. For a two-dimensional, horizontal aquifer, the flow model is based on a solution to Equation (25) or (27). A flow model has only one state variable: the hydraulic head, h(x,y,t). Solutions to the flow model provide predictions of h as a function of space and time. The required input is in the form of boundary and initial conditions on h, the system stresses, and the aquifer properties. The latter may be expressed in terms of hydraulic conductivity, K(x,y), and specific storage, $S_s(x,y)$; or in terms of the transmissivity, T(x,y), and storage coefficient, S(x,y).

Transport models have two components: a flow model and a model of the solute transport process. The first is as described above; the second is based on a solution to Equation (15), perhaps with a retardation factor included. A transport model has two state variables: the hydraulic head, h(x,y,t), and the solute concentration c(x,y,t). In many cases, the flow model is restricted to steady-state flow so that h is a function only of space. Solutions by the transport model provide predictions of c as a function of space and time. Then, the required input is in the form of boundary conditions on h, boundary conditions and initial conditions on c, the system stresses, and the aquifer properties. In addition to T(x,y) and $S_s(x,y)$, the required aquifer properties included the longitudinal and transverse dispersivities, α_L and α_T; the diffusion coefficient, D^*; the effective porosity, n_e; and some measure of retardation for the solute.

Most solute-transport models explicitly consider only a single chemical component. Geochemical speciation models, on the other hand, consider the multicomponent chemical interactions in a static fluid. In general, speciation models calculate the chemical mass distribution between dissolved, adsorbed, and precipitated species. Most speciation models use a combination of equilibrium expressions for each species and mass-balance equations for each chemical component to completely describe the chemical state of the system.

The main differences between chemical speciation computer codes are the selection of input variables and the ease of performance of various tasks, such as reaction-path calculations or free-energy minimization. For example, WATEQ (Truesdell and Jones, 1974) uses the concentrations of aqueous components as input and calculates only the concentrations of aqueous species. PHREEQE (Parkhurst et al., 1980) extends the capabilities of WATEQ by allowing calculations of equilibration with mineral species. MINEQL (Westall et al., 1976) uses "total" (solid and aqueous) concentrations of components and calculates the concentrations of aqueous and solid species of the most stable system.

The processes of groundwater flow, contaminant transport, and chemical reaction occur interactively and not as separate distinct processes. In an effort to address this fact, there is increasing emphasis on the development and application of coupled models that treat these three processes in an integrated fashion. Coupled models solve the governing equations for the flow, transport, and chemical-reaction processes simultaneously. Uncoupled models solve these equations separately or sequentially. Coupled groundwater-flow/solute-transport/ geochemistry models are extremely complicated. The volume and variety of input data required to accurately apply such a model to an actual field problem are staggering. The complexity of coupled models and the required volume of data, combined with the limitations on financial and personnel resources associated with most practical groundwater contamination problems, has limited the use of fully coupled flow/transport/ geochemistry models. Yeh and Tripathi (1989) provide additional discussion of these models in a comprehensive review of coupled "hydrogeochemical transport models."

The application of solute-transport models that couple a flow equation and a solute concentration equation can also be difficult in a field environment. Obtaining dispersivities and retardation factors is difficult. If advection is thought to be the dominant transport process, then a flow model alone can be used to predict the plug-flow advance of a contaminant-plume boundary. This advective approach to transport simulation is widely used in practice. It is particularly well-suited to the use of models in the design of contaminant control systems in high-permeability aquifers, where the pumping process creates high gradients in which advective transport is often the most critical process. In Chapters IV and V, the optimization techniques are presented for advection-based design analysis.

2. Analytical, Semianalytical, and Numerical Solution Techniques

Flow and transport models used in groundwater studies can also be classified according to the method used to solve the governing equations that constitute the model. Solution techniques commonly applied in groundwater problems include analytical, semianalytical, and numerical methods.

Analytical solutions are those which can be found in the form of exact, closed-form solutions to the governing partial differential equations. An analytical solution provides the spatial and temporal distribution of hydraulic head or concentration for a given set of initial and boundary conditions.

Analytical solutions are attractive because the process of interest can be analyzed by simply evaluating the analytical solution using appropriate estimates of the hydrogeological parameters. Analytical solutions are limited to simple problems involving either radial flow or steady uniform flow. In general, analytical solutions are not applicable to flow or transport problems involving highly irregular boundaries or highly heterogeneous or anisotropic porous media. Javandel et al. (1984) review several analytical solutions for groundwater flow and transport.

In contaminant capture design, analytical solutions can be used to develop preliminary conceptual models of the real flow system. Such conceptual models can serve as the basis for collecting additional data and developing a more accurate numerical model for use in the design of the hydraulic control action.

Semianalytical techniques, as defined by Javandel et al. (1984), use the complex velocity potential to mathematically describe the flow system, combined with numerical analysis techniques to produce graphical representations (flow nets) of the flow system. Because this approach relies on the concept of flowlines for the representation of groundwater flow and solute transport, the technique is limited to steady-state two-dimensional flow and transport problems. In practice, this technique is further restricted to problems involving uniform regional flow with non-uniform components in the vicinity of sources and sinks (remedial action wells). Despite these restrictions, semianalytical techniques can provide useful information, especially at the scoping stage of a capture design problem. Javandel et al. (1984) provide a summary of semianalytical solutions for several solute transport and remedial action design problems.

For relatively simple groundwater contamination problems, semianalytical methods are an inexpensive and simple tool for the design of groundwater control actions. For more complicated contamination problems or those requiring more thorough remedial action design, semianalytical techniques can be used to economically generate design alternatives for further analysis with optimization techniques. Chapter IV provides a discussion of the use of semianalytical techniques as a tool for preliminary capture and containment design.

Numerical solution techniques currently have the widest application in groundwater studies. We do not develop the theory here. Briefly, the common factor of all numerical methods is that the differential equations for groundwater flow and solute transport are solved after being transformed into algebraic approximations of the original equations. The algebraic approximations are solved for the variable of interest (head or concentration) at every point of a mesh that encompasses the entire study area. The hydrogeologic parameters involved are assigned unique values for each point in this mesh. This capacity to incorporate realistic spatial and temporal variability and realistic geometry is the major advantage of numerical models. The interested reader is directed to one of the many specialized texts on the subject (Wang and Anderson, 1982; Huyakorn and Pinder, 1983; Bear and Verruijt, 1987). A catalog of numerical models used for groundwater investigations is provided by van der Heijde et al. (1985).

In groundwater remediation problems, numerical models are applied in all phases, from initial development of a conceptual flow model to final design of a remedial system. In Chapter V, numerical models are the vehicles used to apply optimization techniques to groundwater remediation problems.

3. Optimization Models

In the course of designing a capture and containment system, the designer may consider several alternative system designs. In the absence of an objective procedure for selecting the best design, multiple simulations are performed to assess the performance of each design with regard to sometimes vague evaluation criteria. The simulation that best satisfies the evaluation criteria is then chosen as the preferred design.

Such a practice is essentially a process of subjective optimization. The chosen alternative is the one that is in some sense optimal, or "the best," in the designer's mind. Applying linear or quadratic programming techniques in an aquifer management model can make this process less subjective.

Optimization, as applied to the design of pumping systems, involves defining an objective function and then finding the set of decision variables that maximize or minimize the value of this objective function. A set of restrictions, or constraints, specify various conditions that must be satisfied by the optimal solution alternative. The constraints allow the designer to incorporate physical limitations of the flow system and performance criteria for the design into the solution.

Gorelick (1983, 1988) provides comprehensive reviews of optimization techniques applied to groundwater quality-management modelling. Chapter V provides a detailed discussion of optimization theory applied specifically to the task of designing efficient groundwater control systems. Chapters III and IV discuss those aspects of system design that must be addressed prior to simulation-optimization modelling. Chapters VI and VII address those aspects of a design task that are not easily handled by traditional simulation-optimization modelling.

References for Chapter II

Bear, J., *Dynamics of Fluids in Porous Media*, Dover Publications, Inc., New York, NY, 1972.

______, *Hydraulics of Groundwater*, McGraw-Hill, Inc., New York, NY, 1979.

Bear, J., and A. Verruijt, *Modeling Groundwater Flow and Pollution,* Reidel Publ., 1987.

Corey, A. T., *Mechanics of Immiscible Fluids in Porous Media*, Water Resources Publications, 1986.

Davis, S. N., and R. J. M. De Wiest, *Hydrogeology,* John Wiley and Sons, Inc., Wiley, New York, NY, 1966.

de Marsily, G., *Quantitative Hydrogeology, Groundwater Hydrology for Engineers*, Academic Press, Inc., Orlando, FL, 1986.

Domenico and Schwartz, *Physical and Chemical Hydrogeology,* John Wiley and Sons, Inc., 1990, 824 pp.

Freeze, R. A., and J. A. Cherry, *Groundwater*, Prentice-Hall, Inc., Englewood Cliffs, NJ, 1979.

Garrels, R. M., and C. L. Christ, *Solutions, Minerals and Equilibria*, Harper and Row, New York, NY, 1965.

Gillham, R. W., "The Capillary Fringe and Its Effect on Water Table Response," *Journal of Hydrology*, Vol. 67, pp. 307–324, 1984.

Gorelick, S. M., "A Review of Distributed Parameter Groundwater Management Modelling Methods," *Water Resources Research*, Vol. 19, No. 2, 1983.

______, "Incorporating Assurance Into Groundwater Quality Management Models in Groundwater Flow and Quality Modelling," edited by E. Custodio, A. Gurgui, and J. P. Lobo Ferreira, NATA ASI Series, *Mathematical and Physical Sciences*, Vol. 224, pp. 135–150, 1988.

Huyakorn, P. S., and G. F. Pinder, *Computational Methods in Subsurface Flow,* Academic Press, 1983.

Javandel, I., C. Doughty, and C. F. Tsang, *Groundwater Transport: Handbook of Mathematical Models*, American Geophysical Union, Water Resources Monograph 10, 1984.

Mackay, D. M., P. V. Roberts and J. A. Cherry, "Transport of Organic Contaminants in Groundwater," *Environmental Science Technology,* Vol. 19, p. 384– 392, 1985.

McDonald, M. G., and A. W. Harbaugh, *A Modular Three-Dimensional Finite-Difference Ground-Water Flow Model,* U. S. Geological Survey, 1984.

National Institute for Occupational Safety and Health, *Pocket Guide to Chemical Hazards*, Department of Health and Human Services

(NIOSH) Publication No. 85–114, U.S. Government Printing Office, Washington, D.C., 1985.

Parkhurst, D. L., D. C. Thorstenson, and L. N. Plummer, *PHREEQE, A Computer Program for Geochemical Calculations*, U.S. Geological Survey, Water-Resources Investigations Report No. 80–09, 1980.

Stumm, W., and J. J. Morgan, *Aquatic Chemistry,* Wiley, 1970.

Todd, D. K., *Groundwater Hydrology*, Second Edition, John Wiley and Sons, Inc., New York, NY, 1980.

Trescott, P. C., G. F. Pinder, and S. P. Larson, *Finite-Difference Model for Aquifer Simulation in Two Dimensions With Results of Numerical Experiments,* U.S. Geological Survey Techniques of Water Resources Investigations, Book 7, Chapter C1, 1976.

Truesdell, A. H., and B. F. Jones, "WATEQ, A Computer Program for Calculating Chemical Equilibria of Natural Waters," *Journal Research U.S. Geological Survey*, 2(2), 1974.

van der Heijde, P., Y. Bachmat, J. Bredehoeft, B. Andrews, D. Holtz, and S. Sebastian, *Groundwater Management: The Use of Numerical Models*, American Geophysical Union, Water Resources Monograph 5, 1985.

Verschueren, K., *Handbook of Environmental Data on Organic Chemicals*, Second Edition, Van Nostrand Reinhold Company, New York, NY, 1983.

Wang, H. F., and Mary P. Anderson, *Introduction to Groundwater Modeling,* W. H. Freeman, 1982.

Weast, R. C., and M. J. Astle (eds.), *CRC Handbook of Chemistry and Physics*, 60th Edition, CRC Press, Inc., Boca Raton, FL, 1979.

Westall, J. C., J. L. Zachary, and F. M. M. Morel, *MINEQL, A Computer Program for the Calculation of Chemical Equilibrium Composition of Aqueous Systems*, Ralph M. Parsons Laboratory for Water Resources and Environmental Engineering, Department of Civil Engineering, Massachusetts Institute of Technology, Technical Note No. 18, 1976.

Yeh, G. T., and V. S. Tripathi, "A Critical Evaluation of Recent Developments in Hydrogeochemical Transport Models of Reactive Multichemical Components," *Water Resources Research*, Vol. 25, No. 1, pp. 93–108, 1989.

CHAPTER III

Data Needs and Quality Control

Systems for the control of contaminated groundwater are expensive to design, construct, and operate. The success of these systems relies on a thorough understanding of the hydrogeologic environment and how contaminants behave within that environment. Failure to understand the groundwater flow and contaminant transport processes at a site may lead to expensive efforts to redesign ineffective systems. Improperly designed strategies may even worsen a contamination problem. Minimizing the costs and maximizing the probability for success of a groundwater control action require that adequate data of high quality be collected to characterize the hydrogeological system, the contaminant behavior within that system, and the response of the entire system to a given action.

This chapter summarizes the types of data required in the design of capture and containment actions, together with recommendations and references on measurement methods and quality control. The data types are divided into (1) geographical, geological, and geophysical data; (2) hydrogeologic parameters; (3) hydrogeologic variables; (4) hydrochemical parameters; (5) hydrochemical variables; and (6) fluid and contaminant properties. In this classification, *parameters* are time-independent porous-media properties such as hydraulic conductivity, and *variables* are potentially time-dependent indicators of the system state such as hydraulic head or contaminant concentration.

Before the discussion of each data type, we present a general introduction of the objectives of data collection programs.

A. Data Uses, Scales, and Quality

1. Intended Uses

The objectives of collecting field data at a contaminated site are determined mainly by the intended use. Data may be required for (1) site characterization, (2) model development and calibration, (3) remedial systems design, (4) operational performance evaluation, (5) early warning detection of unanticipated contaminant migration, or (6) operation and post-closure regulatory compliance. The first three items in this list serve the needs of remedial design. The last three items in the list serve the needs of monitoring and performance assessment.

Site characterization is an iterative process that proceeds from early screening efforts through increasingly more detailed data collection efforts to a comprehensive characterization of the site that is suitable for remedial design. It usually includes the measurement of physical and chemical entities, state variables, and hydrogeological parameters.

The design of data collection networks for site characterization must go hand-in-hand with the selection of modelling strategies for groundwater control. The field measurement network must be established in such a way that it will deliver data that are compatible with design modelling needs. This is an iterative process with the results of field measurements leading to improvements in the site model, and the results of site modelling providing direction on additional data needs.

In the past, site characterization studies seldom included sufficiently detailed test plans for hydrogeological parameters and key hydrogeologic features. Emphasis was often placed almost totally on plume delineation. As a result, the remedy selection process often went forward without the benefit of information that was vital to the proper evaluation of the potential for success of various possible remedial actions. Delineation of plume extent is important, but undue emphasis on contaminant concentrations alone, at the expense of the physical parameters that control pump-and-treat design, is unwise. Detailed hydraulic testing, bench treatability studies, and pilot operations of individual extraction wells should begin during the site characterization period.

In operational and post-closure data collection efforts, it is proper to place emphasis on the state variables: hydraulic head and contaminant concentration. Measurements of hydraulic heads and hydraulic gradients can be used to assess the performance of pump-and-treat systems and to provide data for active management of the system through increasing or decreasing pumping rates. Measurement of contaminant concentrations can be used to provide early warning detection of anticipated contaminant migration and/or to satisfy regulatory compliance. Further discussions of monitoring network design for performance evaluation are included in Chapter VII.

2. Scale

The relationship between the scale at which a measurement of a particular variable or parameter is made and the scale at which it is to be applied is an important concern (Anderson, 1979; Mercer et al., 1982). Scale of measurement and scale of application are characteristics of virtually any data value used in site characterization or remedial design. From a practical perspective, the greatest concerns regarding scale are commonly associated with hydraulic conductivity and contaminant concentration data. This discussion of scale will be restricted to these two entities. In both cases, one can conceive of various measurement scales, which are representative of volumes up to a few tens of cubic meters; and various application scales, which are normally much larger than the measurement scales and range from site scale up to regional scale.

Let us look at hydraulic conductivity first. The measurement methods outlined in Chapter III.C.1 range from laboratory measurements based on grain-size analysis on cores to multiple-well pumping tests in aquifers. These various test methodologies do not often yield the same result when performed at the same location. There are many reasons for this, but the overriding factor has to do with scale. It has been shown, for example, that laboratory measurements of the hydraulic conductivities of clay and silt samples tend to be two to four orders of magnitude lower than the values determined by field tests. They also tend to have a much greater variability (Chapuis, 1989; Klotz et al., 1980). Not only are such samples compacted during drilling, during recovery from the core barrel, and during insertion into the testing device, but the volume of the sample is too small to guarantee intersection of the few fractures or sand stringers that often conduct most of the flow through the stratum from which the sample was obtained.

Depending on the magnitude of the hydraulic stress created, the value of hydraulic conductivity obtained from a field test may vary considerably. Small-volume tests, such as slug and bail tests, may affect only a few cubic meters of the subsurface. As a result, it is not unusual for the hydraulic-conductivity values obtained from slug tests of closely spaced wells to differ by several orders of magnitude. The results are highly sensitive to minor variations in well-construction details, such as the screen position, filter pack, and grouting. The testing locations may be effectively isolated despite their physical proximity to one another. Conversely, test methods that involve a larger volume of aquifer, such as pumping tests, often yield more consistent results when multiple tests of the same aquifer are conducted.

The most typical application for hydraulic conductivity values is as input to simulation models of the hydrogeologic system. The simulation model may be a simple analytical solution or a numerical model. Simulations may be carried out at the site scale for remedial design purposes or at regional scale for regulatory compliance assessment. In each case, the simulation model will require conductivity input for the domain or for each nodal block or finite element.

The question of how best to scale up small-scale measurement values to large-scale simulation values is not easy to answer. The emerging discipline of geostatistics provides the best framework for doing so, but a detailed development of geostatistical ideas is beyond our scope. An introduction to the concepts can be gleaned from Appendix A. A more complete presentation is available in de Marsily (1986). We will simply note here that hydraulic conductivity values measured in the field are believed to usually exhibit a lognormal distribution, so that the best estimate of a representative site-scale, nodal-block conductivity that can be obtained from a set of slug-test values is probably the mean of the log-conductivity values. The same would be true for obtaining a representative regional-scale nodal-block conductivity from a set of pumping-test values in an aquifer. If the scale of measurement and the scale of application are similar, as might be the case for pumping-test data and a site-scale model, it may be possible to apply the measured values directly to the model nodal blocks. The question then arises as to how to assign values at unmeasured blocks. Geostatistical kriging provides a defensible approach to interpolation (de Marsily, 1986). Other approaches such as Thiessen polygons or trend-surface analysis may also be suitable.

Perhaps of more importance than conductivity values, is the idealization required in a model of the hydrogeological stratigraphy and structure. This, too, is a function of scale, and is in many ways more an art than a science. It is assumed that readers have experience with hydrogeologic model development and may refer to practical texts on modelling, such as Wang and Anderson (1982) or Walton (1985) for advice.

Contaminant concentration measurements are also a function of scale. Measurements made on samples collected from boreholes with large open intervals represent a weighted average of the point concentrations that exist in the individual stratigraphic layers. They are likely to indicate concentrations that underestimate the actual maximum values. On the other hand they may lead a hydrologist to overestimate the vertical extent of contamination. Measurements of this type may be suitable as calibration data for regional-scale plume-migration prediction but must be used with caution for site-scale design.

For detailed site-scale modelling associated with remedial design, it is preferable to have concentration data that more faithfully reflect detailed subsurface contaminant distributions. Sampling from multiple-port piezometers or specific packed-off borehole intervals may be necessary to provide the scale of data needed for design analysis.

Bailers or other low-volume sampling devices are commonly used to obtain samples from fairly small-diameter monitoring wells for organic chemical analyses. However, there may be advantages in sampling discharging wells, such as public supply wells. For such wells, contaminant concentration levels are often a function of the time since pumping began. If this fact is exploited by the collection of successive samples as pumping begins and continues (e.g., chemical time-series sampling), then each successive sample represents groundwater from a part of the aquifer farther from the well than the samples obtained before it (Keely, 1982). In this sense, the overall effect is equivalent to increasing the number of point-sampled monitoring wells. Large gains in efficiency and economy can be realized by sampling in this manner because hundreds of point-sampled wells might be required to sample the same volume of aquifer as is sampled by a single pumping well. The volume of the aquifer sampled is important because sample concentrations are often presumed to represent the average chemical quality of fairly large volumes of the aquifer from which they were withdrawn. This occurs in numerical modelling efforts where grid blocks of several thousand cubic meters are assigned average concentration values. The arrival patterns of contaminants as a function of pumpage can also be examined to yield insights into the locations of contaminant sources and plumes, much as tracer tests are used to determine preferential avenues of flow. Nearby monitoring wells that lie within the cone of influence of a capture and containment system become useful for chemical time-series sampling, analogous to observation wells for drawdown measurements during aquifer tests.

The scale question also influences how one interpolates between measured concentrations and how one extrapolates contaminant trends in time. These decisions impact the selection of sampling locations and sampling frequencies. In a very general sense, this aspect of the scale problem can be addressed by recognizing the existence of three zones at most contaminated sites. These zones include (1) the region within the plume, (2) the region between the plume and a resource that must be protected, and (3) the region associated with the protected resource. Examples of resources requiring protection from a contaminant plume include a municipal well field or a water-supply reservoir.

The nominal plume bounds may encompass an area of a few hundred acres and an average depth of a few tens of feet. Within the plume, a relatively dense monitoring network is appropriate. Each monitoring location may require a cluster of monitoring wells; each monitoring well in the cluster would be constructed to sample a depth-specific zone and allow vertical delineation of the plume composition. By contrast, the areas that would be impacted if a remediation is not implemented at a site are often many thousands of meters downgradient and may encompass many thousands of acres. Occasionally, monitoring wells must be constructed within the potentially impacted areas because the existing wells are not properly located. If the existing wells are located properly, they can serve as monitoring points. Between the contaminant plume and the protected resource areas, the major data-collection objective is to provide an early warning if one fails to contain the plume. The number of monitoring points required to meet this objective decreases with the increasing distance from the plume. This results in a distribution of monitoring wells and hydrogeologic testing locations that is skewed toward a dense pattern immediately downgradient of the plume and a sparse pattern farther downgradient.

3. Quality Versus Quantity

Regardless of the data objectives, there eventually comes a point in the budgetary planning for any data collection where one faces a tradeoff between quality and quantity. Should one install more monitoring wells and settle for run-of-the-mill data, or make do with a lesser number of monitoring wells and perform sophisticated field tests? Despite many attempts to prove otherwise, the absolute rule is: choose quality over quantity. It is far better that excessive care is taken to retrieve a few reliable data values than to have a large number of near-meaningless data values. In groundwater contamination investigations, the descent from high quality to poor quality is made easily; sophisticated field methods are either performed properly or they are not, and data that lack credibility are useless. The economics of bad data collections are readily apparent. Not only can the magnitude and extent of the contamination problem be completely misunderstood, but the selected remedy can be so inappropriate that it actually worsens a contamination problem.

B. Geographical, Geological, and Geophysical Data

1. Geographical Data

Contaminant transport studies can benefit from information on topographic elevations, drainage patterns, and land use.

Topography and drainage are required to place the site within its regional hydrologic context. Water-table configurations frequently exist as a subdued replica of the topography, and in this sense topography acts as a primary control on regional and local groundwater flow systems. The flow systems, in turn, control potential contaminant migration routes. Topographic divides usually constitute upslope boundaries to local flow systems, and valleys and drainage courses often constitute downslope boundaries. Highlands usually represent recharge areas; streams, lakes, marshes and wetlands usually indicate discharge areas. Freeze and Cherry (1979) provide a more detailed description of topographic and hydrologic controls on groundwater flow systems.

Topographic data are most commonly available as elevation contour maps. These are available at large scales (contour intervals of 20 to 100 feet) from the U.S. Geological Survey, U.S. Forest Service, U.S. Army Corps of Engineers, and various state agencies. At the local/site scale, subdivision plans and city planning maps often provide contour intervals as small as 0.25 foot.

Because it is essential to accurately measure all water-level elevations with respect to a common datum, the available topographic maps are not suitable for defining this datum. Differences of only a few feet are comparable to the entire natural drop in water levels across many sites. The elevations of the tops of the monitoring wells must be surveyed by a professional surveyor to an accuracy ± 0.01 foot or better. Failure to achieve this level of accuracy will lead to errors in water-level elevations and the potential for misinterpretation of groundwater flow directions. Surveyed elevations should be checked periodically.

Aerial photographs taken at intervals of 5 to 10 years since the late 1940s are available for most urban areas. Aerial photographs, soils maps, land-use maps, and demographic maps allow the assessment of historic and current land cover, land use, and population density. Such general geographical information can be important in locating sources of contamination and in assessing the constraints that are likely to exist to the siting and construction of remediation facilities.

2. Geological Data

Differentiation between aquifers and aquitards, confined and unconfined aquifers, porous media and fractured rock, and saturated and unsaturated zones, as outlined in Chapter II.A, requires a clear understanding of the site-scale and regional-scale lithology, stratigraphy, and structure. Geologic information that can be used for such interpretation can be obtained both from published sources and from site investigation.

Geologic maps and reports prepared by the U.S. Geological Survey or state agencies are available for all of the United States. Unfortunately, bedrock geology and surficial geology are seldom covered in the same report. In addition, at most locations, more detailed information is available on bedrock geology than is available on surficial geology. At sites where bedrock is exposed at the surface or where contaminants have entered fractured bedrock aquifers, published maps and reports on bedrock geology can aid in delineating the regional stratigraphic and structural setting. In the more usual situation where unconsolidated surficial deposits of fluvial or glacial origin are of primary importance, it may be necessary to place greater interpretive emphasis on available physiographic maps, land-use maps, soils maps, and aerial photographs.

In many states, regulations require water-well drillers to file lithologic logs of drilled wells. In some cases, core repositories are maintained for core recovered from drillholes at larger projects. This type of archival information can be invaluable in interpreting the geologic environment at a contaminated site.

Even more helpful than basic geological information is the possible existence of interpreted hydro-geological reports. Such studies may have been carried out by the U.S. Geological Survey (USGS) and reported in USGS Water Supply Papers, Professional Papers, or Hydrologic Investigations Atlases; or they may reside in USGS Open File Reports. Many areas of the U.S. were the subject of Regional Aquifer Systems Analysis (RASA) studies conducted by the USGS during the 1980s. Many of these USGS sources are catalogued in *New Publications of the U.S. Geological Survey*, which is published monthly by the USGS. Other potential sources may include state geological surveys or water resources departments and university theses and dissertations.

Because it is seldom possible to fully interpret a site from published sources, it is almost always necessary to carry out an exploratory drilling program during site characterization. Lithologic and geophysical logging should be performed on all exploratory drill holes, and representative cores should be recovered from at least some of the holes. Split-barrel or diamond-drill core samples can be used to help define the stratigraphy at a site. Cross-hole correlations allow interpretations to be made of the thickness, extent, and continuity of aquifers and aquitards.

Aller et al. (1989) and Driscoll (1986) provide detailed information on drilling and sampling techniques for groundwater.

3. Geophysical Data

Stratigraphic interpretations can be enhanced by the use of borehole geophysics. In hydrogeological applications, emphasis is usually placed on the two electric logs (spontaneous potential and resistivity). E-logs often provide the most accurate detail for the selection of well-screen intervals and contaminant sampling ports. They can also be used to make quantitative estimates of formation porosity.

Keys (1968) suggests a more complete logging suite for hydrogeological purposes. He recommends a driller's log, a geologic log, a spontaneous potential log, a resistivity log, a natural-gamma log, and a caliper log. Keys and MacCary (1971) and Campbell and Lehr (1973) provide good summaries of these borehole geophysical techniques.

There are also surface geophysical techniques that are used to some extent in hydrogeological studies. The seismic refraction method can provide information on site stratigraphy. The electrical resistivity method, because it reflects formation water salinity, can be used to map plume boundaries if the differences between plume and background salinities are significant. The reader is directed to a standard geophysics text such as Dobrin and Savit (1988) for further information. Ayers (1989) provides a detailed case history of a seismic refraction application in a hydrogeological context.

In recent years, new geophysical techniques such as ground-penetrating radar, and cross-hole tomography have been espoused in hydrogeological studies (Daily and Ramirez, 1989), but they represent emerging technologies that have yet to become routine in engineering applications. Magnetometer surveys can sometimes be used to aid in source location by identifying the presence of buried ferrous metals.

C. Hydrogeologic Parameters

As in the case of geologic data, hydrogeologic parameters can be obtained from both field investigations and previously published site-specific reports. Sources for previously published reports include the USGS and other Federal agencies, state geologic surveys and water resources departments, and university geology and civil engineering departments (theses and dissertations).

1. Hydraulic Conductivity and Transmissivity

The hydraulic conductivity, K, and the transmissivity, T, of an aquifer were defined in Chapter II.B.3. The two are related by

$$T = Kb \tag{37}$$

where b is the aquifer thickness. Assuming that a reasonable estimate is available for b, it is always possible to determine K from T, or T from K.

Aquifer transmissivity or conductivity is required for the assessment of advective contaminant transport and for the selection of locations and pumping rates for pumping wells in a pump-and-treat remedial design. They are fundamental hydrogeologic parameters, and it is necessary to measure or estimate them at virtually every site at which contaminant remediation is under consideration.

Apart from parameter estimation methods, there are five basic approaches to the determination of hydraulic conductivity: (1) estimates from grain-size analysis; (2) laboratory measurements on soil or rock core; (3) in situ, single-hole slug or bail tests; (4) in situ, single-hole packer tests; and (5) in situ, multiple-well pumping tests.

In the early scoping period of an investigation at a site where unconsolidated sands constitute the aquifer of interest, it may be possible to make preliminary estimates of hydraulic conductivity on the basis of grain-size gradation curves. A simple and apparently durable empirical relation, known as the Hazen approximation, predicts a power law relation between K and d_{10}:

$$K = Ad_{10}^2 \tag{38}$$

where d_{10} is the grain-size diameter at which 10 percent by weight of the soil particles are finer and 90 percent are coarser. For K in centimeters per second (cm/s) and d_{10} in millimeters (mm), the coefficient A is equal to 1.0. There are more sophisticated methods available based on similar principles. Freeze and Cherry (1979), quoting Masch and Denny (1966), provide a nomograph that relates K to the mean grain size, d_{50}, and a parameter, σ, that is a measure of the spread of the gradation curve.

Laboratory measurements of hydraulic conductivity can be carried out on repacked columns of unconsolidated sand or on undisturbed cores of sand, clay, or rock obtained from split-spoon sampling or diamond-drill coring of exploratory boreholes. Laboratory measurements are made with one of two types of apparatus, either a constant-head permeameter or a falling-head permeameter.

In a constant-head test, a soil sample of length L and cross-sectional area A is enclosed between two porous plates in a cylindrical tube, and a constant head differential, ΔH, is set up across the sample. The hydraulic conductivity is determined by

$$K = \frac{QL}{A\Delta H} \tag{39}$$

where Q is the volumetric discharge through the system.

In a falling-head test, the head, as measured in a tube of cross-sectional area a, is allowed to fall from H_0 to H_1 during time t. The hydraulic conductivity is calculated from

$$K = \frac{aL}{At} \ln \left(\frac{H_0}{H_1}\right) \tag{40}$$

where A is the cross-sectional area of the sample.

The hydraulic conductivity of clays is best determined in the laboratory from consolidation tests. Testing procedures for both permeameters and consolidometers are included in most hydrogeology or soil mechanics texts (cf. Bear, 1979; Freeze and Cherry, 1979; Todd, 1980; Mitchell, 1976; Craig, 1983).

It is possible to determine in situ hydraulic-conductivity values by simple inexpensive tests carried out in a single piezometer. The Hvorslev (1951) method of analysis is suited to "point" piezometers that are open or screened over only a short interval at their base. The test is initiated by creating an instantaneous change in the water level in the piezometer through a sudden introduction or removal of a known volume of water. The recovery of the water level over time is then measured and recorded. When water is added, the tests are known as slug tests; when water is removed, the tests are called bail tests. It is possible to create the same effect by suddenly introducing or removing a solid cylinder of known volume.

For a bail test in a homogeneous, isotropic, infinite medium, a plot of the recovery data, in the form of $(H - h)/(H - H_0)$, on a logarithmic scale, versus time, t, on an arithmetic scale, should exhibit a straight-line relation. On this plot, H is the hydraulic head (as measured by the water-level elevation in the piezometer) prior to bailing; H_0 is the hydraulic head immediately after bailing and prior to recovery; and h is the hydraulic head at time t during the recovery process. For a piezometer of radius R, and a piezometer intake of similar radius and length L (and assuming a ratio, $L/R > 8$), the hydraulic conductivity is given by

$$K = \frac{R^2 \ln(L/R)}{2LT_0} \tag{41}$$

where T_0, known as the basic time lag, is defined as the value of t for which $(H - h)/(H - H_0)$ is equal to 0.37. This is the simplest test configuration. Hvorslev (1951) and Cedergren (1967) provide formulae for more complex hydrogeological layering and for alternate screen geometry with nonconforming L/R ratios. An alternative but similar interpretation procedure is provided by Bouwer and Rice (1976) and Bouwer (1989).

In situ packer tests are carried out in the packed-off interval of a single borehole. If the packed-off interval corresponds to a specific high-permeability stratigraphic layer bounded above and below by horizontal aquitards, then the confined-aquifer test interpretation procedure of Cooper et al. (1967) and Papadopulos et al. (1973) can be invoked. This test can be run in bail-test or slug-test mode. In bail-test mode, recovery data in the form, $(H - h)/(H - H_0)$, on an arithmetic scale, are plotted against time, t, on a logarithmic scale, and the resulting curve is compared against a set of type curves presented by Papadopulos et al. (1973) to calculate the transmissivity T of the packed-off aquifer. The Papadopulos curves are reproduced in Freeze and Cherry (1979). Packer tests can be carried out on a number of intervals in a borehole, thus providing valuable data on the vertical variations in conductivity values.

In addition to the basic packer test, there are a variety of alternative test geometries and test procedures. These include multiple-packer arrangements and techniques that are based on transient pulse testing. These methods have particular application to sites in fractured rock. Gringarten (1982), Hsieh et al. (1983), and Pickens et al. (1987) provide detailed discussions of the methodologies.

By far the most common method for obtaining site-scale transmissivity values in confined and unconfined aquifers is that based on an interpretation of multiple-well pumping tests. There is no shortage of excellent references on this topic (cf. Walton, 1970; Walton, 1987; Kruseman and de Ridder, 1970; Bear, 1979; Freeze and Cherry, 1979), so only the conceptual rudiments of the methodology are described here.

The simplest test configuration is that for a homogeneous, isotropic, confined aquifer with one fully penetrating pumping well operating at constant discharge, Q. The drawdowns in head in the aquifer are measured at one or more observation wells. The determination of T from a pumping test involves a direct application of the formulae developed by Theis (1935) for the transient, radial flow of groundwater toward a pumping well, see Equations (31) and (32) in Chapter II.G.1. Drawdowns, $h_0 - h$, versus time, t, are plotted on a log-log scale and compared with the type curve, U(u), also plotted on a log-log scale, that represents the Theis solution. The two curves are superimposed on one another and matchpoint values of $h_0 - h$, t, u, and U(u) are read from the axes of the two graphs. Transmissivity is calculated as

$$T = \frac{QU(u)}{4\pi(h_0 - h)}. \qquad (42)$$

One can also determine T in a similar manner from the head recovery data following the cessation of pumping. It is also possible to determine T from observation-well drawdown or recovery data using the Cooper and Jacob (1946) semi-log plotting procedure as described in each of the earlier referenced texts. These texts also describe the interpretation procedures and provide the pertinent formulae for the determination of T for leaky confined aquifers and unconfined aquifers. The latter is of particular importance at many contaminated sites involving surficial unconsolidated aquifers. The type curves for an unconfined aquifer (after Neuman, 1975) are more complex than those for a confined aquifer, but the concepts of the curve-matching procedure are similar.

Multiple-well pumping tests are more expensive than single-well slug tests and the question arises as to the appropriate trade-off strategy between these two approaches to the determination of hydraulic conductivity. Is it better to have a few large-volume values from pumping tests or a large number of point values from slug tests? For the purposes of contaminant migration assessment under natural (non-pumping) flow conditions, the large number of point values probably has greater value. Advective transport may be concentrated in high-K microstratigraphic stringers and a slug-testing or packer-testing program is more likely to locate such features. However, pumping tests are justified for wells that will be used as part of the pump-and-treat remediation. Pumping tests in this case provide the necessary site-scale information to allow the design of a suitable pumping strategy to contain or extract the plume of contaminated water.

2. Storativity and Specific Yield

The storativity, S, of a confined aquifer, and the specific yield, S_y, of an unconfined aquifer were defined in Chapter II.B.3. In a confined aquifer, the storage term is a function of aquifer porosity, aquifer compressibility, and fluid compressibility; in an unconfined aquifer it is almost solely a function of porosity.

The storativity of a confined aquifer can be determined along with the transmissivity from a multiple-well pumping test. For a homogeneous, isotropic, confined aquifer, and a test configuration that meets the Theis assumptions, the storativity can be calculated from the log-log matchpoint values of $h_0 - h$, t, u, U(u) as

$$S = \frac{4uTt}{r^2} \tag{43}$$

where T is the transmissivity calculated from Equation (6) and r is the distance between the pumping well and the observation well.

The specific yield of an unconfined aquifer can be determined from the interpretation procedures for an unconfined pumping test. As noted in Chapter II.B.3, this determination is complicated by the difficulties associated with the basic concept of specific yield.

Specific yield can also be determined from disturbed or undisturbed aquifer samples using drainage experiments in the laboratory (Todd, 1980). As with the case of laboratory hydraulic-conductivity measurements, the representativeness of laboratory specific-yield measurements is subject to doubt.

3. Porosity

Porosity was defined in Chapter II.B.1. Differentiation was made there between total porosity, n_t, and effective porosity, n_e. It is easier to measure n_t, yet n_e is the parameter of true interest in most contaminant transport studies. Effective porosity is required to calculate the average linear velocity of groundwater flow, which is the pertinent velocity to use in travel time calculations for advective contaminant transport.

Porosity can be measured in five ways: (1) laboratory drying experiments, (2) laboratory drainage experiments, (3) in situ unconfined multiple-well pumping tests, (4) in situ single-well borehole-dilution tests, and (5) in situ two-well tracer tests.

In laboratory drying experiments (Vomocil, 1965), total porosity is calculated from

$$n_t = 1 - \frac{\rho_b}{\rho_s} \tag{44}$$

where ρ_b is the bulk mass density of the sample and ρ_s is the particle mass density. The bulk density is the oven-dried mass of the sample divided by its field volume. A particle density of 2.65 grams per cubic centimeter (g/cm^3) is typical for alluvial material.

In a laboratory drainage experiment (Todd, 1980), gravity drainage of a known volume of saturated subsurface material produces a volume of water that, when divided by the total volume of the sample, constitutes the specific yield, S_y (Chapter III.C.2). The waters that remain on the grains by wetting tension constitute the specific retention, S_r, of the sample. The following relations apply:

$$n_t = S_r + S_y, \quad (45)$$
$$S_y \leq n_e \leq S_y + S_r.$$

If there is reason to believe that the specific retention includes only dead-end pores or otherwise isolated pores, then one can approximate n_e by S_y. If n_e can be approximated by S_y, then a determination of S_y from a multiple-well pumping test in an unconfined aquifer can be used as an estimate of n_e.

Single-well borehole dilution tests (Drost et al., 1968; Halevy et al., 1967; Grisak et al., 1977) provide a measure of the average linear velocity, V, of the natural, horizontal, steady-state groundwater flow rate past a borehole. In this test, a nonreactive tracer is introduced instantaneously at a concentration, c_0, into a packed-off borehole interval of volume, ω, and vertical cross-sectional area, A (perpendicular to the horizontal flow across the well bore). The concentration, c, declines with time, t, due to the mixing of the tracer with the natural groundwater flow. The average linear velocity is calculated from

$$V = -\frac{\omega}{\beta A t} \ln\left(\frac{c}{c_0}\right) \quad (46)$$

where β is a factor that depends on the geometry of the well screen and associated gravel pack. It has values in the range 0.5 to 4.0 (Drost et al. 1968).

Calculation of the effective porosity from a borehole dilution test requires independent knowledge of the hydraulic gradient of the flow and the hydraulic conductivity of the aquifer. The porosity is calculated from a rearrangement of Equations (6) and (8) from Chapter II.C.2:

$$n_e = -\frac{K\Delta h}{V\Delta l}. \quad (47)$$

Single-well tracer tests can be also be conducted by injection and subsequent withdrawal of a solution of tracer chemicals in a single well through a zone that has been isolated by dual packers straddling the injection interval. The advantage of this method lies in the ability to recover the injected tracer chemicals. Its shortcoming lies in the small volume of aquifer affected and the extreme sensitivity of the results to well construction details.

In situ, two-well tracer tests avoid some of the shortcomings of the single-well tests. However, they too require independent knowledge of K and $\Delta h/\Delta l$, and the use of Equation (47), to determine n_e from the tracer-test measurement of V.

Two basic kinds of two-well tests are conducted: the natural-gradient test and the forced-gradient test. In the natural-gradient two-well tracer test, a small volume of concentrated tracer solution is released instantaneously from one well, and a second well directly downgradient is monitored to time the arrival of the leading edge, peak (or centroid), and trailing edge of the tracer plume. In the forced-gradient two-well tracer test, dilute tracer solution is continuously pumped into the first well, while the other well downgradient withdraws groundwater at an identical rate. In either case, the average linear velocity (of the natural or stressed system) is determined as the distance between the two wells divided by the time to arrival of the tracer plume.

Many types of nonreactive chemicals have found application in tracer tests. They range from simple salts such as NaCl or $CaCl_2$, to radioisotopes such as ^3H, ^{131}I, or ^{29}Br. Downhole ion-specific electrodes for rapid measurement of ion concentrations simplify test procedures.

D. Hydrogeologic Variables

1. Aquifer Recharge and Discharge

Recharge of unconfined surficial aquifers takes place as infiltration through the unsaturated zone from precipitation, snowmelt events, and leakage through the bottom of surface water bodies. Recharge to deeper confined aquifers takes place by saturated leakage through confining beds. Recharge can also take place in arid environments by stream-bed losses from intermittent influent streams.

Discharge from unconfined surficial aquifers occurs as springs and seeps, as baseflow contributions to effluent streams, and as evapotranspiration from the land surface. Discharge from deeper confined aquifers takes place as upward saturated leakage across confining beds to overlying surficial aquifers. Discharge from both types of aquifers occurs through the pumping of water-supply wells.

Aquifer recharge and discharge reflect the existing groundwater flow system in a basin. Their rates are controlled by available precipitation and by the topography and geology of the groundwater basin. Under natural conditions, on an average annual basis, recharge must equal discharge. The average annual position of the water table will be such that the groundwater flow delivered from recharge areas to discharge areas by the flow system meets the needs of the average annual steady-state water balance for the basin. Water-table fluctuations reflect daily, seasonal, and annual variations in precipitation. Freeze and Cherry (1979) review the concepts of regional groundwater flow systems, infiltration, recharge, and water-table fluctuations.

Recharge and discharge are extremely difficult to measure directly. The most commonly used method to estimate their value arises from the calibration of steady-state or transient models of regional groundwater flow. This topic is discussed in Chapter III.E.

In addition to indirect climatic water balance or soil-moisture budget calculations, direct methods of estimating recharge include (1) analyses of infiltration across the unsaturated zone from precipitation events and (2) interpretations of water-table fluctuations.

Infiltration analysis requires field measurement of volumetric soil-moisture content and/or soil-moisture pressure head in a vertical profile, together with in situ or laboratory determination of the unsaturated hydrogeologic properties of the soil. These latter properties consist of the two primary hysteretic curves of moisture content versus pressure head and unsaturated hydraulic conductivity versus pressure head. Field measurement techniques for moisture content include gravimetric sampling, resistivity cells, and neutron probes. Field measurement techniques for pressure head usually utilize soil tensiometers. Laboratory measurements are carried out with pressure-plate apparatus and/or unsaturated soil-cell permeameters. Field instruments are difficult to calibrate, and the representativeness of laboratory measurements of soil properties raises large questions. It has not been shown that this type of intensive measurement program can yield estimates of recharge that are any more reliable than that achieved by use of precipitation records and a water-balance calibration of a groundwater flow model.

Methods have been suggested whereby water-table fluctuations in recharge areas can be used to estimate recharge rates. Unfortunately, water-table fluctuations are influenced by many factors other than recharge rates, including evapotranspiration, air-entrapment effects, barometric-pressure effects, agricultural irrigation and drainage, groundwater pumping, tidal influences near oceans, bank-storage effects near streams, and time-lag due to piezometer or well construction.

It is almost impossible to successfully unravel the individual effects of the many contributing factors.

Direct methods of estimating discharge include (1) calculations of evapotranspiration, (2) base flow separation of streamflow, and (3) interpretations of water-table fluctuations in discharge areas.

2. Hydraulic Head, Fluid Pressure, and Water-Table Elevation

Hydraulic head, h, and fluid pressure, p, were defined in Chapter II.C.2. They are related by

$$h = z + (p/\rho g). \qquad (48)$$

Given this relationship, it is always possible to determine h from p or p from h.

The water-table represents the locus of points where the fluid pressure is atmospheric (gage pressure equals zero) and $h = z$, with z representing the water-table elevation measured with respect to the datum.

Water levels measured in deeper wells or piezometers that are open only at one specific screened interval are representative of the hydraulic head at the open interval. Water levels measured in shallower wells that are open all along their length are representative of the water-table elevation.

Values of hydraulic head and/or water-table elevation are required to calculate the directions of groundwater flow. They may also be required as calibration variables in models of groundwater flow. In addition, measurements of drawdown in hydraulic head are utilized in pumping tests to determine aquifer properties. Hydraulic head measurements are also used in performance evaluations to assess whether gradient control has been achieved by a pumping system. Hydraulic-head values are among the most important measurements taken at a contaminated site.

Water-level measurements in wells and piezometers can be taken with (1) chalked tape, (2) float-type automatic recorders, (3) pressure transducers, (4) acoustic probes, (5) electrical sensors, and (6) air lines (Aller et al., 1989). The traditional methods employ chalked tape for spot measurements and float-type drum-chart recorders for continuous records. These methods are simple, reliable, and accurate.

Electrical sensors are also used for the spot measurement of water-level elevations and usually consist of a spool of double-strand electrical wire, the ends of which are fastened to the top and bottom ends of a nonconductive probe. When the top of the probe makes contact with the water surface, the electrical circuit is completed, illuminating an indicator light or sounding a tone at the above-ground spool. Electrical probes may be used for rapid water-level measurements during drawdown studies because the probe can be lowered successively without being raised between measurements.

A rather common water-level measuring device that is built into most municipal and industrial supply wells is the air line. A small diameter pipe (e.g., 0.5-inch diameter copper tubing) is mated to the casing of the well and truncates, open-ended, several feet below the expected maximum drawdown elevation in the well at its maximum pumping rate. An air pump is used to force air into the tube until all of the water is expelled and the air begins to bubble out. A pressure gage on the air line will read its maximum value at this time. That value represents the pressure exerted at the bottom of the air line by the column of water that lies above it. The pressure reading can easily be converted to a hydraulic-head value with Equation (48).

Pressure transducers used for continuous water-level measurements contain a strain gage and amplification circuits that generate an electrical signal which reflects the fluid pressure at the strain gage. This signal is converted into a pressure measurement at a data logger. Most pressure-transducer designs also incorporate an internal atmospheric venting tube inside the electrical cable, so that the pressure measurements are immune to barometric pressure fluctuations. Pressure transducers are reliable (calibration drift is typically less than one-tenth of a foot over months of continuous operation), fairly rugged (though sensitive to overpressuring—the strain gage can be crushed if its maximum range is exceeded), and commonly accurate to within 0.01 foot. Most importantly, these measurements are easily automated so that no repositioning is needed to acquire successive measurements. This feature permits the rapid collection of measurements (greater than one per second). Finally, the measurements acquired are readily stored in digital form that simplifies data handling and analysis.

Hydraulic-head measurements may be required to obtain an understanding of spatial gradients, temporal changes, or both. Spatial coverage may be required horizontally across the extent of a single aquifer, or vertically across an aquifer-aquitard sequence. Vertical gradients can be measured in (1) individual wells installed in separate boreholes and screened at different intervals; (2) individual wells, nested in a single borehole and screened at different intervals; or (3) a single multiple-port well screened at several different intervals. The first method is the safest but it is also the most expensive. The reliability of multiple-port well technology has constantly improved in recent years.

Important considerations regarding hydraulic-head measurements include the following:

a. Active hydrogeologic settings may produce significant fluctuations in water levels over time scales of a few days or even a few hours. Such fluctuations can result from changes in stage of a river or reservoir, operation of nearby water-supply wells, or operation of nearby remedial action systems. The influence of such factors can be detected and quantified by collecting automated water-level data at short time intervals.
b. The construction of the wells from which water-level measurements are obtained should be kept in mind when comparing data. The length and position of the screened interval on a well may influence the water level observed in the well. Short, screened intervals provide depth-specific hydraulic-head measurements for the depth at which the screen is placed. Long screened intervals provide an integrated average of the hydraulic heads over the screened portion of the aquifer system.
c. It is incorrect to group together water-level measurements from wells that are screened at significantly different vertical positions within an aquifer. Piezometric contour maps drawn from the resulting data will present a picture that is an undefinable mixture of the horizontal and vertical distribution of hydraulic head.
d. In cases where significant variations in groundwater density are found, hydraulic-head measurements should be corrected to reflect a common density before comparing data or calculating gradients (Jorgensen et al., 1982). Methods of calculating equivalent fresh-water heads from data from wells with brackish or saline water are provided by Pickens et al. (1987).

Water-supply wells, irrigation wells, and industrial-supply wells have traditionally been constructed with long screened intervals. Consequently, measurements of their static water-level elevations are integrated averages of vertical-head differences over a large section of the aquifer. These water-level elevations may be plotted to produce contour maps of the horizontal component of the regional flow system, but they should not be combined with site-scale measurements from properly constructed observation wells to produce contour maps because the two sets of data are not compatible.

Water-level measurements that have been obtained from water-supply wells during operation are affected by the pressure drop that occurs as groundwater passes through the entrance openings of the well screen. The magnitude of the effect is a function of the well construction details and the flow rate. Hence, it is very difficult to produce accurate estimates of the water-level elevation immediately outside the well screen.

3. Velocity and Flux

The average linear groundwater velocity, V, was defined in Chapter II.C.2. It is the appropriate velocity to use when calculating advective travel times of dissolved contaminants. When calculating flux toward a pumping well or across an external compliance boundary, the correct velocity to use is the specific discharge, q, which was also defined in Chapter II.C.2.

Groundwater velocity can be measured directly with tracer tests or can be calculated indirectly from measurements of hydraulic gradient, hydraulic conductivity, and porosity.

For direct measurement, one of three types of tracer test can be used: (1) single-well borehole dilution test, (2) natural-gradient two-well tracer test, or (3) forced-gradient two-well tracer test. These tests have been briefly described in Chapter III.C.3 in the discussion of porosity determination. The value of V can be determined from a single-borehole dilution test using Equation (46). In a two-well tracer test, V is calculated as the distance between the two wells divided by the time of arrival of the peak of the tracer plume.

If independent measurements of the magnitude of the gradient, $\Delta h/\Delta l$, hydraulic conductivity, K, and effective porosity, n_e, are available, V can be determined from Darcy's law:

$$V = -\frac{K}{n_e}\frac{\Delta h}{\Delta l}. \tag{49}$$

The contaminant mass flux, Q_c, at concentration, c, through a cross-sectional area, A, is given by

$$Q_c = Vn_eAc. \tag{50}$$

E. Data Estimates From Model Calibration

1. Estimates of Hydraulic Conductivity and Recharge From Steady-State Simulations

Simulations of groundwater flow conditions can be carried out with finite-difference or finite-element computer models of the type described in Chapter II.H.

In steady-state form, there are two different two-dimensional geometries in which such simulations might be carried out. The first is a two-dimensional vertical cross section along a line parallel to the direction of groundwater flow across the site. This geometry allows one to investigate the influence of topography and stratigraphy on the vertical and horizontal components of groundwater flow under natural conditions. The second geometry is a two-dimensional horizontal representation of a particular confined or unconfined aquifer. Simulations can be carried out under pre-development steady flow, and under the steady-state conditions that develop some time after a pumping/containment system has been installed at the site. In the latter case, the model would be based on Equation (27).

Steady-state simulations of groundwater flow in either geometry require as input (1) the distribution of hydraulic conductivity, $K(x,z)$, or transmissivity, $T(x,y)$; and (2) the recharge/discharge rate, W'. If W' is known, the simulations produce as output the distribution of hydraulic head, $h(x,z)$, or $h(x,y)$. If the hydraulic head configuration is known, a simulation model can be used to produce the distribution of W'.

The model calibration process involves a comparison of the predicted values of hydraulic head with observed measurements at particular wells or piezometers. Trial-and-error variations in the hydraulic-conductivity distribution and/or the recharge rates are then invoked in an attempt to improve the fit. These variations must of course be faithful to those values of hydraulic conductivity that have been measured at specific locations at the site. It is seldom that there are measured values of recharge, but calibrated recharge values must be faithful to the range of climatic conditions known to exist at the site.

This trial-and-error approach is the simplest and most widely used method of model calibration. However, automated calibration procedures, based on inverse theory and utilizing statistical criteria to satisfy specific objective functions with respect to fit, are becoming more common (Yeh, 1986; Bear and Verruijt, 1987).

It must be recognized that calibrated or inverse solutions are not unique. There may be many possible combinations of hydraulic-conductivity distribution and recharge rate that will reproduce in a general way sparse suite of measured hydraulic-head and hydraulic-conductivity values. Calibration remains an art in which the judgment of the modeller, which is invoked on the basis of knowledge of the hydrogeological conditions at the site, plays an important role. However, the statistical foundation for modern inverse methods provides an important means of assessing model reliability.

Despite these limitations, model calibration provides one of the best available method for estimating aquifer recharge and discharge.

2. Estimates of Transmissivity, Storativity, and Recharge From Transient Simulations

Transient simulations of groundwater flow are most often utilized in a two-dimensional horizontal representation of a confined or unconfined aquifer in order to predict the temporal and spatial distribution of drawdown in hydraulic head under the influence of pumping from the aquifer. Equation (26) is the one that is solved in such models.

In predictive mode, the required input includes the transmissivity, T(x,y); the storativity, S(x,y); and the stress and recharge rates, W′(x,y,t). Output consists of the time-space distribution of hydraulic head, h(x,y,t). The calibration process involves comparing predicted time series of hydraulic heads with measured values of head at particular well locations at particular times. Such observations may span a period of natural fluctuations in groundwater conditions, or they may include a period when the aquifer has been stressed by a pumping test or by pumped production wells. Automated or trial-and-error variations in the transmissivity, storativity, or recharge distributions are invoked to obtain a better fit between predicted and observed values of head. As with the steady-state calibrations, transient calibrations are not unique. Nevertheless, the calibration process usually results in a more defensible and more compatible combination of aquifer properties and recharge rates than can be gleaned from a noncalibrated interpretation of available measurements.

F. Hydrochemical Parameters

1. Dispersivity and Matrix Diffusion

Hydrodynamic dispersion of dissolved solutes was discussed in a qualitative manner in Chapter II.F.2. Little emphasis was placed there on a quantitative description because the design methodologies presented in Chapter IV and V for the design of capture and containment systems are based on the assumption that advective components of contaminant transport overwhelm dispersive components in most practical applications. For the same reason, our discussion in this subsection on measurement technologies for the determination of dispersivity will be brief and qualitative.

Dispersivity is commonly assumed to have two major components. Longitudinal dispersivity describes contaminant spreading parallel to the direction of flow; transverse dispersivity describes the spreading perpendicular to it. Transverse dispersivities are generally much smaller than longitudinal dispersivities (Anderson, 1979), so emphasis in this subsection will be on the longitudinal component.

The longitudinal dispersivity of aquifer materials can be determined from (1) laboratory tracer tests, (2) field tracer tests, (3) back analysis of plume extent, (4) calibration of transient contaminant transport models, and (5) estimates based on the statistical distribution of hydraulic-conductivity values.

Tracer tests in laboratory columns are generally viewed as providing little indication of in situ dispersivity of geologic materials. Borehole-size samples have little relevance in the analysis of transport problems at the site or regional scale, and in the analysis of spacial variability of velocity at the field scale.

Two types of field tracer tests can be used for longitudinal dispersivity determinations: (1) two-well natural-gradient tracer tests and (2) two-well forced-gradient tracer tests. The methodology has already been discussed in Chapters III.C.2 and III.D.3 with respect to porosity and velocity determinations from such tests. In both types of test, tracer chemicals are introduced in one well and tracer concentrations are measured as a function of time in the other well. Natural-gradient tests use the existing gradient to carry the contaminants; forced-gradient tests attempt to speed up the process by using an injection-well/ extraction-well pair. The distribution over time of the tracer concentrations at the downgradient well can be used to calculate the dispersivity. Fried (1975) presents an outline of the test methods and their analysis. While both natural-gradient and forced-gradient two-well tests can yield the same information, they differ in their sensitivity to dispersive forces because of the different velocities involved. Natural-gradient tracer tests tend to be biased toward producing large dispersivities, presumably because the primary hydraulic influence is regional and thereby subject to the widest range of hydraulic conductivities. Forced-gradient tracer tests tend to be biased toward smaller dispersivities, presumably because the primary hydraulic influence is highly localized.

There are many excellent recent case histories of field tracer tests that investigated in situ contaminant-transport properties. Interested readers are directed to the papers describing test sites in Arizona (Hsieh et al., 1983), Alabama (Molz et al., 1986, 1988), and Ontario (Roberts et al., 1986; Freyberg, 1986; Mackay et al., 1986a, 1986b).

It is also possible to determine site-scale dispersivities by back analysis of measured plume extent at contaminated sites. The analysis can be carried out using an analytical inverse technique or by calibration of transient contaminant-transport models (Huyakorn and Pinder, 1983; Wagner and Gorelick, 1987).

Dispersion is caused by the presence of heterogeneities in the hydraulic conductivity of an aquifer. Gelhar and his co-workers as well as Dagan (summarized in Gelhar, 1986; Dagan, 1986) have developed predictive analytical expressions that allow one to estimate dispersivity, given a measure of the variance of hydraulic conductivity in an aquifer. Such a measure can be obtained from a geostatistical interpretation of a set of point measurements of hydraulic conductivity from slug-test data.

One particularly common style of aquifer heterogeneity is that of horizontally layered microstratification. In such environments, Feenstra et al. (1984) observed dispersion to be largely the result of advection and matrix diffusion processes, whereby contaminants are carried along advective flow paths of variable permeability and diffuse laterally into the matrix of low-permeability layers and lenses. Diffusion occurs *into* the matrix at the front end of a contaminant pulse and *out* again at the back end. Sudicky et al. (1985), Starr et al. (1985), and Guven and Molz (1986) describe tracer-test interpretations in stratified media.

2. Retardation Factor

Retardation of dissolved organic contaminants was discussed in Chapter II.F.3. A primary mechanism leading to retardation is sorption of dissolved solutes on organic matter. This partitioning of contaminant between the liquid and the solids causes the average velocity of the solute to be reduced from that of the groundwater flow by a factor, R_f, where R_f is known as the retardation factor. The retardation factor is contaminant- and media-dependent. Retardation is the simplest means of representing complex chemical transport properties. One should always use retardation factors with caution. Cherry et al. (1984) and Mackay et al. (1985) provide excellent summaries of the concepts of contaminant retardation.

The retardation factor for a particular contaminant in a particular media can be determined from (1) laboratory column experiments, (2) laboratory batch experiments, (3) field tracer tests, (4) calibration of transient contaminant-transport models, and (5) estimates based on knowledge of the octanol-water partition coefficient.

Laboratory column experiments and batch tests suffer from the usual limitations with respect to the representativeness of the laboratory results in comparison to the true but unknown in situ values. Such limitations plague all laboratory tests on small samples.

Two-well tracer tests can be used to estimate retardation along with dispersion, but independent measurements of the hydraulic gradient, hydraulic conductivity, and effective porosity are required if unique solutions to the tests are desired.

Transient contaminant-transport models can be calibrated jointly for dispersion and retardation, given reliable measurements of plume concentration patterns and an independent estimate of the hydraulic parameters.

A common method used to estimate retardation factors involves application of empirical formulae relating R_f to the octanol-water partition coefficient, K_{ow}, of the contaminant in question and the organic carbon content, f_{oc}, of the aquifer material. Several possible relationships between K_{ow}, f_{oc}, and R_f are reported by Mackay et al. (1985). Verschueren (1983) and Montgomery and Welkom (1990) provide tables of K_{ow} values for a wide range of organic contaminants. Organic carbon contents must be measured on soil or rock samples collected from the contaminated aquifer under study.

G. Hydrogeochemical Variables: Concentration

Reliable measurements of contaminant concentrations in the subsurface are required for a variety of purposes associated with the design of a pump-and-treat remedial system:

1. The design of the extraction-well network requires knowledge of the horizontal and vertical extent of the contaminant plume.
2. The design of the treatment facilities requires knowledge of the concentrations of the influent water that will be delivered by the extraction-well network.
3. Concentration values may be needed to calibrate predictive models of contaminant transport used in the design of the remedial system and in the performance evaluation of the system during operation.
4. Concentration values will be required for samples taken from monitoring wells for early warning detection or for regulatory compliance.

There are three essential elements to the measurement of contaminant concentrations: (1) monitoring-well design, (2) sampling protocol, and (3) analytical issues. These elements are treated in turn in the following three subsections. The question of sample representativeness was discussed in Chapter III.A.2 as an issue of scale, and those arguments are not repeated here.

The art of reliable sampling and analysis of groundwaters to meet the needs of performance evaluation and regulatory compliance is a difficult one, given the extremely low concentrations that must be detected and the extremely high potential for sample contamination. The treatment here is brief. More detailed presentations on these important topics are provided by Barcelona et al. (1985), Aller et al. (1989), and by the many detailed articles that appear quarterly in the journal *Groundwater Monitoring Review*.

1. Monitoring-Well Design

Campbell and Lehr (1973) and Driscoll (1986) describe well-drilling technology in general, and Aller et al. (1989) assess the particular merits of each technology with specific reference to monitoring wells that will be used to obtain samples for chemical analysis. The primary limitation on drilling methodology for monitoring wells is the desire to avoid the introduction of nonnative drilling fluids into the borehole. This leads one to favor either air-rotary drilling, in which the circulation fluid is air rather than water or drilling mud; or hollow-stem auguring, in which no circulation fluid is used. However, air-rotary drilling is restricted to consolidated or semiconsolidated formations, and drillers often want to enhance their penetration rates by adding small quantities of foam surfactants that are not chemically inert. If foam or direct mud-rotary drilling is used, then all foam and mud must be removed by the well-development process, or the chemical interaction between the mud and the formation fluids may prevent collection of a sample that is representative of the in situ groundwater quality.

Monitoring-well construction issues revolve around (1) well diameter, (2) casing and screen materials, (3) annular seals, and (4) well development.

Monitoring-well diameters must be selected to accommodate borehole geophysical tools, water-level measuring devices, groundwater-sampling devices, and aquifer testing equipment. Generally this favors large rather than small borehole diameters, usually in the range 6 to 12 inches. Often, diameters are smaller.

Casing materials must be selected such that they do not adsorb chemical constituents from the formation water or leach chemical constituents into it. Adsorption and leaching are believed to be minimal for stainless steel casings. Many of the popular polyvinyl chloride (PVC) thermoplastic casings are susceptible to chemical attack by high concentrations of certain organic solvents, including low-molecular-weight ketones, aldehydes, amines, and chlorinated alkenes and alkanes (Barcelona et al.,

1983). Fluoropolymer casings such as Teflon® are thought to exhibit less leaching than PVC (Aller et al., 1989).

If sample contamination by casing leaching occurs, it will likely be concentrated at the well screen where surface areas are largest and flow rates are highest. Stainless steel or Teflon® well screens are recommended.

The annular opening above the filter pack that surrounds the screened interval of a monitoring well must be sealed to prevent vertical migration of contaminated fluids between aquifers. The most common seal materials are pelletized, granular, or powdered bentonite and neat cement grout. Care must be taken to ensure that shrinkage, bridging, or poor hydration does not allow leakage through the annular seal.

Well-development techniques are designed to eliminate the effects of drilling from the monitoring zone. Development methods include bailing, surging, and pumping with backwashing. It is imperative to avoid introducing additional fluids into the monitoring zone during the development process. Care should be taken to minimize the degree of vertical communication and induced contamination created by monitoring wells.

2. Sampling Protocol

The design of a sampling program for collection of groundwater samples includes consideration of (1) sampling devices, (2) purging strategy, (3) sample-collection protocols, and (4) sampling frequency.

Aller et al. (1989) list the available sampling devices as bailers, grab samplers, syringe samplers, suction-lift pumps, gas-drive samplers, bladder pumps, and submersible pumps. Barcelona et al. (1985) recommend bladder pumps as the most reliable. Conventional bailers and grab samplers require very careful operation. In all cases, sample integrity is best guaranteed by employing dedicated samplers and pumps and careful decontamination procedures.

The number of well volumes to be purged from a monitoring well before collection of a water sample must be tailored to the (1) hydraulic properties of the geologic materials being monitored, (2) well-construction parameters, (3) desired pumping rate, and (4) sampling methodology to be employed. Barcelona et al. (1985) provide examples of well-purging calculations for specific cases. They argue that rule-of-thumb guidelines for purging that involve 3-, 5-, or 10-well volumes are outmoded. In addition, purging rates should be kept below development rates to avoid well damage. If high contaminant concentrations are expected in the purged water, it may be necessary to dispose of the water in a permitted facility. In these circumstances, it will be desirable to minimize the total volume of purged water.

A sample-collection protocol, as outlined by Barcelona et al. (1985) involves specific procedures for sample collection, sample filtration, sample preservation, and sample storage and transport. Sample filtration may be necessary in some cases and unacceptable in others. The question of when and whether to filter field samples is a contentious issue; expert advice should be sought. Barcelona et al. (1985) also recommend field measurement of parameters, such as pH, that may be affected by storage. They also recommend liberal use of field blanks and standards. These can help to identify erroneous data and permit quantitative correction for bias. Blanks and standards also provide an independent check on the performance of the analytical laboratory. The development of a reliable sampling protocol is a complex process that must be designed to meet the specific goals of a particular monitoring effort. Readers are reminded of the point made in Chapter III.A.3—even the most precise and accurate laboratory analysis of a contaminant concentration is worthless if the reliability of the sampling procedure by which the sample was obtained is in doubt.

Sampling frequency for early-warning-detection monitoring wells or regulatory-compliance monitoring wells has traditionally been established by regulation rather than reason. Barcelona et al. (1985) argue for frequency calculations that are hydrogeologically based. They provide a nomograph that allows determination of a reasonable sampling frequency given estimates of hydraulic conductivity and effective porosity of the aquifer, the distance from the source to the monitoring well, and the hydraulic gradient across this distance. Sampling frequencies of 1, 10, 100, or 1,000 days could result from reasonable sets of input values.

3. Analytical Issues

A laboratory analysis program to measure chemical concentrations in groundwater samples requires decisions on (1) the constituents to be measured and (2) the methods of analysis to be employed. The second of these topics is outside the scope of this book, and the first will only be discussed in very general terms.

The suite of chemical constituents included as analytes for a sampling run must depend on the objectives of the monitoring program. One might differentiate, as Barcelona et al. (1985) do, between detection monitoring, which simply strives to identify the presence of a plume, and assessment monitoring, which is intended to assess the type, extent, and dynamics of a contaminant plume, perhaps as a part of the site characterization or performance-evaluation process. The first approach might require only general chemical parameters such as pH, total dissolved solids (TDS), total organic carbon (TOC), and total organic halogen (TOX). The second approach would require specific analysis of particular metals or organic compounds that constitute the suspected contaminants at the site. In specifying analytical requirements at a site, one might group the required measurements into a threefold classification: (1) properties (such as pH and Eh) and constituents (such as Cl^-, $SO_4^=$, $NaCl^+$, Ca^{++}, etc.) that provide a description of the background groundwater quality, (2) constituents (such as metals and organics on the EPA Priority Pollutant List) that provide an indication of the type and level of contamination that exists at the site, and (3) constituents (such as arsenic, PCBs, or pesticides) that might be of specific interest at a particular location.

H. Fluid and Contaminant Properties

The intrinsic properties of groundwater that are of interest in contaminant studies are density, compressibility, and viscosity. They were defined and discussed in Chapter II.B.2.

The intrinsic properties of liquid contaminants that are of interest are density, viscosity and solubility. They were defined and discussed in Chapter II.E. In addition, the octanol-water partition coefficient is useful in making retardation factor estimates, as discussed in Chapter III.F.2. It is not common to measure any of these fluid properties as part of the remedial investigations at a particular site. Rather, values are commonly taken from standard references.

Values for the compressibility of water are available in Weast (1985) or in many groundwater texts. Information on the viscosity of various fluids can also be found in Weast (1985). Information on the density, solubility, and octanol-water partition coefficient for a wide range of organic contaminants can be found in several sources including Verschueren (1983), Montgomery and Welkom (1990), and the National Institute for Occupational Safety and Health (1985).

I. Microbiological Data

Microbiological transformations of organic contaminants can take place in the subsurface (Borden et al. 1989; Bouwer and Wright, 1988; Jensen et al., 1988; Major et al., 1988; Roberts et al., 1989). With current site-assessment procedures, this fate mechanism is seldom analyzed quantitatively. Where such mechanisms are thought to be operative, qualitative investigations of in situ microbial populations and nutrient conditions may be carried out.

Prior to the 1970s, it was presumed that the subsurface was largely uninfluenced by bacteria. This was shown to be untrue in the early 1970s, when various microbial populations were found in subsurface-sediment samples. Since then, researchers have attempted to identify organisms that might contribute to the degradation of specific organic contaminants. It is unlikely that an indicator organism will be found that can be used to benchmark the potential for biodegradation of contaminants in general.

The population dynamics of subsurface biota depend on a number of factors including the availability of nutrients such as methane, oxygen, nitrogen, and phosphate. In addition, the ability of microorganisms to degrade an organic contaminant depends on the structure of the contaminant and its concentration. Threshold concentration effects have been observed where an adequate supply of nutrients is available and yet little degradation occurs because contaminant concentrations are too low. It is possible that low dissolved carbon concentrations may not provide enough energy to stimulate biotic activity. On the other end of the contaminant concentration scale, the contaminant may be present in certain portions of the plume at concentrations that are toxic to indigenous biota.

It is outside the scope of this book to discuss subsurface-sediment sampling methodology for microbiological investigations. The interested reader is directed to Bitton and Gerba (1984).

J. Quality Assurance and Quality Control

A quality-assurance (QA) program is a system of written procedures used to ensure that field and laboratory measurement systems produce reliable data. A quality-control (QC) program is a system of records and documented checks that audits the effectiveness of the QA procedures.

In groundwater contamination studies, QA/QC procedures have been routinely applied to the sampling and analysis of contaminant concentrations. It is equally important that they be applied to the collection of hydrogeological parameters such as hydraulic conductivity and hydrogeological state variables such as hydraulic head. QA/QC procedures should be used to verify that field and laboratory measurement systems operate within acceptable limits, and these limits should be set during the design of data acquisition programs for each measurement that the program requires.

A quality-assurance program must ensure that measurements are both accurate and precise. The accuracy of a measurement is the degree of closeness of a measurement to its true value. For most hydrogeological measurements, the true value is unknown and accuracy is reported as bias. Bias is established by making measurements of field or laboratory standards. Proof of measurement bias indicates systematic errors in field measurements that must be corrected. The precision of a measurement is a measure of its reproducibility. Precision is often expressed as the standard error of the mean value of a set of replicate determinations.

Examples of the more common quality-assurance issues are provided by considering measurements of hydraulic head and contaminant concentration. With respect to hydraulic-head measurements it is important that:

1. Well-casing elevations are accurately surveyed.
2. Fluid density effects are taken into account for saline waters or non-aqueous-phase liquids.
3. The depth of the screened interval is correctly documented.
4. Large, open screened intervals that cross more than one hydrostratigraphic unit are avoided.
5. Leakage around annular seals in multiport installations is prevented.
6. Downhole fluid pressure instrumentation is properly calibrated.

In the measurement of contaminant concentrations, it is important that:

1. All drilling fluids are removed from monitoring wells by well development procedures.
2. Screens and casing materials are made of nonreactive materials.
3. Grouts and joint-sealing material are not allowed to introduce spurious chemical constituents into the well bore.
4. Downhole equipment is decontaminated before installation and use.
5. Leakage around annular seals is avoided.
6. Adequate well purging is carried out prior to sampling.
7. Unbiased sampling protocols are used.
8. All downhole sampling equipment is in proper working order.
9. All field measurement devices are properly calibrated.
10. Proper labeling and identification of samples is carried out.

The recordkeeping requirements of a quality-control program will not be discussed in detail here. It is sufficient to note that all measurements should be fully documented. Most groundwater measurements are associated with a specific well or piezometer; a complete record of drilling history, well-completion data, and well-pumping history should be maintained for each well. Aller et al. (1989) and Barcelona et al. (1985) provide examples of well-documentation records.

References for Chapter III

Aller, L., T. W. Bennett, G. Hackett, R. J. Petty, J. H. Lehr, H. Sedoris, D. M. Nielsen, and J. E. Denne, *Handbook of Suggested Practices for the Design and Installation of Ground-Water Monitoring Wells*, National Water Well Association, 1989.

Anderson, M. P., "Using Models to Simulate the Movement of Contaminants Through Groundwater Flow Systems," *Critical Reviews in Environmental Controls*, Vol. 9, pp. 97–156, 1979.

Ayers, J. D., "Application and Comparison of Shallow Seismic Methods in the Study of an Alluvial Aquifer," *Ground Water*, Vol. 27, pp. 550–563, 1989.

Barcelona, M. J., J. P. Gibb, and R. Miller, "A Guide to the Selection of Materials for Monitoring-Well Construction and Groundwater Sampling," *Illinois State Water Survey*, SWS Contract Report 327, 1983.

Barcelona, M. J., J. P. Gibb, J. A. Helfrich, and E. E. Garske, "Practical Guide for Groundwater Sampling," *Illinois State Water Survey*, SWS Contract Report 374, 1985.

Bear, J., *Hydraulics of Groundwater*, McGraw-Hill, Inc., 1979.

Bear, J., and A. Verruijt, *Modelling Groundwater Flow and Pollution*, D. Reidel Publishing Company, 1987.

Bitton, G., and C. P. Gerba (editors), *Groundwater Pollution Microbiology*, Wiley Interscience/John Wiley & Sons Inc., 1984.

Borden, R., M. Lee, J. M. Thomas, P. Bedient, and C. H. Ward, "In Situ Measurement and Numerical Simulation of Oxygen Limited Biotransformation," *Ground Water Monitoring Review*, Vol. 9, pp. 83–91,1989.

Bouwer, H., The Bouwer and Rice Slug Test: An Update, *Ground Water*, Vol. 27, pp. 304–309, 1989.

Bouwer H., and R. C. Rice, "A Slug Test for Determining Hydraulic Conductivity of Unconfined Aquifers With Completely or Partially Penetrating Wells," *Water Resources Research*, Vol. 12, pp. 423–428, 1976.

Bouwer, E. J., and J. P. Wright, "Transformations of Trace Halogenate Aliphatics in Anoxic Biofilm Columns," *Journal of Contaminant Hydrology*, Vol. 2, pp. 155–169, 1988.

Campbell, M. D., and J. H. Lehr, *Water Well Technology*, McGraw-Hill, 1973.

Cedergren, H. R., *Seepage, Drainage, and Flow Nets,* 3rd Ed., John Wiley & Sons, Inc., 1967.

Chapuis, R. P., "Shape Factors for Permeability Tests in Boreholes and Piezometers," *Ground Water*, Vol. 27, pp. 647–654, 1989.

Cherry, J. A., R. W. Gillham, and J. F. Barker, "Contaminants in Groundwater: Chemical Processes," in *Ground-Water Contamination, Studies in Geophysics*, National Academy Press, 1984.

Cooper, H. H., Jr., and C. E. Jacob, "A Generalized Graphical Method for Evaluating Formation Constants and Summarizing Well Field History," *Transactions of the American Geophysical Union*, Vol. 27, pp. 526–534, 1946.

Cooper, H. H., J. D. Bredehoeft, and S. S. Papadopulos, "Response of a Finite-Diameter Well to an Instantaneous Charge of Water," *Water Resources Research*, Vol. 3, pp. 263–269,1967.

Craig, R. F., *Soil Mechanics*, 3rd Ed., van Nostrand Reinhold, 1983.

Dagan, G., "Statistical Theory of Groundwater Flow and Transport: Pore to Laboratory, Laboratory to Formation, and Formation to Regional Scale," *Water Resources Research*, Vol. 22, No. 9, pp. 120S–134S, 1986.

Daily, W., and A. Ramirez, "Evaluation of Electromagnetic Tomography to Map In-situ Water in Heated Welded Tuffs," *Water Resources Research*, Vol. 25, pp. 1083–1096, 1989.

de Marsily, G., *Quantitative Hydrogeology: Groundwater Hydrology for Engineers*, Academic Press Inc., 1986.

Dobrin, M. B., and C. H. Savitt, *Introduction to Geophysical Prospecting*, McGraw-Hill, 1988.

Driscoll, F. G., *Ground Water and Wells*, Johnson Division, 1986.

Drost W., D. Klotz, A. Koch, M. Moser, F. Neumaier, and W. Rauert, "Point Dilution Methods of Investigating Groundwater Flow by Means of Radioisotopes," *Water Resources Research*, Vol. 4, pp. 125–146, 1968.

Feenstra, S., J. A. Cherry, E. A. Sudicky, and Z. Haq, "Matrix Diffusion Effects on Contaminant Migration From an Injection Well in Fractured Sandstone," *Ground Water*, Vol. 22, pp. 307–316, 1984.

Freeze, R. A., and J. A. Cherry, *Groundwater*, Prentice Hall, Inc., 1979.

Freyberg, D., "A Natural Gradient Experiment on Solute Transport in a Sand Aquifer. 2. Spatial Moments of the Advection and Dispersion of Nonreactive Tracers," *Water Resources Research*, Vol. 22, pp. 2031–2046, 1986.

Fried, J. J., *Groundwater Pollution,*, Elsevier, 1975.

Gelhar, L. W., "Stochastic Subsurface Hydrology From Theory to Applications," *Water Resources Research*, Vol. 22, pp. 135S–145S, 1986.

Goltz, M. N., and P. V. Roberts, "Interpreting Organic Solute Transport Data From a Field Experiment Using Physical Nonequilibrium Models," *Journal of Contaminant Hydrology*, Vol. 1, pp. 77–94, 1986.

______, "Simulations of Physical Nonequilibrium Solute Transport Models: Application to a Large-Scale Field Experiment," *Journal of Contaminant Hydrology*, Vol. 3, pp. 37–64, 1988.

Gringarten, A. C., "Flow Test Evaluation of Fractured Reservoirs," in *Recent Trends in Hydrogeology*, T.N. Narasimhan, ed., Geological Society of America Special Paper 189, pp. 237–264, 1982.

Grisak, G. E., W. F. Merritt, and D. W. Williams, "A Fluoride Borehole Dilution Apparatus for Groundwater Velocity Measurements," *Canadian Geotechnical Journal*, Vol. 14, pp. 554–561, 1977.

Guven, O., and F. J. Molz, "Deterministic and Stochastic Analyses of Dispersion in an Unbounded Porous Medium," *Water Resources Research*, Vol. 22, pp. 1565–1574, 1986.

Halevy, E., H. Moser, O. Zellhofer, and A. Zuber, "Borehole Dilution Techniques: A Critical Review," in *Isotopes in Hydrology*, IAEA, Vienna, Austria, pp. 531–564, 1967.

Hsieh, P. A., S. P. Neuman, and E. S. Simpson, "Pressure Testing of Fractured Rocks: A Methodology Employing Three-Dimensional Cross-Hole Tests," *U.S. Nuclear Regulatory Commission Topical Report*, NUREG CR-3213, 176 pp., 1983.

Huyakorn, P. S., and G. F. Pinder, *Computational Methods in Subsurface Flow*, Academic Press, 1983.

Hvorslev, M. J., "Time Lag and Soil Permeability in Groundwater Observations," *U.S.Army Corps of Engineers Waterways Experiment Station Bulletin 36*, Vicksburg, MS, 1951.

Jensen, B. K., E. Arvin, and A.T. Gundersen, "Biodegration of Nitrogen- and Oxygen-Containing Aromatic Compounds in Groundwater From an Oil-Contaminated Aquifer," *Journal of Contaminant Hydrology*, Vol. 3, pp. 65–76, 1988.

Jorgensen, D. G., T. Gogel, and D. C. Signor, "Determination of Flow in Aquifers Containing Variable-Density Water," *Ground Water Monitoring Review*, Vol. 2, pp. 40–45, 1982.

Keely, J. F., "Chemical Time-Series Sampling," *Ground Water Monitoring Review*, Vol. 2, pp. 29–38, 1982.

Keys, W. S., "Well Logging in Groundwater Hydrology," *Ground Water*, Vol. 6, pp. 10–18, 1968.

Keys, W. S., and L. M. MacCary, "Application of Borehole Geophysics to Water Resources Investigations," in *Techniques of Water Resources Investigations*, U.S. Geological Survey, Book 2, Chapter E1, 1971.

Klotz, D., K. P. Seiler, H. Moser, and F. Neumaier, "Dispersivity and Velocity Relationship From Laboratory and Field Experiments," *Journal of Hydrology*, Vol. 45, pp. 169–184, 1980.

Kruseman, G. P., and N. A. de Ridder, "Analysis and Evaluation of Pumping Test Data," *International Institute of Land Reclamation and Improvement Bulletin 11*, Wageningen, The Netherlands, 1970.

Mackay, D. M., W. P. Ball, and M. G. Durant, "Variability of Aquifer Sorption Properties in a Field Experiment on Groundwater Transport of Organic Solutes: Methods and Preliminary Results," *Journal of Contaminant Hydrology*, Vol. 1, pp. 119–132, 1986a.

Mackay, D. M., D. L. Freyberg, P. V. Roberts, and J. A. Cherry, "A Natural Gradient Experiment on Solute Transport in a Sand Aquifer: 1. Approach and Overview of Plume Movement," *Water Resources Research*, Vol. 22, pp. 2017–2029, 1986b.

Mackay, D. M., P. V. Roberts, and J. A. Cherry, "Transport of Organic Contaminants in Groundwater," *Environmental Science and Technology*, Vol. 19, pp. 384–392, 1985.

Major, D. W., C. I. Mayfield, and J. F. Barker, "Biotransformation of Benzene by Denitrification in Aquifer Sand," *Ground Water*, Vol. 26, pp. 8–14, 1988.

Masch, F. D., and K. J. Denny, "Grain-Size Distribution and its Effect on the Permeability of Unconsolidated Sands," *Water Resources Research*, Vol. 2, pp. 665–677, 1966.

Mercer, J. W., S. D. Thomas, and B. Ross, *Parameters and Variables Appearing in Repository Siting Models*, U.S. Nuclear Regulatory Commission, NUREG/CR-3066, 1982.

Mitchell, J. K., *Fundamentals of Soil Behaviour*, Wiley, 1976.

Molz, F. J., O. Guven, J. G. Melville, and J. F. Keely, *Performance and Analysis of Aquifer Tracer Tests with Implications for Contaminant Transport Modelling*, U.S. EPA Office of Research and Development, EPA/600/2-86-062, 1986.

Molz, F. J., O. Guven, J. G. Melville, J. S. Nohrstedt, and J. K. Overholtzer, "Forced-Gradient Tracer Tests and Inferred Hydraulic Conductivity Distributions at the Mobile Field Site," *Ground Water*, Vol. 26, pp. 570–579, 1988.

Montgomery, J. H., and L. M. Welkom, *Groundwater Chemicals Desk Reference*, Lewis Publ., 1990.

National Institute for Occupational Safety and Health, *NIOSH Pocket Guide to Chemical Hazards*, U.S. Department of Health and Human Services, U.S. Government Printing Office, Washington, D.C., 1985.

Neuman, S. P., "Analysis of Pumping Test Data From Anisotropic Unconfined Aquifers Considering Delayed Gravity Response," *Water Resources Research*, Vol. 11, pp. 329–342, 1975.

Papadopulos, S. S., J. D. Bredehoeft, and H. H. Cooper, "On the Analysis of Slug Test Data," *Water Resources Research*, Vol. 9, pp. 1087–1089, 1973.

Pickens, J. F., G. E. Grisak, J. D. Avis, D. W. Belanger, and M. Thury, "Analysis and Interpretation of Borehole Hydraulic Tests in Deep Boreholes: Principles, Model Development and Applications," *Water Resources Research*, Vol. 23, pp. 1341–1376, 1987.

Roberts, P. V., M. N. Goltz, and D. M. Mackay, "A Natural Gradient Experiment on Solute Transport in a Sand Aquifer: 3. Retardation Estimates and Mass Balances for Organic Solutes," *Water Resources Research*, Vol. 22, pp. 2047–2058, 1986.

Starr, R. C., R. W. Gillham, and E. A. Sudicky, "Experimental Investigation of Solute Transport in Stratified Porous Media: 2. The Reactive Case," *Water Resources Research*, Vol. 21, pp. 1043–1050, 1985.

Sudicky, E. A., R. W. Gillham, and E. O. Frind, "Experimental Investigation of Solute Transport in Stratified Porous Media: 1. The Nonreactive Case," *Water Resources Research*, Vol. 21, pp. 1035–1041, 1985.

Theis, C. V., "The Relation Between the Lowering of the Piezometric Surface and the Rate and Duration of Discharge of a Well Using Groundwater Storage," *Transaction of the American Geophysical Union*, Vol. 2, pp. 519–524, 1935.

Todd, D. K., *Groundwater Hydrology*, 2nd Ed., John Wiley & Sons Inc., New York, NY, 1980.

Verschueren, K., *Handbook of Environmental Data on Organic Chemicals*, 2nd Ed., Van Nostrand Reinhold, 1983.

Vomocil, J. A., "Porosity," in *Methods of Soil Analysis*, Pt. I., C. A. Black, ed., American Society of Agronomy, Madison, WI, pp. 299–314, 1965.

Wagner, B. J., and S. M. Gorelick, "Optimal Groundwater Quality Management Under Parameter Uncertainty," *Water Resources Research*, Vol. 23, No. 7, pp. 1162–1174, 1987.

Walton, W. C., *Groundwater Resource Evaluation*, McGraw-Hill, 1970.

______, *Practical Aspects of Ground Water Modeling*, 2nd Ed., National Water Well Association, 1985.

______, *Groundwater Pumping Tests: Design and Analysis*, Lewis Publishers, Inc., 1987.

Wang, H., and M. P. Anderson, *Introduction to Groundwater Modeling: Finite Difference and Finite Element Methods*, W. H. Freeman and Company, 1982.

Weast, R. C. (editor), *Handbook of Chemistry and Physics*, 66th Ed., CRC Press Inc., 1985.

Yeh, W. W.-G., "Review of Parameter Identification Procedures in Groundwater Hydrology: The Inverse Problem," *Water Resources Research*, Vol. 22, No. 2, pp. 95–108, 1986.

CHAPTER IV

Strategies for Design of Capture and Containment Remedial Systems

A. Introduction

The first three chapters reviewed the principles of groundwater contamination and summarized the types of field data needed for a proper analysis of a contamination event. We are now ready to turn to the central focus, that of design of pumping systems for capture and containment of groundwater contamination. In this chapter we look at several alternative strategies for the design of such systems. Chapter V presents a detailed design methodology, and Chapter VI provides qualitative guidelines for design in complex settings.

1. Modes of Occurrence of Groundwater Contamination

The remedial strategy to be employed at a site will depend on the location and mode of occurrence of groundwater contamination. Foremost, it is important to know whether contamination is located in unconsolidated surficial materials or in fractured bedrock. It is also important to know whether the contamination occurs in the unsaturated zone above the water table or in the saturated zone below the water table, or both. Finally, it is important to know whether the contaminants are miscible or immiscible, and if immiscible, whether they are LNAPLs or DNAPLs. Summarizing these issues, it is apparent that contamination at a site may occur in one or more of the four following modes: (1) as residual immiscible contamination in the unsaturated zone, (2) as pools of immiscible LNAPL contamination floating on the water table, (3) as pools of immiscible DNAPL contamination at depth in the saturated zone, and (4) as a dissolved-solute plume of miscible contamination in the saturated zone.

Because the emphasis of this book is on capture and containment systems, it should be noted that such systems are best-suited for the remediation of dissolved-solute plumes of miscible contamination in the saturated zone of an unconsolidated aquifer. The complexities discussed in Chapter VI are primarily those introduced by the presence of fractured rock, NAPL pools, and unsaturated-zone contamination. The existence of any of these complexities, especially if they occur in combination, may be sufficiently adverse to call into question our ability to successfully remediate a site.

2. Remedial Strategies and Remedial Technologies

It is important that capture and containment technology be seen within the context of a wider suite of available remedial technologies. First, let us differentiate, as the EPA does, between individual *remedial technologies*, such as capping or pumping, and a *remedial strategy*, which involves the grouping of one or more of these technologies into an overall remedial plan.

Within this framework, three broad remedial strategies could be developed for any particular site: (1) no action, (2) monitoring alone, and (3) monitoring and remedial action.

Given the availability of these three alternative strategies, a five-step process must be carried out for any site: (1) selection of the appropriate remedial strategy; (2) defense of the selected remedial strategy to the appropriate regulatory agencies involved; (3) design of the monitoring network for those sites requiring monitoring; (4) design of the component technologies of the remedial strategy for those sites requiring remedial action; and (5) construction and operation of the remedial systems.

The process used to determine whether remedial action is required at a site involves four major steps: (1) determining if contaminants are present in the groundwater, (2) identifying potential receptors of contaminated groundwater, (3) calculating potential future contaminant concentrations at the receptor points, and (4) determining if the concentrations at the receptor points are acceptable. In the highly regulated environment that has developed in the United States in the past decade, receptor points usually take the form of regulatory compliance points, and acceptable concentrations usually take the form of maximum concentration limits established by legislation and enforced by State or Federal regulatory agencies.

If remedial action is required, a large number of remedial technologies should be considered. Table 4 provides a list of some of the most commonly considered options. Remedial activities at "Superfund" sites, authorized under CERCLA over the past few years, provide a growing body of knowledge about the feasibility of these various technologies in different hydrogeological environments.

Table 4 shows that remedial action at most Superfund sites has included a requirement for some soil excavation. Capping has been widely

Table 4. Remedial Technologies

Objective	Remedial Technology
Source Removal	Excavation of Soils
Source Containment	Capping Cutoff Walls
Contaminant Removal: Unsaturated Zone	Soil-Vapor Extraction
Contaminant Removal Saturated Zone (cleanup)	LNAPL Bailing or Skimming Enhanced Recovery Flushing Steam DIsplacement Surfactants Enhanced Biodegradation Extraction
Contaminant Containment: Saturated Zone (migration control)	Extraction and Injection Wells Slurry Walls Sheet Piling

used at facilities where sources are of limited areal extent and source boundaries are well established. Cutoff walls have also been used at a limited number of sites. Soil-vapor extraction has been widely used but usually in combination with one or more of the other technologies. Removal of LNAPL from the water table by bailing or skimming has become relatively common. Methods based on enhanced biodegradation, or enhanced recovery through flushing, steam displacement, or the introduction of surfactants, have been investigated in the laboratory and, in some cases, at pilot-plant scale but are not yet sufficiently proven to constitute common alternatives for commercial remediation.

With this brief summary complete, the capture and containment technology can be placed in context. It has been the most common remedial technology to date, and it is likely to remain so. In some cases it may be coupled with excavation, capping, or soil-gas collection, but in most cases it will be the primary component of the remedial strategy, especially at sites where dissolved-contaminant plumes have developed in the saturated zone in unconsolidated surficial aquifers. We again note that capture and containment systems alone will usually not result in complete aquifer clean up. If clean up is achieved, it will often take decades or longer.

3. Objectives of Capture and Containment Systems

As indicated in Table 4 (and discussed earlier in Chapter I.B.4) there are two very different reasons why one might use pumping wells as a component of a remedial strategy. On the one hand, the goal may be contaminant removal from the plume; the purpose is often *hot-spot cleanup*. On the other hand, the goal may be to eliminate contaminant-plume migration; in this case, the purpose is *containment through migration control*.

For the *hot-spot cleanup* option, the primary component of a capture and containment technology will involve extraction wells located within the plume in the zones of highest concentration. The contaminant-removal option is most effective for high-solubility contaminants forming a solution with a density close to that of water and little predilection for sorption or matrix diffusion. Unfortunately, few contaminants meet these specifications. Extraction wells cannot be expected to be efficient in removing contaminants that are sorbed on aquifer materials, contaminants that have diffused into the matrix of low-permeability materials, or contaminants that occur as LNAPL or DNAPL pools. The desorption and matrix-diffusion processes are likely to be slow, and pumping of many aquifer volumes, more or less in perpetuity, would be required to attain complete cleanup. LNAPL or DNAPL pools (or even residual globules, fingers, and ganglia) are equally hard to remove. After all, petroleum engineers only achieve partial recovery of oil and gas from their reservoirs, even with secondary-recovery waterflooding techniques. It is unlikely that remedial action DNAPL recovery efforts will be complete. The usual observation in pump-and-treat extraction systems is that contaminant concentrations decline over time to some low non-zero value. At that point, large volumes of water are being treated to remove small quantities of contaminants. When pumping is stopped, concentrations in the groundwater often rise again. We are forced to conclude that once a subsurface volume of aquifer has been contaminated it is difficult, if not impossible, to return the aquifer to its pristine condition. Even after significant contaminant removal, it is likely that many groundwater samples taken on the site will still fail to meet maximum concentration limits.

If we accept this discouraging scenario with respect to contaminant removal, we must turn to the other potential objective, that of migration control. Here, the potential for meeting remedial objectives is much more encouraging. For the migration-control option, it would be common to use both extraction wells and injection wells. Their purpose is to control hydraulic gradients in such a way that the advective plume-front velocity is reduced to zero, and the contaminant plume is contained within the volume of aquifer material already contaminated prior to the instigation of remedial action. Some contaminant-removal wells might also be included in the design, but with the recognition that complete cleanup cannot be expected. It is noteworthy that by providing optimal containment, enhanced remediation methods such as surfactant injection or bioremediation may be used.

Our emphasis will be on pump-and-treat systems that are designed with the objective of hydraulic containment of dissolved contaminant plumes through migration control.

4. Components of Capture and Containment Systems

Capture and containment systems usually involve several components, including (1) extraction wells, (2) injection wells, (3) pipeline networks, and (4) treatment facilities. Extracted water is treated to remove contaminants to a level that meets regulatory standards. Depending on the situation, the treated water may be reinjected into the aquifer by means of injection wells, made available for water-supply use, or released to surface water. The simplest injection/extraction systems to design are those that do not involve water-supply or surface-water release, in which case total injection rates can be set equal to total extraction rates, and local hydraulic gradient control can be achieved without regional water-level declines. However, it is often found that design injection rates are hard to sustain due to the clogging of well screens and well-pack materials.

Potential treatment systems may be based on physical separation through carbon adsorption or air-stripping, chemical treatment such as oxidation, or biological treatment involving activated sludge. Carbon adsorption systems and air-stripping towers have been widely used in the Superfund program. Potential system designs span the range from individual treatment facilities associated with each extraction-injection well pair to large central treatment facilities connected to the wells by complex pipeline networks. In urbanized areas, the addition of a pipeline network to an already-complex network of roads, sewers, and service lines may not be a simple matter. There may be trade-offs between the costs of complex pipeline networks and the benefits of large central treatment facilities. Our mandate is limited to the design issues associated with extraction and injection well networks.

Information on costs of all components of a pump-and-treat system can be obtained from the CORA software developed by the EPA and is available through their contractor: CH_2M–Hill, Mid-Atlantic Office, P.O. Box 4400, Reston, VA 22090.

B. A Framework for Design

In this subsection we begin to address the question of how to design a network of extraction wells and injection wells for the purpose of plume-migration control.

In a more general context, the process of engineering design involves a sequence of decisions between *alternatives*. Alternatives are established so that they meet the technical *objectives* of the project. In most projects it is necessary to meet these objectives within a set of technical, legal, political, or economic *constraints*. Engineering alternatives are differentiated from one another on the basis of their technical components. The variables that can be used to define and differentiate alternatives are known as *decision variables*. Decision variables may take on discrete values, giving rise to discrete alternatives; or they may be continuous functions, giving rise to a continuous range of alternatives.

Designers base their decisions on an economic analysis of the alternatives. The design framework must provide a link between the economic milieu in which decisions are made and the results of the technical analyses on which decisions are based. A discussion of the various methods that can be used to determine which alternative is "best" is postponed until Chapter IV.C. Let us first clarify the concepts associated with alternatives, objectives, constraints, and decision variables for the case at hand.

1. Decision Variables, State Variables, and Hydrogeological Parameters

The alternatives that can be identified for a capture and containment system revolve around the well-network geometry. The *decision variables* include (1) the number and location of extraction wells, (2) the number and location of injection wells, (3) the pumping and/or injection rates for each well, and (4) the pumping and/or injection schedules for each well. These are the variables that can be specified, managed, or controlled by the design engineer. The purpose of the design process is to identify the best combination of these decision variables. If we specify the pumping rate, Q, as a function of space and time (positive for extraction, negative for injection), then all four of the decision variables identified above are essentially collected into a single decision variable, Q(x,y,t).

All the design approaches described in this chapter involve the use of a simulation model of groundwater flow and transport. In its most general form, the simulation model has a hydraulic component based on the flow equation and a contaminant component based on the transport equation. The *state variables* in this context are the hydraulic head, which is the dependent variable in the flow equation, and the concentration, which is the dependent variable in the transport equation. In steady-state simulations, the state variables are functions of position; in transient simulations, they are functions of position and time. Many contaminant simulations couple a steady-state hydraulic model with a transient transport model.

Hydrogeological parameters include all media properties such as porosity, hydraulic conductivity, transmissivity, storativity, dispersivity, and the like. In heterogeneous media these parameters may vary through space, but they usually do not vary with time.

The input to a transient transport simulation must include information on the spatial distribution of the hydrogeological parameters, together with information on the initial conditions and boundary conditions. Output from the simulation takes the form of predicted changes in the state variables through space and time. In some cases, output includes calculations of *auxiliary variables*, which are additional, conveniently defined output quantities such as velocities, gradients, travel times, or capture zone dimensions.

2. Objective Functions

For this discussion, assume that the goal of our pumping scheme is the containment of a contaminant plume such that no further downgradient migration of the plume front occurs. There may be a number of alternative well networks that can meet this goal within the types of constraints discussed in the following subsection. Such feasible alternatives are compared with one another on the basis of an economic objective function.

From the perspective of the engineer (or the owner-operator he or she represents), we can define an objective function as the net present value of the expected stream of remedial costs, taken over an engineering planning horizon, and discounted at the market interest rate. If an objective function, Z_j, is defined for each $j = 1 \ldots N$ alternatives, then the goal is to minimize Z_j, where

$$Z_j = \sum_{t=0}^{T} \left[\frac{1}{(1 + i)^t} \right] [C_j (t)] \tag{51}$$

and Z_j = objective function for alternative j [$],
T = planning horizon [years],
i = annual discount rate [decimal fraction], and
$C_j(t)$ = costs of alternative j in year t [$].

For a remedial pump-and-treat scheme, the costs include the *capital costs* associated with site investigation, well installation, and treatment-facility construction; and the *operational costs* associated with pumping and treatment over the life of the project. If the alternatives all have similar total pumping rates, then the costs associated with treatment will be more-or-less common for all alternatives and they can be removed from the comparative analysis. Furthermore, the differences in capital costs between alternatives are often small, and the relative merits of the alternative pump-and-treat network designs are decided on the basis of their relative pumping costs (where the term *pumping* includes both *extraction* and *injection*).

In Chapter V, the optimization procedures are carried out with an objective function that emphasizes total pumping. In fact, if treatment costs are the dominant expense and are assumed to depend only on pumping rates, then the total pumping rate is an economic surrogate for the total cost in the objective function. As shown in Chapter V.D.3, this leads to a linear objective function that can be solved with a linear-programming algorithm. If, on the other hand, pumping costs are the dominant expense and are assumed to be a function of both the pumping rate and the total lift to bring water from the well bore to the surface, a quadratic objective function results and a quadratic-programming approach is needed.

There are many other technical objectives, other than those associated with minimizing pumping, that could be formulated in an objective function. For example, the objective function might reflect an attempt to minimize the maximum drawdown, maximize the minimum hydraulic head, or minimize the sum of squared deviations from target heads, drawdowns, gradients, or velocities. There are also alternative economic criteria to that of maximizing net present value. There are criteria based on minimizing maximum regret, where regret is defined as the opportunity loss suffered by making a non-optimal decision. There are criteria that give greater weight to alternatives that exhibit robustness over a wide range of potential technical or economic conditions. It is also possible to formulate multiobjective problems in the same framework as the one we have described for a single objective, but of course solution methodologies are more complex. Having drawn attention to some of the complexities, we proceed now along the simpler and more conventional track based on a single-objective, cost-minimization objective function.

3. Constraints

The objective must be met, and alternatives compared, within a set of constraints derived from technical, economic, legal, or political conditions associated with the remedial project. There may be constraints on decision variables, state variables, or auxiliary variables. They may take the form of either equalities (e.g., drawdown must equal 10 meters) or inequalities (e.g., drawdown must not exceed 10 meters).

Constraints on decision variables may involve the number of wells or their pumping rates. With respect to pumping rates, it may be necessary to meet a certain demand or, on the contrary, it may be necessary not to exceed a certain capacity. There may be limitations on rates for individual wells or on total well field pumping. There may be limitations on the rates themselves or possibly on the changes in rates that are acceptable. In extraction/injection scenarios, there may be a requirement that the two be balanced or that a particular imbalance be maintained.

Constraints on the state variables might include requirements that hydraulic heads be maintained above a certain level or below a certain level or that contaminant concentrations not exceed regulatory standards at a compliance point.

Constraints on auxiliary variables could include limitations on the magnitudes of drawdowns, gradients, or velocities; or the restrictions of gradients or velocities to certain directions. In the design of pump-and-treat systems for migration control, the overall technical objective of attaining containment of a contaminant plume may often be replaced by a series of gradient-control constraints.

The linear- and quadratic-programming optimization techniques described in Chapter V are capable of identifying the alternative that minimizes cost while simultaneously satisfying all of the constraints.

4. Program Integration

The design process for remediation of contaminated groundwater involves a sequence of at least three steps: (1) design of a site investigation program, (2) design of the remedial well network, and (3) design of a monitoring network. Each step involves a decision among alternatives. How many holes will be drilled during site investigation? How many wells are needed in the pump-and-treat network? What monitoring-well spacing is required?

In this book, the design framework is limited to the design of the remedial well network itself. However, it is important to emphasize that a successful remedial action is dependent on (1) a proper site investigation prior to design of the remedial action and (2) a thorough performance assessment program during and after construction of the remedial system. We have discussed data needs in Chapter III, and we address performance assessment in Chapter VII. The point to be made here is that the same design framework used for the remedial network, with decision variables, objectives, and constraints, can also be used for the design of site investigation programs and monitoring networks.

Better yet, the framework might be expanded to allow an integrated design process that allows the engineer to assess economic trade-offs between the various steps. Would it be better, for example, to use minimal site investigation and conservative design; or would it be better to carry out a detailed site investigation in the hopes of buying reduced construction costs. The owner-operator would like to know how to partition his or her resources among the competing requirements of site investigation, remedial action, and monitoring. While this type of expanded and integrated design process is desirable, it has not yet been developed into on-the-shelf technology and, therefore, a detailed methodology is not presented here.

5. Deterministic and Stochastic Analysis

The process of engineering design involves making decisions under conditions of uncertainty. This is particularly so in engineering projects that require a knowledge of the hydrogeological environment, where uncertainty as to the system's properties and expected conditions is far greater than in most traditional engineering practice. There is uncertainty associated with the parameter values needed for design calculations and with the very geometry of the system being analyzed. The uncertainties of lithology, stratigraphy, and structure introduce a level of complexity to hydrogeological analysis that is completely unknown in most other engineering disciplines.

Recognition of these uncertainties has led hydrogeological researchers to adapt geostatistical techniques, first developed in the mineral exploration field, to a hydrogeological context. Geostatistical interpretations of field data can be used to generate probabilistic interpretations, whereby our uncertainty as to the geometry of the geologic system and the values of hydrogeological parameters within the geologic system can be placed in a quantitative framework.

This leads to two possible approaches to the simulation component of the design framework: deterministic analysis and stochastic analysis. In a *deterministic analysis*, all initial conditions, boundary conditions, and hydrogeological parameter values are assumed to be known with certainty. In a *stochastic analysis*, one or more of these features is represented as having a distribution in probability.

The classical approach to groundwater modelling is deterministic. The modeller estimates the most likely parameter values and then makes a single simulation to estimate the most likely output values. Deterministic modelling is often carried out in conjunction with sensitivity analysis, whereby a set of simulations are run to investigate the influence of changes in input parameters on output variables. This provides a quantitative assessment of the impact of changes in parameter values across the range of uncertainty, but it does not associate a probability with each of the possible outcomes.

In one type of *stochastic analysis*, geostatistical methodology is invoked to generate a set of equally likely realizations of the hydrogeological environment at a site. The simulation model is applied to each realization.

With this approach, the final answer is not merely a set of single-valued output variables. Instead, each variable at each location at each point in time has a probability density function associated with it. The mean value can be interpreted as the most likely value; it should be equal to the value calculated using a deterministic simulation. The variance can be interpreted as a measure of the uncertainty in the output variable generated by the uncertainty in input parameters.

There are a variety of possible methods of stochastic analysis, but the most commonly used one is that of Monte Carlo analysis as described above. This technique involves multiple runs with the exact same simulation model that would require only one run in a deterministic analysis. Stochastic analysis is thus much more computer-intensive than deterministic analysis.

We introduce these stochastic concepts in preparation for the next part of this chapter, wherein the differentiation between decision analysis and optimization analysis rests in part on an understanding of these concepts. However, when it comes to our recommended design procedures, both the simulation-analysis approach in Chapter IV.D and the optimization approach in Chapter V are carried out in a deterministic rather than a stochastic framework.

C. Alternative Approaches to Design

Figure 4 summarizes the various options that have been considered in our discussion of a design framework for pump-and-treat remedial systems. The asterisks identify the decisions we have reached with respect to the various options, at least insofar as they apply to the quantitative design methods presented in this chapter and the next. These design methods are limited to *saturated* conditions and are best suited to *unconsolidated* aquifer materials or bedrock aquifers that are porous and/or sufficiently fractured to be treated as an equivalent porous medium. They can be applied to both uniform and non-uniform *steady flow* and under transient flow conditions. It is assumed that the objective of the remedial system is migration control of a *dissolved-solute plume*. Both analytical and numerical simulation methods are utilized, but they are limited to a *hydraulic treatment* of the problem that assumes the primacy of *advective transport*. We assume a *deterministic framework* that does not take uncertainty into account, except possibly by means of a sensitivity analysis. Our objective function is one that *minimizes cost*.

With these conditions in mind, four alternative approaches to design are identified at the bottom of Figure 4: (1) simulation, (2) simulation plus optimization, (3) simulation plus decision analysis, and (4) qualitative guidelines. These alternatives are more clearly laid out on Figure 5, where we differentiate between simple systems, optimizable systems, and non-optimizable systems. For simple systems, application of a simulation model alone may suffice as an approach to design. Optimizable systems make use of a simulation model coupled to an optimization procedure or a decision-analysis procedure. Non-optimizable systems are too complex to satisfy the assumptions required by the simulation model, the optimization procedure, or the decision-analysis procedure; in such cases, we must revert to a design process based on qualitative guidelines.

Figure 5 shows three approaches for optimizable systems. These require further discussion. They differ from one another in three ways: (1) the upper approach uses a deterministic simulation model; the middle and lower approaches are stochastic; (2) the upper and middle approaches use an optimization model; the lower uses decision analysis; and (3) the upper optimization model uses linear and quadratic programming; the middle one uses nonlinear programming. The difference between deterministic and stochastic analysis has been discussed above. The difference between linear and nonlinear programming will be described in Chapter V. The difference between optimization and decision analysis is discussed next.

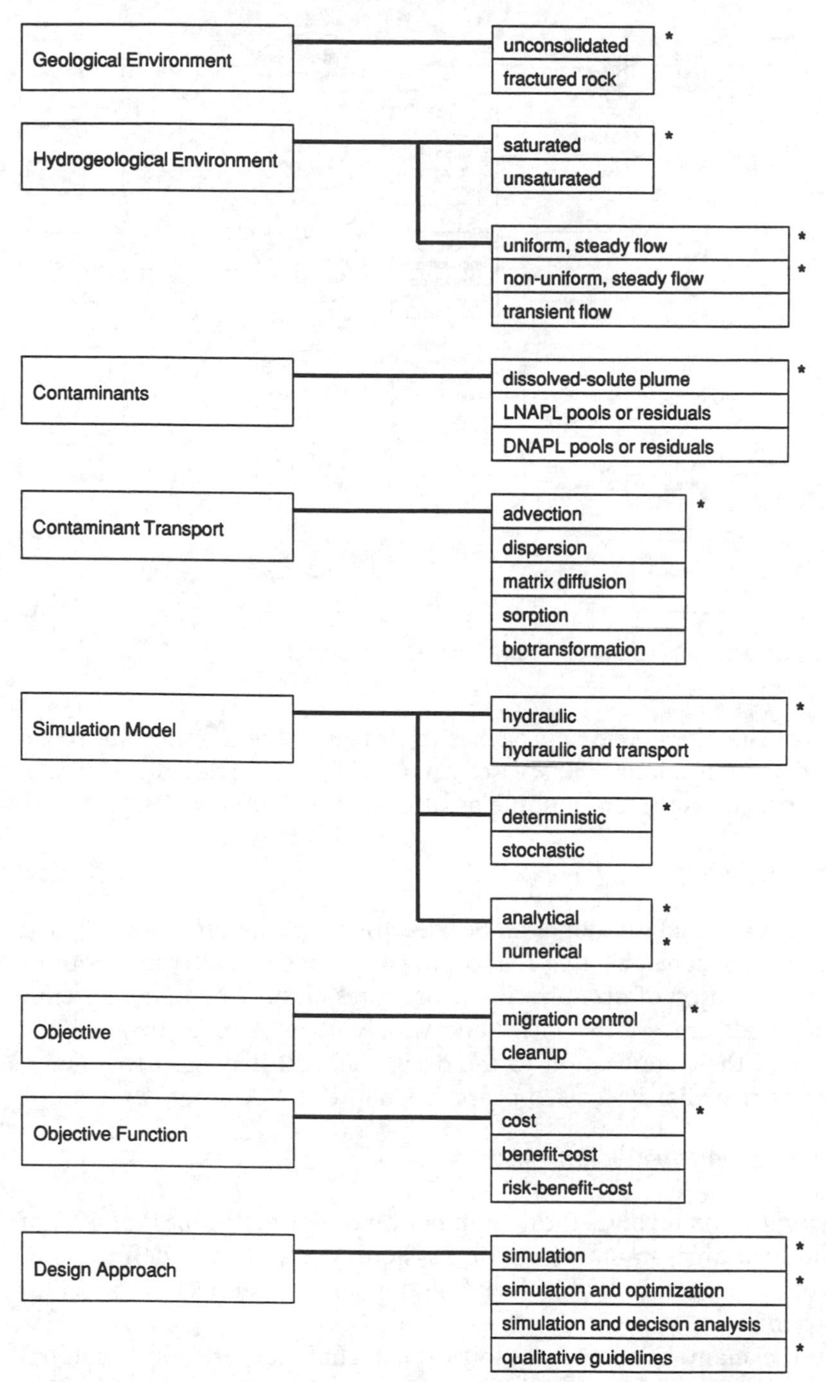

Figure 4. Summary of design options for pump-and-treat remedial systems; asterisks indicate conditions best suited to pump-and-treat remediation.

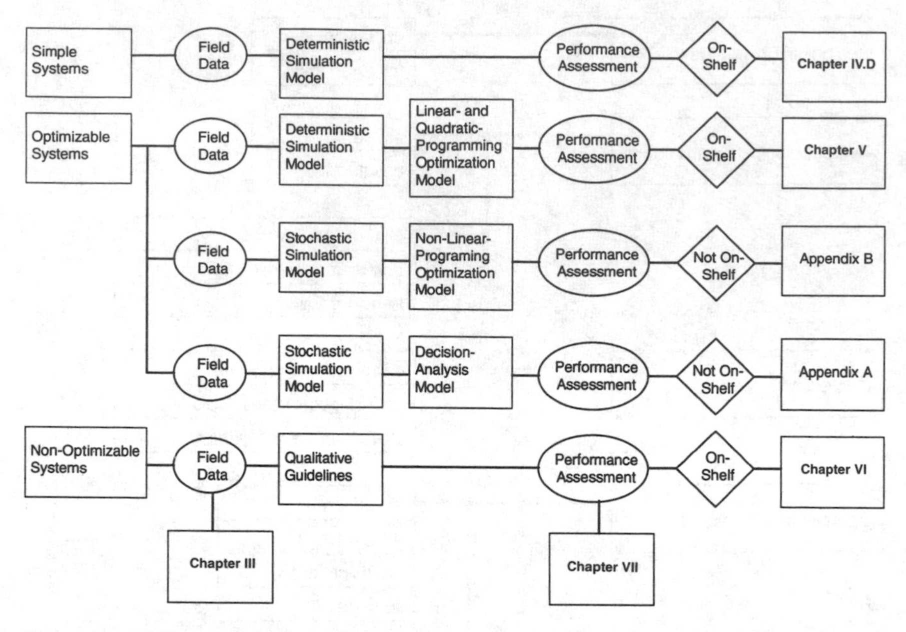

Figure 5. Differentiation of alternative remedial approaches.

All of the alternative approaches to design in Figure 5 are based on field data collection as discussed in Chapter III, and all require a performance-assessment module as discussed in Chapter VII.

1. Simulation

Simulation analysis alone can be used for simple systems as suggested in Figure 5, or it can be used for scoping and screening alternatives prior to the application of optimization procedures in more complex systems. The methods are "on-the-shelf" and widely used. A more detailed discussion of their application to the design of well networks for pump-and-treat remedial systems appears in Chapter IV.D.

2. Simulation Plus Optimization

Optimization involves the determination of optimal values for a set of decision variables in an engineering system. Optimality is defined with respect to a specified objective function and is subject to a set of constraints.

Of the many available optimization techniques, the one that has proven most popular and tractable for coupling with groundwater simulation models is linear programming. It requires a linear objective function, linear constraints, and linear flow equations in the simulation model. If the latter two linearities are retained, it is possible to move from a linear to a quadratic objective function and remain in the same optimization framework. Gorelick (1983) provides a review of linear and quadratic optimization techniques applied to groundwater problems. Lefkoff and Gorelick (1987) provide a user's manual for their linear and quadratic optimization program, AQMAN, which is the program described and recommended in Chapter V. This program is "on-the-shelf" and can be applied directly to the design of extraction-well/injection-well networks. Gorelick and Wagner (1986) and Lefkoff and Gorelick (1986) report applications to aquifer remediation.

The requirement for linear flow equations in the simulation model limits applications to treating systems as confined aquifers. The methods presented in Chapter V are applicable to unconfined aquifers only if the drawdowns and buildups in hydraulic head created by the extraction and injection wells are small in comparison with the total saturated thickness of the aquifer. Under these conditions, the confined-aquifer equations can be applied to an unconfined aquifer with little loss in accuracy. There are also simple iterative methods whereby the nonlinear equations can be linearized (cf. Danskin and Gorelick, 1985). This technique is also demonstrated in Chapter V.

At the research level, restrictions with respect to nonlinearity have been fully removed (Ahlfeld et al., 1988a, 1988b) . Gorelick et al. (1984) discuss aquifer reclamation design with a simulation/optimization methodology that allows either linear or nonlinear objective functions, constraints, and flow and transport equations. Unfortunately, documented manuals for the programs used in that study are not yet available.

It has been traditional to apply simulation/optimization techniques in a deterministic framework, and this is the most common framework utilized in "on-the-shelf" optimization packages. Researchers have investigated the effects of uncertainty on optimization problems through stochastic analysis (cf. Gorelick, 1987; Ward and Peralta, 1990). There are two ways that a linear program can be applied in a stochastic framework. *Stochastic linear programming* treats the coefficients of the objective function and/or the constraints as random variables, but the constraints hold with probability equal to one. *Chance-constrained programming* states the constraints probabilistically. The coefficients are treated deterministically but the constraints are only satisfied on an expected value basis. Wagner and Gorelick (1987) present a chance-constrained nonlinear optimization solution to the plume-capture problem. Tung (1986) describes a chance-constrained model in a groundwater management context. Ward and Peralta (1990) present an "on-the-shelf" code that is capable of solving chance-constrained or deterministic optimization problems for short-term emergency plume containment.

3. Simulation Plus Decision Analysis

Decision analysis involves the determination of the best alternative (that is, the best values for a set of decision variables) from a discrete set of specific alternatives. For example, we might wish to decide between a particular three-well extraction system and a particular five-well extraction system. Decision analysis is less general than optimization in that optimization provides the optimal alternative from the set of all possible alternatives, whereas decision analysis provides only the best alternative from a specified set of alternatives. On the other hand, it is less limited with respect to linearity than are linear-programming optimization techniques.

There is a fundamental difference in the treatment of objectives and constraints between optimization and decision analysis. In an optimization framework, the objective function involves only the costs [Equation (51)], or in some other applications, the costs and benefits. The performance requirements on the engineered system appear as constraints. In a decision-analysis framework, the potential failure to meet performance requirements produces risks, and the risks are given dollar value and included with the benefits and costs in the objective function. The goal becomes to maximize Z_j over j = 1. . . ..N alternatives, where

$$Z_j = \sum_{t=0}^{T} \left[\frac{1}{(1 + i)^t}\right] [B_j(t) - C_j(t) - R_j(t)] \tag{52}$$

and $B_j(t)$ = benefits of alternative j in year t [$],
$C_j(t)$ = costs of alternative j in year t [$], and
$R_j(t)$ = risks of alternative j in year t [$].

The risks, $R_j(t)$, associated with alternative j in year t are defined as

$$R_j(t) = [Pf_j(t)]\ [Cf_j(t)] \tag{53}$$

where $Pf_j(t)$ = probability of failure of alternative j in year t [decimal fraction], and
$Cf_j(t)$ = cost associated with a failure of alternative j in year t [$].

For a remedial pump-and-treat scheme designed to provide migration control for a contaminant plume, failure would be defined by the spread of contamination into previously uncontaminated areas of the aquifer. This could occur by contaminants slipping downstream through the well network or by lateral migration of contaminants across the presumed capture-zone boundaries. A failure of the system would be associated with a failure to meet constraints on the magnitude and/or direction of drawdowns, gradients, or velocities. Whatever the reasons, there will be expected costs associated with potential failures. These could take the form of regulatory penalties, loss of goodwill in the community, possible facility closure, and/or the costs of further remedial action.

In many risk-analysis textbooks (cf. Crouch and Wilson, 1982), Equation (53) has a third factor on the right-hand side. It is a term that allows one to take into account the risk-averse nature of some decision makers. We will not address this issue.

The primary point to be made here is that decision analysis with a risk-cost-benefit objective function requires a stochastic analysis. The probability-of-failure term in the objective function can be determined with a simulation model operating in Monte Carlo mode. The Monte Carlo simulations are carried out on a set of geostatistically generated realizations of the hydrogeological regime that reflects our uncertainty as to the geological system, hydrogeological parameter values, and/or initial plume distribution. It is this input uncertainty that creates output uncertainty and the output uncertainty that creates risk. There is risk associated even with respect to the "best" design geometry we can produce for the network of remedial wells.

Risk can be reduced by a commitment to additional costs, either for additional site investigation, which reduces input uncertainty, or for an increased number of pumping wells, which provides a more conservative remedial design. The "best" design from the owner-operator's perspective is the one that maximizes Z_j, not one that satisfies any predetermined acceptable level of risk. One might contrast this approach with a chance-constrained optimization scheme, which is also a stochastic approach. There, the probability of failure to meet a constraint is not usually coupled with the cost of failure; and the optimal solution is usually determined for a specified probability of failure, Pf, (or reliability, 1 – Pf), a process that is similar in principle to setting an *a priori* acceptable risk.

In summary, decision analysis is less general than optimization, and it is well-suited to a risk-based philosophy of engineering design. Its greatest weakness lies in the difficulties associated with quantifying the anticipated cost of failure. The coupling of a stochastic simulation model and a decision-analysis model is one of the possible design approaches for an optimizable system. This approach has recently been espoused by Massmann and Freeze (1987), but thus far their applications have been directed toward the design of new waste-management facilities rather than remedial action, and documented computer programs are not yet "on-the-shelf." We cannot recommend the approach as an alternative to optimization at this time. Nevertheless, future developments in this area are worth watching for, and a more detailed outline of the philosophy and methodology is included in Appendix A of this book.

D. Simulation Analysis

In very simple hydrogeologic settings or for scoping and initial screening of alternatives, simulation analysis applied alone, without subsequent optimization, can be an instructive design tool. It requires, as a start, an understanding of capture zones.

1. Capture Zones

The design framework for contaminant-plume migration control with a pump-and-treat remedial system is based on the concept of capture zones.

Capture zones are best explained for steady-state flow conditions in a horizontal, confined aquifer. Consider a small portion of such an aquifer, with a regional hydraulic gradient as shown in Figure 6a, and orthogonal regional flow directions as shown in Figure 6b. Now assume that it has been pumped for a sufficiently long time to attain steady-state conditions. A drawdown cone will have developed in the hydraulic head field as shown in Figure 6c, and the associated flow lines will be as shown in Figure 6d. Water will be drawn into the well from the stippled capture zone. It is located primarily on the (preoperational) upgradient side of the well but includes a small region of the downgradient side as well. Point B is a stagnation point.

The *capture zone* associated with an extraction well is defined as that portion of the aquifer that contains groundwater that will eventually be captured and discharged by the well. It does not include the entire area of perturbed heads, unless the velocity of the preoperational, natural flow system is zero.

Use of the capture-zone concept in remedial design should be clear from Figure 6d. If a contaminant plume exists within the stippled region, it will not migrate outside the capture-zone boundaries of Well A.

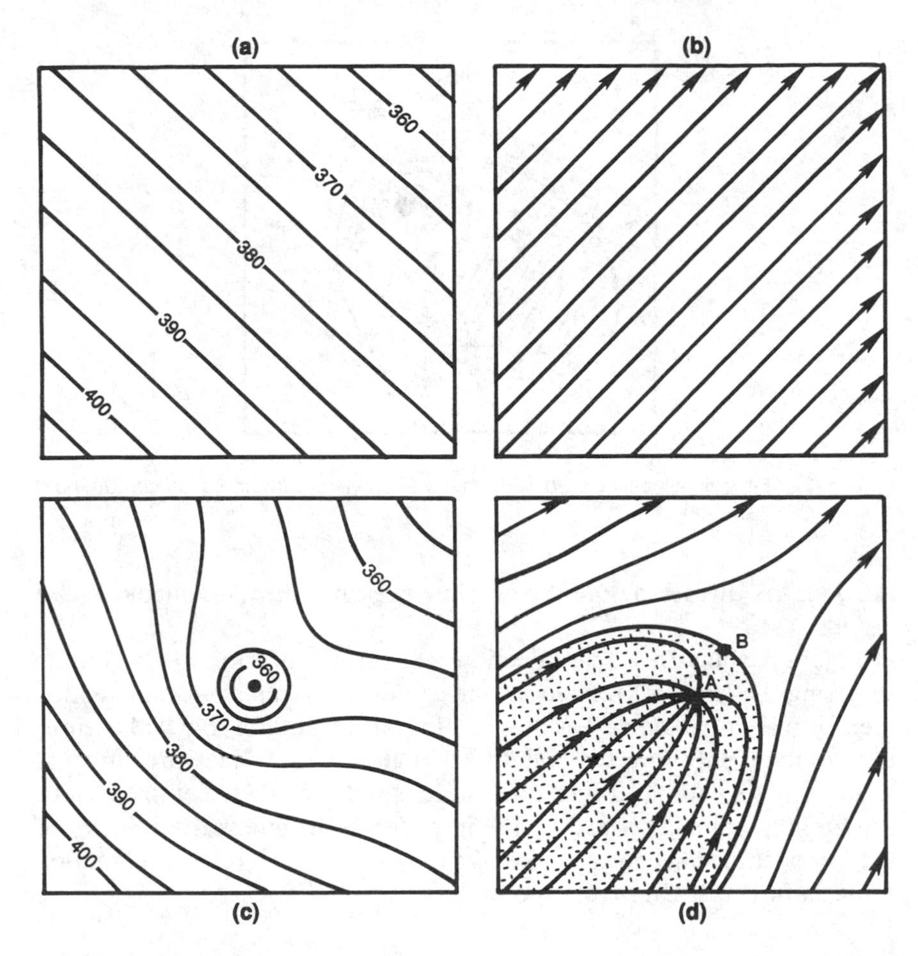

Figure 6. Capture-zone concept.

Given a contaminant-plume geometry, the design goal for a migration-control system is to establish a well network that will create a capture zone that encompasses the entire plume.

It must be emphasized that capture zones are a purely hydraulic concept. They therefore address only the advective component of contaminant transport. For this component, one can define time lines as shown in Figure 7. A plug of water inside the 1-year time line will be captured by Well A within 1 year. If a plume were totally encompassed by the 20-year time line as shown in Figure 7, one might be tempted to think that all contamination would be drawn into the well in 20 years. However, as noted earlier in our discussion of the limitations to total contaminant removal, this is not likely to be the case due to the influences of sorp-

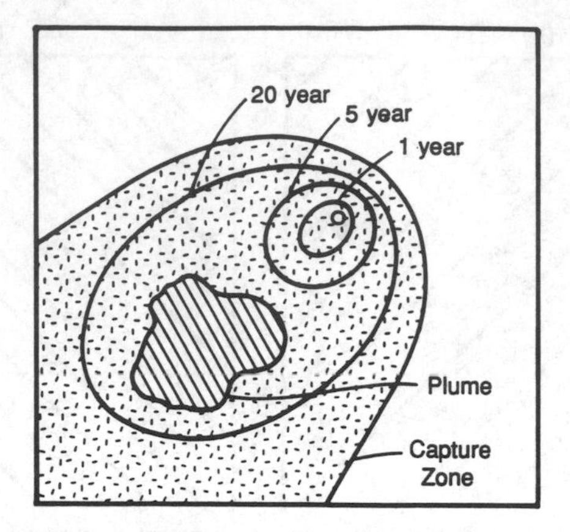

Figure 7. Time lines associated with the advective components of contaminant transport.

tion, matrix diffusion, and the possible presence of non-aqueous phase liquids.

Figure 8 shows a single injection well in a uniform flow field. By reversing the frame of reference, one can define a *rejection zone* associated with an injection well as that portion of the aquifer that will eventually contain only injected water. All regional flow lines are diverted around it. As with a capture zone, it does not include the entire area of the buildup cone generated by the injection well, unless the velocity of the preoperational, natural flow system is zero. Here the stagnation point is just upstream from the well.

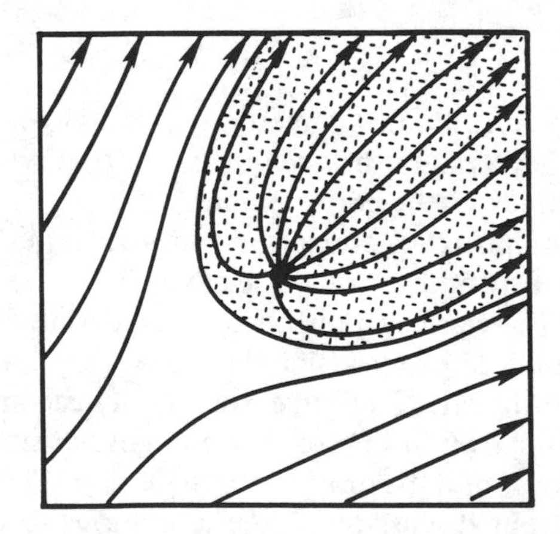

Figure 8. Single injection well in a uniform flow field.

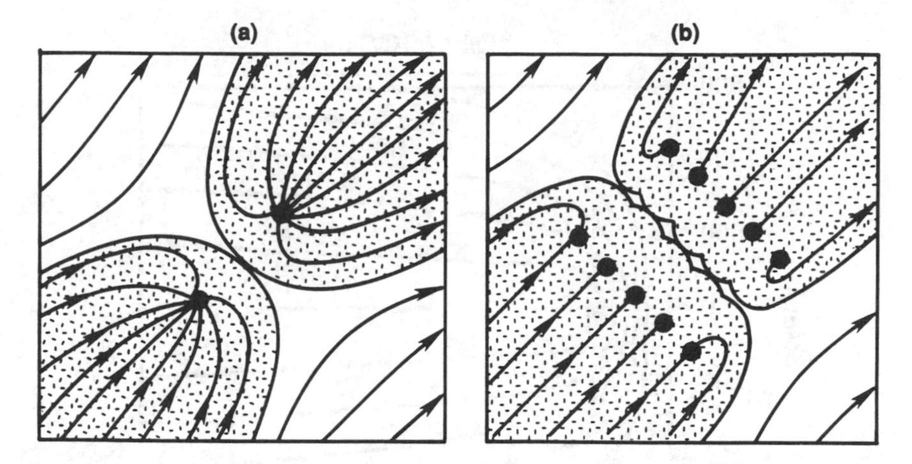

Figure 9. Flow regimes for (a) extraction/injection well pair and (b) paired line of extraction and injection wells.

Figure 9a shows the flow regime for an extraction/injection well pair, and Figure 9b shows a paired line of extraction and injection wells. In the latter case, the stagnant point becomes a stagnant zone. In both cases, migration control is achieved with little impact on regional flow.

The concept of a capture zone was introduced into the groundwater literature by Keely and Tsang (1983) and has been popularized through the widely used monograph of Javandel et al. (1984) and the paper based on it by Javandel and Tsang (1986). These latter papers provide general analytical solutions for capture-zone geometry for a two-dimensional representation of a homogeneous, isotropic confined aquifer under uniform, steady flow.

Under such conditions, capture-zone width is directly proportional to the pumping rate, Q [L^3/T], and inversely proportional to the product of aquifer thickness, b [L], and the regional specific discharge, q [L/T]. Javandel et al. (1984) provide capture-zone type curves for single- and multiple-well extraction systems. Figure 10 reproduces a set of their type curves showing capture zones for a four-well extraction system for several values of the parameter Q/bq. Note that because $q = KI$ and $T = Kb$, the parameter of Q/bq can also be expressed as Q/TI where T is aquifer transmissivity and I is the magnitude of the regional flow gradient.

The limitations on the analytical expressions for capture-zone geometry are quite severe, but most of them can be removed by using a numerical model to calculate the postoperational flow net. With numerical models, one can treat heterogeneous and anisotropic systems, uncon-

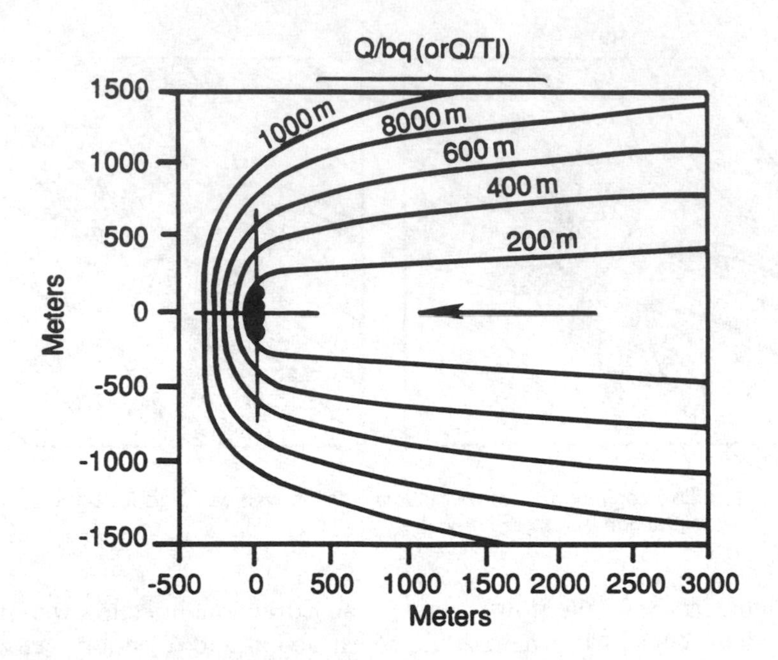

Figure 10. Capture zones for a four-well extraction system.

fined aquifers, nonuniform preoperational flow, and complex extraction/injection well networks. Shafer (1987a, 1987b) provides examples of such an approach. Figure 11a shows the hydraulic-conductivity pattern for one of his examples. Figure 11b shows the hydraulic-head distribution that results from the placement of a single pumping well in such an aquifer, and Figure 11c shows the 20-year capture zone for this well.

With the concepts associated with capture zones clearly in hand, we can proceed to show how they can be used in a simulation-based design procedure.

We will describe two approaches, one that applies to uniform-flow capture-zone analysis and one that applies to nonuniform-flow capture-zone analysis. The first uses analytical solutions and the programs RESSQ (Javandel et al., 1984) or DREAM (Rounds and Bonn, 1989). The second uses numerical solutions based on MODFLOW (McDonald and Harbaugh, 1984) and the programs GWPATH (Shafer, 1987b) or MODPATH (Pollock, 1989).

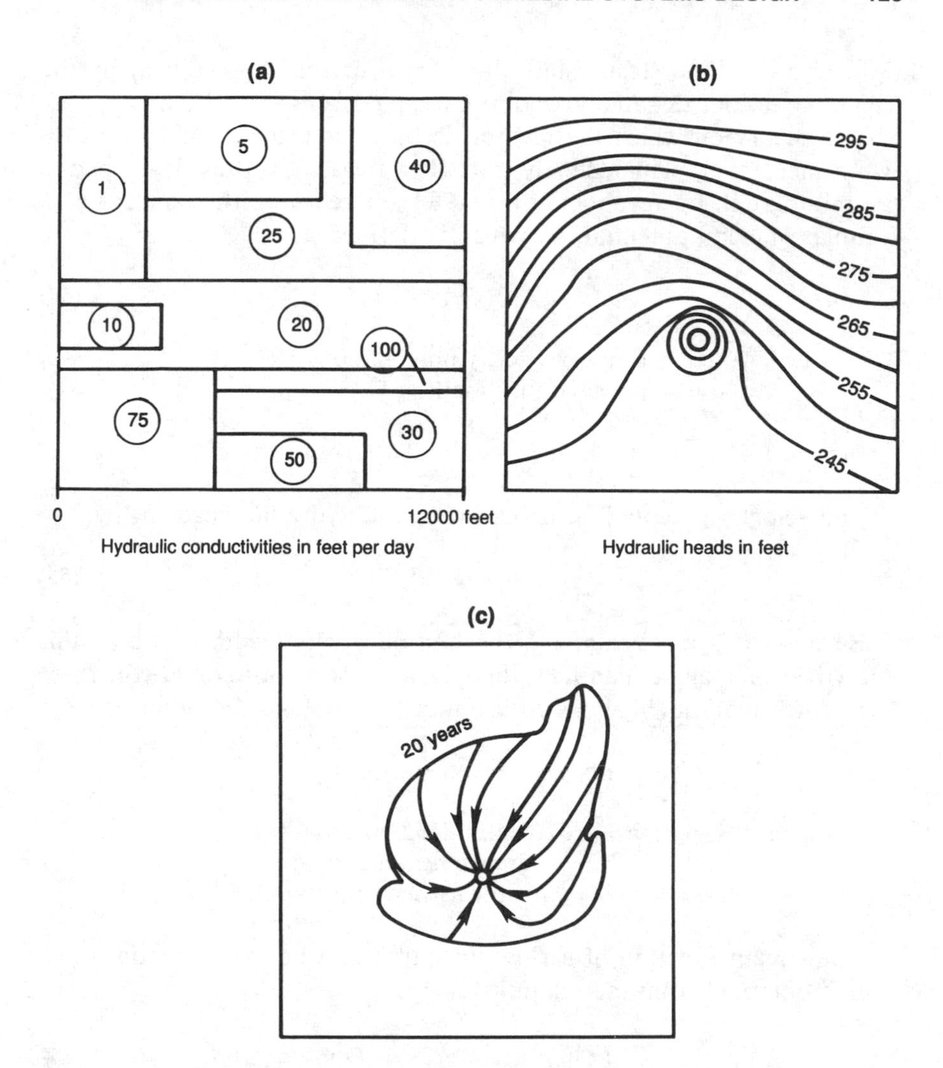

Figure 11. Capture-zone geometry using numerical models.

2. Uniform-Flow Capture-Zone Analysis

This subsection will address the design of a pumping-well network in a homogeneous, isotropic, horizontal confined aquifer, in which the preoperational, natural flow field is one of uniform, steady flow, like that shown earlier in this chapter in Figures 6a and 6b. In this simple system, the transient drawdown from an extraction well, or the transient buildup from an injection well, can be calculated using the Theis equation, and the final steady-state drawdown or buildup is given by the Thiem equation, as summarized in Chapter II.G of this book.

The theory on which the method is based is usually carried out on two-dimensional, planar, steady-state flow fields. It is fully developed in the monograph by Javandel et al. (1984). It relies on the concept of a complex velocity potential, Φ, which is defined as

$$\Phi(x,y) = \phi(x,y) + i\psi(x,y), \tag{54}$$

where $\Phi(x,y)$ = complex velocity potential $[L^2/T]$,
$\phi(x,y)$ = potential function $[L^2/T]$,
$i = \sqrt{-1}$, and
$\psi(x,y)$ = stream function $[L^2/T]$.

The velocity potential Φ, is related to the hydraulic head, h, by

$$\Phi = Kh. \tag{55}$$

Because K is a constant for the homogeneous media to which this analytical development applies, there is no reason not to divide the three terms of Equation (54) by K to produce a complex *head* potential:

$$P(x,y) = h(x,y) + iw(x,y) \tag{56}$$

where $P(x,y) = \Phi/K$ = complex head potential [L],
$h(x,y) = \phi/K$ = hydraulic head [L], and
$w(x,y) = \psi/K$ = modified stream function [L].

The stream function of a flow system with a known potential function is obtained from the relationships:

$$\left(\frac{\partial w}{\partial x}\right) = -\left(\frac{\partial h}{\partial y}\right) \qquad \left(\frac{\partial w}{\partial y}\right) = \left(\frac{\partial h}{\partial x}\right) \tag{57}$$

which simply state that h and w are everywhere orthogonal.

The components of specific discharge are given directly by Darcy's law:

$$q_x = -K\left(\frac{\partial h}{\partial x}\right) \qquad q_y = -K\left(\frac{\partial h}{\partial y}\right). \tag{58}$$

For a flow system that is influenced both by uniform regional gradients, and a number of extraction or injection wells, the equations that define h and w can be developed on the basis of the superposition principle. They are well-known from classical potential theory (Javandel et al., 1984):

$$h(x,y) = -I\,(x\cos\alpha + y\sin\alpha) - \sum_{i=1}^{n} \left(\frac{Q_i}{4\pi T}\right) \ln\left[(x - x_i)^2 + (y - yi)^2\right] \qquad (59)$$

$$w(x,y) = -I\,(y\cos\alpha - x\sin\alpha) - \sum_{i=1}^{n} \left(\frac{Q_i}{2\pi T}\right) \tan^{-1}\left(\frac{y - y_i}{x - x_i}\right) \qquad (60)$$

where I = preoperational, regional hydraulic gradient [decimal fraction]; note that I is always positive;
α = angle between regional flow direction and positive x axis [degrees or radians];
(x,y) = coordinates of point at which h and w are being evaluated [L];
Q_i = pumping rate of ith well [L^3/T], positive for injection, negative for extraction;
n = number of wells;
(x_i,y_i) = coordinates of ith well [L]; and
T = transmissivity of aquifer [L^2/T].

The first term in Equations (59) and (60) is due to the regional gradient. The second term is due to the pumping wells; the similarities in (59) to the Thiem equation (34) should be evident.

With Equations (59) and (60) it is possible to calculate h(x,y) and w(x,y) for a large number of locations, (x,y), and hence to map the two functions for any particular set of n pumping wells at locations, (x_i, y_i), with pumping rates, Q_i. Figure 6c, presented earlier in this chapter, is an example of an h(x,y) plot determined from Equation (59). Figures 6d, 8, and 9b are examples of w(x,y) plots determined from Equation (60).

Differentiating (59) with respect to x and y and multiplying through by -K, as indicated by Equation (58), provides analytical expressions for the components of specific discharge, q_x and q_y:

$$q_x(x,y) = KI\cos\alpha + \sum_{i=1}^{n} \left\{\frac{Q_i}{2\pi b}\,\frac{(x - x_i)}{[(x - x_i)^2 + (y - y_i)^2]}\right\} \qquad (61)$$

$$q_y(x,y) = KI \sin\alpha + \sum_{i=1}^{n} \left\{ \frac{Q_i}{2\pi b} \frac{(y - y_i)}{[(x - x_i)^2 + (y - y_i)^2]} \right\} \tag{62}$$

The components of the average linear velocity, V_x and V_y, are then given by

$$V_x = q_x/n_e \qquad V_y = q_y/n_e \tag{63}$$

where n_e is the effective porosity of the aquifer. This is the velocity at which contaminants move through the aquifer toward an extraction well during remediation. Calculations of V(x,y) can be used to develop plots of remedial time lines like those shown earlier in Figure 7.

For contaminants that are retarded, the velocity components are given by

$$V_x = q_x/n_eR_f \qquad V_y = q_y/n_eR_f \tag{64}$$

where R_f is the retardation factor, as defined in Chapter II.F. No comparably simple method is available to take into account dispersion or matrix diffusion in an analytic, hydraulics-based model.

The analytical methodology presented in this chapter can also be used to determine capture-zone geometry. Javandel and Tsang (1986) use Equation (60) to develop an equation for the dividing streamlines that separate the capture zone of a single well, pumping at Q, from the rest of the aquifer (Figure 12). For $\alpha = 0$, it is given by

$$y = \pm \left(\frac{Q}{2TI}\right) - \left(\frac{Q}{2\pi TI}\right)\tan^{-1}\left(\frac{y}{x}\right). \tag{65}$$

Solving this equation for $x = 0$ and $x = \infty$ allows one to calculate the distance between the dividing streamlines at the line of wells and far upstream from the wells. One can also calculate the downstream distance from the well to the stagnation point by solving for x at $y = 0$. For a single extraction well, these distances are given by Q/2TI, Q/TI, and Q/2πTI. Javandel and Tsang (1986) calculate these values for one-, two- and three-well extraction systems; Table 5 records their results. They also provide capture-zone type curves of the kind shown earlier in Figure 10 for one-, two-, three-, and four-well extraction systems.

Coming finally to the point of well-network design, Javandel and Tsang (1986) use their analysis to calculate the maximum distance that can exist between multiple wells such that capture zones are continuous

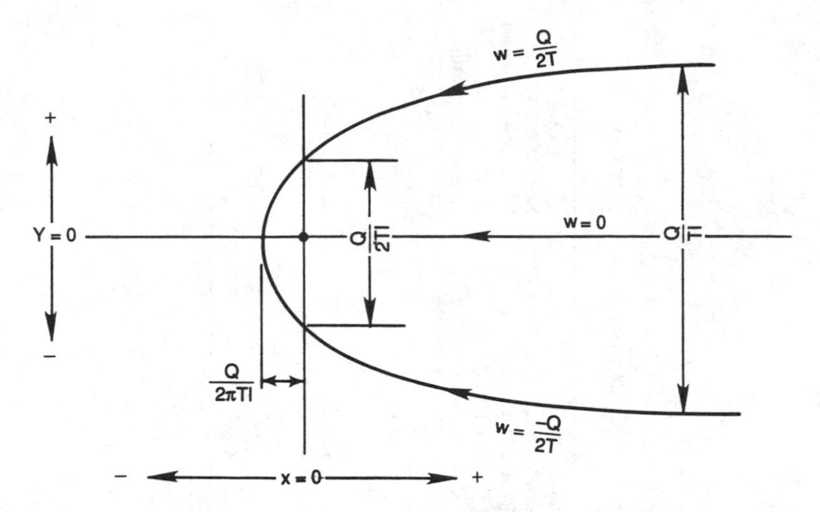

Figure 12. Equation for the dividing streamlines separating the capture zone of a single well from the rest of an aquifer.

and no flow tubes (or contaminants) can slip between the extraction wells. For two or three equally spaced wells, located along a line perpendicular to the regional gradient, and all pumping at the same rate, Javandel and Tsang provide the recommended spacings listed in the right-hand column of Table 5.

The design methodology for a one-, two-, or three-well extraction system using Table 5 involves a trial-and-error procedure with a set of alternative well networks. One tries to identify the lowest cost network that will meet the following specifications, given measured values for aquifer transmissivity, T, and regional hydraulic gradient, I:

1. The capture-zone geometry, as indicated by the values given in Table 5 for the distance between dividing streamlines, must be adequate to encompass the known boundaries of the contaminant plume.
2. The pumping rate, Q, to be applied at each of the wells, must not create drawdowns in excess of any constraints on the available drawdown at the wells.
3. The distances between the wells must be equal to or less than the recommended distances given in Table 5.

It must be emphasized that use of Table 5 to design remedial well networks will *not* lead to an optimal design. The limitations on the analytical solutions on which the table is based are too severe. It will provide a design that works for a pre-specified number of wells, all on a

Table 5. Parameters for Design of Remedial Well Fields Based on Javandel and Tsang (1986) Capture-Zone Theory. For multiple-well systems, Q is the constant pumping rate applied to each well.

Number of Wells	Distance Between Dividing Streamlines at Line of Wells	Distance Between Dividing Streamlines Far Upstream From Wells	Downstream Distance to Stagnation Point at Center Point of Capture Zone	Recommended Distance Between Each Pair of Extraction Wells
1	$\frac{Q}{2TI}$	$\frac{Q}{TI}$	$\frac{Q}{2\pi TI}$	—
2	$\frac{Q}{TI}$	$\frac{2Q}{TI}$	$\frac{Q}{2\pi TI}$	$\frac{Q}{\pi TI}$
3	$\frac{3Q}{2TI}$	$\frac{3Q}{TI}$	$\frac{3Q}{4\pi TI}$	$\frac{^3\sqrt{2}Q}{\pi TI}$

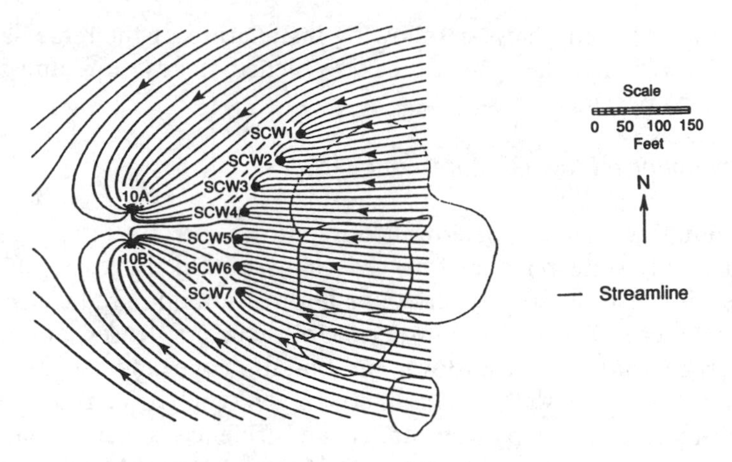

Figure 13. Predicted streamlines for the design network.

line, and all pumping at the same rate. The optimal solution might involve irregular spacings and/or pumping rates.

The concepts and methodology of this chapter have been embodied in a computer code named RESSQ. It is fully documented by Javandel et al. (1984). Given a homogeneous, isotropic, confined aquifer and a system of regional flow, which together with a set of extraction and/or injection wells, creates a steady-state flow field, RESSQ calculates and plots the streamline pattern in the aquifer. With a trial-and-error approach, one can examine the capture-zone geometry of any set of irregularly spaced wells, pumping at any desired rates. RESSQ has been included as the simulation component of a nonlinear optimization model (Greenwald and Gorelick, 1989).

RESSQ is not the only available code for producing streamline patterns. Rounds and Bonn (1989) describe a similar program called DREAM that is a user-friendly, menu-driven program designed specifically for personal computer application. They describe in more detail than Javandel et al. (1984) the complications that arise in the programming and plotting procedures due to the multivalued $\tan^{-1}$ function that appears in Equation (60). RESSQ is available through the International Ground Water Modelling Center at the Colorado School of Mines, Golden, CO 80401. DREAM is available from Lewis Publishers, 121 South Main Street, P.O. Box 519, Chelsea, MI 48118.

Cosgrave et al. (1989) have used a RESSQ-type code in the remedial design of a gradient-control well network at a coal-tar refinery in Illinois. Figure 13 shows the predicted streamlines for the design network. Routine monthly measurements of water-level elevations in a monitoring network during the year following installation of the remedial program showed that the system quickly reached steady state and is performing as designed.

3. Nonuniform-Flow Capture-Zone Analysis

In aquifers that are heterogeneous and anisotropic, the preoperational steady-state regional flow system is not likely to be uniform. Rather, it will show spatial variations in direction and gradient induced by the pattern of high- and low-permeability layers, lenses, and trends. Similarly, steady-state drawdown and buildup cones induced by extraction and injection wells are unlikely to be symmetric; they too will exhibit more complex patterns under the influence of permeability distributions. In these circumstances, RESSQ and DREAM are no longer suitable. The assumptions underlying their analytical basis are no longer satisfied.

It is, however, still possible to utilize deterministic simulation analysis to aid in the design of remedial well networks. It will be necessary to use a finite-difference or finite-element model that is capable of producing steady head distributions in heterogeneous media. There are many such flow-net simulators available. Perhaps the most widely used is MODFLOW, the USGS finite-difference model. This program is capable of transient simulation in three dimensions, and it has many options that allow consideration of wells, drains, streams, recharge, and evapotranspiration. However, it is written in a modular style that is well-suited to invoking the simpler option package we require: steady-state flow in two dimensions with wells. McDonald and Harbaugh (1984) provide easy-to-follow user instructions in their program book. MODFLOW can be run either on a mainframe computer or on a personal computer.

The output from MODFLOW is in the form of hydraulic-head values at finite-difference nodal points; the program does not provide pathline output. To obtain such output, one must use a post-processor program that constructs pathlines from potential-field output. Shafer (1987b) has developed such a program in his GWPATH code. It is an interactive software package for calculating pathlines and travel times in a two-dimensional planar flow field. It allows either forward or reverse pathline tracking, capture-zone analysis, and multiple-pathline capture detection mechanisms. The USGS program, MODPATH (Pollock, 1988; 1989), uses a particle-tracking scheme to develop pathlines for three-dimensional outflow from MODFLOW.

MODFLOW and GWPATH or MODPATH can be used in a trial-and-error format for well-network design in nonuniform flow fields in heterogeneous, anisotropic aquifers in the same way that RESSQ or DREAM is used in uniform flow fields in homogeneous, isotropic aquifers. Given the patterns of aquifer transmissivity, T(x,y), and regional gradients, I(x,y), the well locations, (x_i, y_i), and pumping rates, Q_i, are established such that drawdowns are acceptable and the plume is completely contained within the capture-zone geometry.

Once again, our caveat must be stated. Simulation alone using MODFLOW and GWPATH or MODPATH does not produce an optimized design, only one that is designed to work. In order to produce a design that minimizes a cost-based objective function, while satisfying all constraints, the simulation model must be coupled to an optimization procedure, as described in Chapter V.

References for Chapter IV

Ahlfeld, D. P., J. M. Mulvey, G. F. Pinder, and E. F. Wood, "Contaminated Groundwater Remediation Design Using Simulation, Optimization, and Sensitivity Theory 1, Model Development," *Water Resources Research*, Vol. 24, No. 3, pp. 431–442 1988a.

______, "Contaminated Groundwater Remediation Design Using Simulation, Optimization, and Sensitivity Theory 2, Analysis of a Field Site," *Water Resource Research*, Vol. 24, No. 3, pp. 443–452, 1988b.

Cosgrave, T., P. Shanahan, J. C. Craun and M. Haney, "Gradient Control Wells for Aquifer Remediation: A Modelling And Field Case Study," *Proceedings of National Water Well Association Fourth International Conference on Solving Ground Water Problems with Models*, pp. 1083–1107, 1989.

Crouch, E. A. C., and R. Wilson, *Risk/Benefit Analysis,* Ballinger, 1982.

Danskin, W. R., and S. M. Gorelick, "A Policy Evaluation Tool: Management of a Multiaquifer System Using Controlled Stream Recharge," *Water Resources Research*, Vol. 21, pp. 1731–1747, 1985.

Gorelick, S. M., "A Review of Distributed Parameter Groundwater Management Modelling Methods," *Water Resources Research*, Vol. 19, pp. 305–319, 1983.

______, "Sensitivity Analysis of Optimal Groundwater Contaminant Capture Curves: Spatial Variability and Robust Solutions," *Proceedings of National Water Well Association Conference on Solving Groundwater Problems with Models*, 1987.

Gorelick, S. M., and B. J. Wagner, *Evaluating Strategies for Ground-*

Water Contaminant Plume Stabilization and Removal, U.S. Geological Survey Water Supply Paper 2290, pp. 81–89, 1986.

Gorelick, S. M., C. I. Voss, P. E. Gill, W. Murrary, M. A. Saunders, and M.H. Wright, "Aquifer Reclamation Design: The Use Of Contaminant Transport Simulation Combined With Nonlinear Programming," *Water Resources Research*, Vol. 20, pp. 415–427, 1984.

Greenwald, R. M., and S. M. Gorelick, "Particle Travel Times for Contaminants Incorporated Into a Planning Model for Groundwater Plume Capture," *Journal of Hydrology,* Vol. 107, pp. 73–98, 1989.

Javandel, I., and C. F. Tsang, "Capture-Zone Type Curves: A Tool for Aquifer Cleanup," *Ground Water*, Vol. 24, pp. 616–625, 1986.

Javandel, I., C. Doughty, and C. F. Tsang, *Groundwater Transport: Handbook of Mathematical Models*, American Geophysical Union, Water Resources Monograph 10, 1984.

Keely, J. F., and C. F. Tsang, "Velocity Plots and Capture Zones of Pumping Centres for Groundwater Investigations," *Ground Water*, Vol. 21, pp. 701–714, 1983.

Lefkoff, L. J., and S. M. Gorelick, "Design and Cost Analysis of Rapid Aquifer Restoration Systems Using Flow Simulation and Quadratic Programming," *Ground Water*, Vol. 24, pp. 777–790, 1986.

______, *AQMAN: Linear and Quadratic Programming Matrix Generator Using Two-Dimensional Ground-Water Flow Simulation for Aquifer Management Modelling*, U.S. Geological Survey, Water-Resources Investigations Report 87–4061, 1987.

Massmann, J., and R. A. Freeze, "Groundwater Contamination From Waste Management Sites: The Interaction Between Risk-Based Engineering Design and Regulatory Policy. 1. Methodology," *Water Resources Research*, Vol. 23, pp. 351–367, 1987.

McDonald, M. G., and A. W. Harbaugh, *A Modular Three-Dimensional Finite-Difference Groundwater Flow Model*, U.S. Geological Survey, 1984.

Pollock, D. W., "A Semi-Analytic Integration Scheme for Computing Pathlines for Block-Centered Finite-Difference Ground-Water Flow Models," *Ground Water,* Vol. 26, pp. 743–750, 1988.

______, *Documentation of Computer Programs to Compute and Display Pathlines Using Results From the USGS Modular Three-Dimensional Finite-Difference Groundwater Flow Model*, USGS Open File Report, 89–381, 1989.

Rounds, S. A., and B. A. Bonn, "DREAM: A Menu-Driven Program That Calculates Drawdown, Streamlines, Velocity, and Water Level Elevation," *Proceedings of National Water Well Association Fourth*

International Conference on Solving Groundwater Problems with Models, pp. 329–350, 1989.

Shafer, J. M., "Reverse Pathline Calculation of Time-Related Capture Zones In Nonuniform Flow," *Ground Water*, Vol. 25, pp. 283–289, 1987a.

______, *GWPATH: Interactive Groundwater Flow Path Analysis*, Illinois State Water Survey Bulletin 69, 1987b.

Tung, Y. K., "Groundwater Management by Chance-Constrained Model," *Journal of Water Resources Planning and Management, American Society of Civil Engineers*, Vol. 112, 1986.

Wagner, B. J., and S. M. Gorelick, "Optimal Groundwater Quality Management Under Parameter Uncertainty," *Water Resources Research*, Vol. 23, pp. 1162 -1174, 1987.

Ward, R. L., and R. C. Peralta, *"EXEIS" Expert Screening and Optimal Extraction/Injection Pumping Systems for Short-Term Plume Immobilization*, Air Force Engineering and Services Center, Environics Division, Report No. ESL–TR–89–57, 1990.

CHAPTER V

Design Optimization for Capture and Containment Systems

A. Introduction

1. Aquifer Management Modelling

Under certain circumstances it is possible to use a combination of optimization techniques and simulation models to help in aquifer remediation design. In this chapter, we discuss this combination, its benefits, and its limitations. It must be noted that this report, and particularly this chapter, is intentionally limited in scope. We discuss only so called "on-the-shelf" technology. That is, those techniques for which the mathematical and physical theory exists and for which implementations have been achieved through documented and readily available computer programs. The reader should be aware that several other more complex techniques have been presented in the literature during the past 6 years. Because these methods do not meet the criteria of being "on-the-shelf" technology, yet are important future tools, their description appears in Appendix B rather than in the body of this report.

Aquifer management modelling extends simulation methods to embrace optimization techniques. This combination provides a tool to help determine design strategies for problems of water use and groundwater contamination. As discussed in the previous chapter, simulation alone is valuable for aquifer remediation design. It allows the hydrogeologist to experiment with different management alternatives. The development of a simulation model is a necessary step in any modern analysis of a groundwater management problem. However, simulation alone is often not enough because the problems of aquifer management do not involve prediction alone. Rather, they involve both simulation, for prediction, and optimization to develop the best operating policy for a particular objective taking into account those restrictions that exist on a site-specific basis.

Combined simulation-optimization techniques can be thought of as organized and methodical trial-and-error methods. However, in contrast to most trial-and-error approaches to problem solving, the objective, constraints, and solution search strategies are clearly specified. This means that, given a proper formulation, an optimal engineering design can be identified that minimizes or maximizes an objective function subject to physical, economic, and logistical constraints. In the absence of such a formal statement and solution strategy, "optimization" is a nebulous concept. It is important to note that if design constraints are added or removed or the objective is altered, different "optimal solutions" may result. In this sense, the combined simulation-optimization technique is just a tool to help the hydrologist. The best use of this tool is to develop a family of so-called "optimal solutions" under a broad and varied menu of design considerations. In the context of this book, a variety of "optimal" capture and containment design strategies for aquifer remediation are presented at the end of this chapter. To further clarify the role of simulation in these management models let us state simply that simulation is the best means of including our understanding of the physical system in the design strategy. The union of simulation with optimization yields the advantages of simulation alone but adds the elements of precise management formulation and optimal design.

The identification of optimal operating schemes is the realm of the field of operations research. It is an area of applied mathematics that had initial applications during World War II aimed at the optimal allocation of military resources. Today, operations research is a tremendously important discipline that is concerned with the optimal allocation of any scarce resources. It is widely used in industry and engineering for a variety of allocation, scheduling, switching, mixing, and design problems.

The first application of operations research to the management of a subsurface fluid reservoir was by Lee and Aronofsky (1958). Their formulation sought to maximize profits from petroleum production. An important component of their problem formulation was a simulation model in the form of a simple analytic solution that determined the pressure response to pumping. After a few more published applications appeared in the petroleum engineering literature, Deninger (1970) brought the early methods to bear on a groundwater problem. He considered maximization of water production from a well field subject to restrictions on drawdowns and well facilities. Again, a simple simulation model was included as a key element of the problem statement. Important early contributions to hydraulic management modelling methods were made by Wattenbarger (1970), Maddock (1972a, 1972b), Rosenwald and Green (1974), Aguado and Remson (1974), Schwarz (1976), and Willis and Newman (1977). These studies and nearly 100 others are discussed and critically analyzed in the reviews of Gorelick (1983, 1988).

The latter paper discusses recent methods that handle highly nonlinear problems and attempt to incorporate notions of risk and uncertainty into simulation-optimization design formulations for groundwater-quality management.

Groundwater hydrologists typically are familiar with methods for simulating aquifer response but may be unacquainted with optimization methods. If a primary goal in aquifer remediation design is to select the best well locations and determine the optimal pumping rates, then optimization methods cannot remain a mystery. Many optimization methods appearing in the literature seem complicated. In fact, they are not. Once a basic understanding is developed of these methods, problems can be readily formulated and solved using very robust and easy to use computer programs. In this chapter we will discuss groundwater management problem formulation and stress the combination of simulation and optimization methods for hydraulic gradient control. We will also provide a practical overview of the most basic optimization method, linear programming, to help the reader obtain a working knowledge of optimization methods for designing capture and containment systems. Although we will not cover the general class of nonlinear optimization methods, we will cover a particular type of nonlinear method known as quadratic programming. This method is needed if the goal of a particular remediation design involves minimizing energy costs associated with pumping for contaminant removal.

2. *Preliminary Statement — Overview of Simulation-Optimization Approach*

Let us consider a typical problem of aquifer remediation design. Figure 14 shows a contaminant plume that is migrating toward the water supply wells shown near the bottom of the figure. Without remediation the water supply will become contaminated. To prevent this problem we can utilize a series of remediation wells (cleanup wells) to remove the contaminated water and prevent the continued downgradient migration of the plume. In addition, we want to pump the least quantity possible so that the expense of treatment and disposal can be kept to a minimum. Such a scheme is shown in cross section in Figure 15.

The two most important questions that must be addressed in the design of a pump-and-treat system are

1. Where should the remediation wells be located?
2. How much should be pumped from each well?

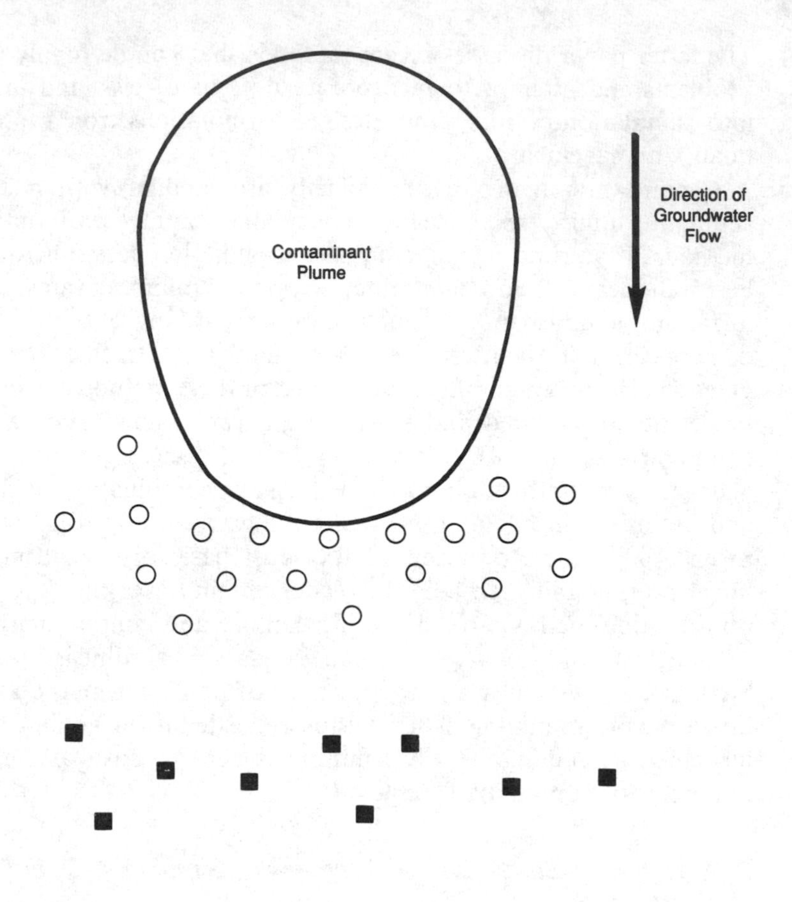

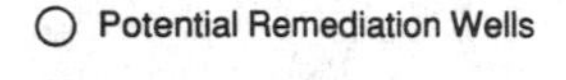

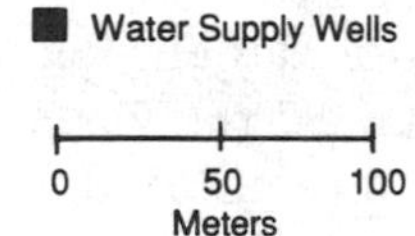

Figure 14. Typical plan view of a contaminant plume migrating toward a series of water-supply wells.

Additional questions that might be posed are

1. Should the number of wells be restricted or specified?
2. Once the contaminated water is removed from the ground and treated, would it be useful to reinject this clean water to assist in plume stabilization and to accelerate the process of flushing out the contaminants?

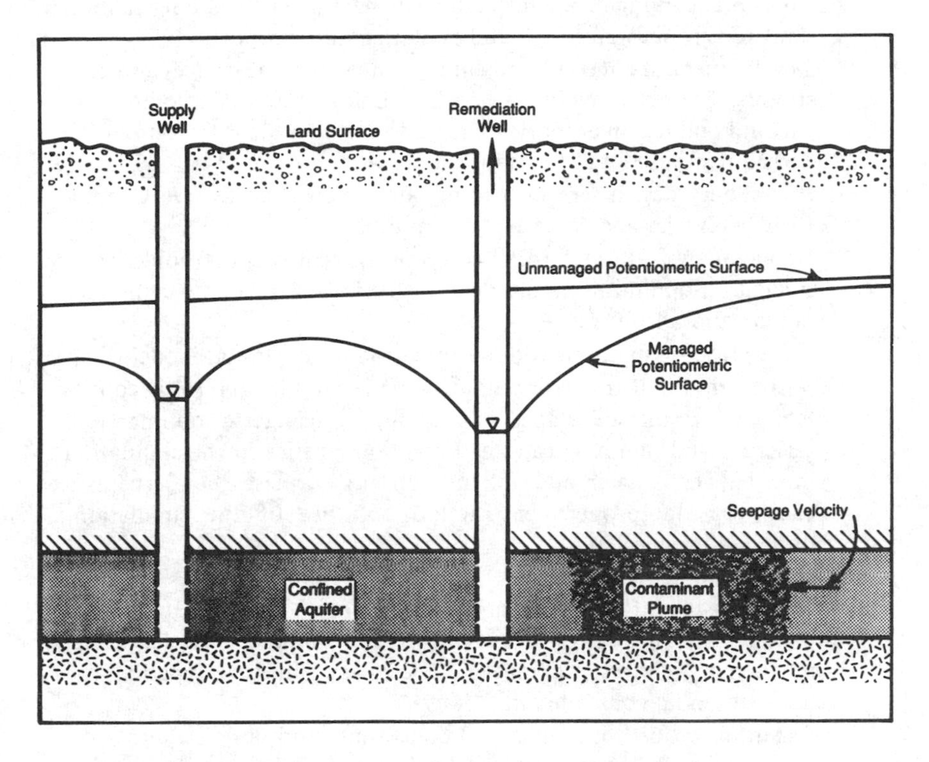

Figure 15. Cross section showing wells and flow directions in a pump-and-treat remediation system.

3. Should the wells be pumped at constant rates or at variable or even cyclic rates?
4. Is it necessary to try to remove all of the contaminant from the groundwater, or can some remain and be left to migrate because it is expected to be at a "safe level," below a water-quality standard or a detection limit?
5. Should the capture and containment system be overdesigned to account for uncertainty regarding the natural hydrogeological environment or the functioning of the engineered system?

The above questions can be answered using an aquifer management model. Ten general steps should be followed:

1. Data collection and field and laboratory analyses for site characterization, including the definition of the hydrogeology and the determination of the nature and extent of the contamination problem.

2. Conceptual and mathematical model development (including calibration) for groundwater flow and contaminant transport.
3. Development of a menu of potential management objectives and constraints. Both can involve physical, technical, economic, logistic, legal, and political matters. This step also includes identification of the decision variables and the state variables.
4. Preliminary design screening using simulation analysis (see Chapter IV.E) with trial-and-error experimentation.
5. Design analysis through formulation of the remediation problem as a simulation-optimization problem or series of related simulation-optimization problems.
6. Generation of solutions with a simulation-optimization model.
7. Verification that the design solution generated in step (6) is consistent with results developed in step (4) by using the optimal well selection and pumping rates as known quantities in the simulation model. (This step should not turn up any surprises but serve as a check on the proper construction and use of the simulation-optimization model).
8. Conducting sensitivity studies of the design solutions using simulation to inspect the influences of model uncertainty (e.g., estimated model parameters) and design uncertainty (e.g., optimal pumping rates, number of wells, etc.) upon the value of the objective function and upon potential constraint violations.
9. Experimentation with additional constraints and design alternatives (e.g., pump-treat versus pump-treat-inject) based upon the menu of design considerations developed in step (3) above. This involves solving several related optimization problems.
10. Analysis and comparison of the costs and benefits associated with each design alternative.

We will first address the two important questions posed at the beginning of this chapter, leaving the others for later discussion. This initial presentation will not be detailed, nor will we provide detailed definitions of the "jargon" that accompanies optimization problems. These aspects will be covered in Chapter V.C.

We can formulate the problem of selecting well locations and pumping rates as one involving simulation and optimization. The formulation is meant as a general instructive introduction and, in fact, its solution will be beyond the scope of this book. However, the concepts and formulation strategy for remediation design are useful in the simpler formulations that follow. Our initial formulation consists of a statement of the objective and a listing of a series of constraints. In words and in mathematical form the formulation is

OBJECTIVE:

Minimize the sum of the pumping rates	Minimize $Z = \sum_{i=1}^{n} q_i$

CONSTRAINTS:

The plume must be contained so that concentrations do not exceed specified values (e.g., water quality standard) at j critical checkpoint locations (e.g., water supply wells).	$C_{sim_j} \leq C^*_j$
The pumping rate for any well must not exceed a specified maximum value.	$q_k \leq q^*_k$
The simulated hydraulic head at each pumping well must be at least a specified minimum value.	$h_{sim_k} \geq h^*_k$
The values of the concentrations, heads, and pumping rates must not become negative.	$C_{sim} \geq 0$ for all values $h_{sim} \geq 0$ for all values $q \geq 0$ for all values

The **decision variables** are the pumping rates, q_i at the n potential pumping wells. Given many possible well sites, the optimal solution will select the appropriate wells and determine their minimal pumping rates. Of the potential well sites, only those values with non-zero pumping rates in the final solution will be active. Those wells represent the optimal selection of wells. The **state variables** are the hydraulic heads and concentrations. Their values are given by the simulation model. It is evident from this formulation that the simulation model is required to provide values of hydraulic head and concentrations appearing in the constraints.

The above problem is one of nonlinear optimization. The nonlinearities stem from the fact that simulated concentrations are a nonlinear function of pumping rates (the decision variables). This is not the case if changes in hydraulic heads are a linear function of pumping. The nature of the nonlinearity associated with simulated concentrations is easily seen. Neglecting details and complexity, note that in the simulation-management formulation above that the pumping rates are the unknown variables of interest. Because the pumping rates are unknown, the simulated hydraulic heads are unknown. Since the hydraulic heads are unknown, the groundwater velocities must be unknown. The simulation of contaminant transport is naturally dependent upon detailed knowledge of the groundwater velocities. In the equation describing contaminant transport (see Chapter II.F), the advective terms contain the products of groundwater velocities and contaminant concentrations. Similarly, the dispersive terms contain products of the dispersion coefficients, which are functions of the groundwater velocities, and the concentration gradients. So in both advective and dispersive terms there are products of velocities and concentrations or concentration gradients. The products are the nonlinearities. The nonlinearities appear as the result of products of unknown concentrations times unknown velocity components in the advective-dispersive equation.

This formulation is straightforward and instructive in terms of aquifer remediation design considerations, but it is a nonlinear problem and its solution involves nonlinear optimization methods. Unfortunately, combined models of contaminant transport simulation and nonlinear optimization have not been "packaged" into "on-the-shelf" methodology. While solution techniques have been presented in the literature (Gorelick et al., 1984; Ahlfeld et al., 1988; Wagner and Gorelick, 1987, 1989), the methodology is still relatively new and codes have not been distributed for common use. Solution of the above simulation-optimization problem is beyond our scope.

Fortunately, hydrodynamic dispersion is often a transport mechanism of secondary importance in aquifer remediation. The high gradients and convergent flow paths developed around pumping wells favor advective transport. Methods for the design of capture and containment systems can be developed using linear simulation management models if we ignore hydrodynamic dispersion. These simpler models can be used to plan hydraulic gradient control schemes that force groundwater to flow in certain desirable directions. They allow us to contain a contaminated zone and optimize removal of the contaminated water.

The formulations and methodology presented in the remainder of Chapter V are for problems closely related to the one above, but they deal exclusively with linear systems, advective transport, and hydraulic gradient control. The goal is to best control the magnitude and direction of the groundwater flow so that all contaminated water within a contaminated zone is contained and removed. Solute concentrations are not explicitly considered in these simplified management models. The linear formulations will follow the methodologies presented by Colarullo et al. (1984), Atwood and Gorelick (1985), Gorelick and Wagner (1986), and Lefkoff and Gorelick (1986, 1987).

B. Background to Linear Systems Management

The simultaneous management and simulation of groundwater for linear systems can be accomplished using two different methods. In both methods, a groundwater simulator is included as part of the constraint set of a linear optimization problem. The difference between them stems from the manner in which the flow simulator is represented. In this brief background section the two methods are defined. In the subsections that follow they are developed in detail.

In the **embedding method** the governing flow equation is discretized using finite differences or finite elements, and the resulting algebraic equations are treated as constraints in a linear programming problem. The discretized flow equation is "embedded" in the optimization problem along with other constraints that restrict local hydraulic heads, gradients, and pumping or recharge rates. The solution to such a problem gives the optimal pumping and injection rates plus the simulated hydraulic heads throughout the aquifer at every finite difference or finite element node. The fact that every hydraulic head in the discretized domain is necessarily a decision variable in the planning model means that the linear programming problem can be huge when using this method. As a result, the method is usually limited to steady-state problems.

In the **response matrix approach** the governing equation is not included in the constraint set, but rather unit solutions to the flow equation are developed and linearly superposed (added together) to simulate any configuration of pumping or recharge. Each unit solution describes the response in terms of the change in head at selected observation points due to a pulse of pumping at a particular location. The responses are assembled into a **response matrix** that is then included in the management model as a set of constraints. Because the responses are only recorded at certain key locations (those of critical interest in managing the system), the response matrix is a highly compact simulator. Compared to the embedding method, the response matrix approach gives less information about simulated heads throughout the entire domain. However, because it only considers key locations, the linear programming problem becomes more manageable and much larger steady-state and transient systems can be dealt with as management problems. The response matrix approach is by far the more general of the two methods. It is also more complicated to explain and requires a complete understanding of linear systems theory. As such the concepts underlying this approach will be explained first. A description of the embedding approach is postponed.

1. Groundwater Flow as a Linear System

The key to combining groundwater flow simulation with any linear optimization method is the notion that the total hydraulic response is a linear function of the system flow stresses and the initial and boundary conditions. Most hydrogeologists are accustomed to applying a simulation model for two-dimensional confined flow that solves the following equation:

$$\frac{\partial}{\partial x_i}\left(T_{ij}\frac{\partial H}{\partial x_j}\right) = S'\frac{\partial H}{\partial t} + W' \qquad i,j = 1,2 \tag{66}$$

where H is head; W′ is pumping, injection, or recharge; T is transmissivity; S′ is storativity; and x and t refer to space and time, respectively.

The solution requires that values have been determined for T and S′, the initial conditions and boundary conditions are prescribed, and a set of flow stresses, W′, (e.g., pumping from 10 pumping wells, injection from 3 wells, recharge from a leaky river channel) are defined. By considering this properly posed *complete problem*, the hydraulic heads over space and time are given by the solution.

The above approach, wherein we consider the *complete problem*, is one way to predict hydraulic heads over space and time. However, there is another approach. The problem is decomposed, solutions are obtained for each piece, and then these component solutions are assembled to give a solution that is identical to that given by the complete problem. This is a property of any linear system. In the flow equation the dependent variables H(x,y,t) do not interact multiplicatively and no coefficients are functions of H(x,y,t). Therefore the solution, consisting of the heads over space and time, can be assembled from the following solution components:

1. Hydraulic heads resulting from initial conditions.
2. Changes in hydraulic heads resulting from changing boundary conditions.
3. Changes in hydraulic heads resulting from the hydraulic stresses, each considered separately. For example, head changes that are
   - due to the 10 pumping wells,
   - due to the 3 injection wells, and
   - due to the recharge along the leaky river channel.

If one adds up the component influences on hydraulic heads due to initial conditions, boundary conditions, and all of the hydraulic stresses, then the behavior of the entire system is reproduced. The process of combining the individual solutions is known as linear superposition.

2. The Response Matrix

We can take the idea of linear superposition further to develop an extremely compact groundwater simulator. This simulator is known as a **response matrix**. It exploits the linearity of Equation (66) to provide the hydraulic heads at any selected times and spatial locations due to the influences of any selected flow stresses.

At this point, let us focus upon the influence on drawdowns solely due to pumping. It is convenient to adopt the idea of a **unit well**. A unit well is a flow-stress location (e.g., pumping well, injection well, portion of a leaky river) with an arbitrary but known **unit stress** or "unit flow rate." The unit stress is a pulse source. It has a fixed constant value for a specified period and has a value of zero thereafter. A unit stress gives rise to a unit response. The important concept here is that any set of flow stresses may be simulated by superposing the unit responses from the unit wells.

The response matrix concept is demonstrated with the following example. Figure 16 shows a map of a portion of a homogeneous aquifer with three wells. We want to predict the total drawdown after 30 days at five observation points A through E in an excavation area shown on the figure. The drawdown responses, due to pumping of a unit well at each of the three locations, are given by solving Equation (66). Each well is treated as a unit well where we select a unit stress of 1 liter per second (L/s). The drawdowns due to each unit well are shown in Table 6. This table is a response matrix. Each element in the matrix is a **response coefficient**. Each column corresponds to one of the three pumping wells. Each row corresponds to one of the five observation points.

Assume that the following pumping pattern is desired for 30 days:

$$Q_1 = 0.004 \text{ m}^3/\text{s}, \; Q_2 = 0.005 \text{ m}^3/\text{s}, \; Q_3 = 0.006 \text{ m}^3/\text{s}$$

It is easy to calculate the pumping rates in terms of the unit stress of 1 L/s = 0.001 cubic meter per second (m^3/s); they are 4.0, 5.0, and 6.0. That is, the rate as a multiple of the unit stress equals the pumping rate times the unit stress conversion factor, in this case 1,000. Total drawdown at an observation point is then readily given by multiplying each row element of the matrix by the pumping rate of each well (expressed in terms of the unit stress) and then summing these values.

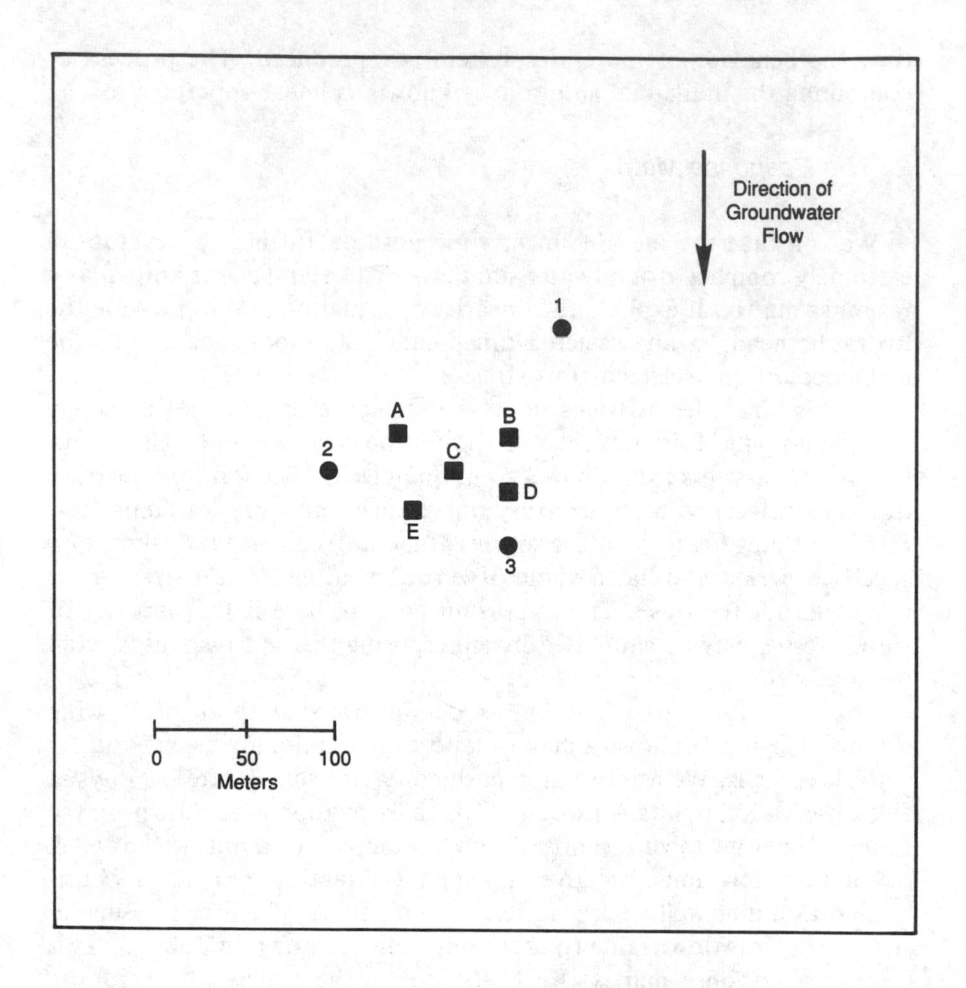

Figure 16. Map of a portion of an aquifer with three pumping wells and five observation points.

The drawdown at observation point A (the first row in the matrix) is

[(0.1698)(4.0)] +	[(0.4177)(5.0)] +	[(0.2321)(6.0)] =	4.160 m
Drawdown due to Q_1	Drawdown due to Q_2	Drawdown due to Q_3	Drawdown at observation point A due to pumping

Table 6. Simple Drawdown-Response Matrix After 30 Days Given Pumping During One 30-Day Period

Observation Point	Drawdown Influence Due to Unit Pumping at		
	Well 1	Well 2	Well 3
A	0.1698	0.4177	0.2321
B	0.2978	0.1848	0.3311
C	0.1893	0.2865	0.3835
D	0.2031	0.1885	0.5450
E	0.1307	0.3673	0.3623

In general the drawdown at any location i can be obtained as

$$\sum_{j=1}^{3} (r_{ij} * Q_j) = S_i \tag{67}$$

Drawdown Due to Each Unit Well	Pumping Rate (in terms of unit stress) of Each Well	Drawdown at Location i.

Written in terms of matrix notation, a single row representing drawdown at a single location (at a specified time) is

$$\mathbf{r_i^t Q} = S_i \tag{68}$$

where $\mathbf{r_i^t}$ = row vector of drawdown-response coefficients showing the influence of a unit well on drawdown at location i (the transposed symbol **t** is needed because we are defining a row of coefficients rather than a column of coefficients),
$\mathbf{Q}$ = column vector of pumping rates (Q_1, Q_2, Q_3), and
S_i = drawdown at location i.

We now must consider the influences on drawdown over time due to pumping schedule changes. We can represent the effects of hydraulic head recovery as a drawdown response. This information is given again by solution to the flow equation and is displayed in Table 7, where the additional rows (compared with Table 6) give the drawdown at the five observation points at the end of two more periods, at 60 days and 90 days. The response coefficients show that the drawdown induced by pumping persists even though the pumping ceases after 30 days. The water levels recover quite rapidly, so after 90 days the drawdown-response coefficients are quite small.

Now we can enlarge the response matrix to account for continued pumping during the next two 30-day periods. Here there is a trick that

Table 7. Simple Drawdown-Response Matrix After 90 Days Given Pumping During One 30-Day Period

Observation Point	Well 1	Drawdown Influence Due to Unit Pumping at Well 2	Well 3	
A	(After 30 days)[a]	0.1698	0.4177	0.2321
B		0.2978	0.1848	0.3311
C		0.1893	0.2865	0.3835
D		0.2031	0.1885	0.5450
E		0.1307	0.3673	0.3623
A	(After 60 days)[b]	0.0936	0.1141	0.1042
B		0.1088	0.0988	0.1108
C		0.0992	0.1085	0.1129
D		0.1010	0.0994	0.1160
E		0.0893	0.1124	0.1121
A	(After 90 days)[c]	0.0592	0.0660	0.0625
B		0.0631	0.0609	0.0645
C		0.0603	0.0642	0.0651
D		0.0610	0.0611	0.0660
E		0.0571	0.0655	0.0649

[a]Drawdown occurs during first 30 days. Note that the greatest drawdown occurs during the first 30 days and drawdown values decline over time due to natural recovery.
[b]Recovery 30 days after pumping ceases.
[c]Recovery 60 days after pumping ceases.

makes this expansion very economical. The idea again relies on linear superposition and is best explained with a simple example. Suppose Well 1 is pumped at a unit flow rate and the drawdown-response is recorded at observation point C at the end of 30 days. Now suppose that pumping continues for another 30 days. At the end of both periods the total drawdown is due to pumping during the first 30 days plus pumping during the second 30 days. The trick is to note that the effects of a unit well for the second 30 days are the same as the effects of that unit well during the first 30-day period but the responses are delayed by 30 days. In other words, the response to a unit pumping impulse is independent of when pumping occurs; in a linear system the influence of a unit of pumping on drawdown can be translated through time. To expand the response matrix to account for pumping over time, one copies the original drawdown-response information to additional columns of the matrix. The first set of columns correspond to pumping during the first 30-day period and additional columns correspond to the second 30-day period. The responses must be staggered or stair-stepped for 30 days; the responses are delayed due to the delay in pumping for sequential periods.

For this example, a transient response matrix is displayed and explained in Table 8. Note the key point of the table is to show that a unit of pumping which occurs between 60 and 90 days gives a certain drawdown response that is identical to that due to a unit of pumping during 0 and 30 days, except the response occurs 60 days later. The upper triangle of the response matrix is zero because a well that has not yet started to pump, say between 60 and 90 days, cannot cause a drawdown response from day 0 to 60. Finally, it is important to note that this trick of stair-stepping the responses due to pumping requires that *all pumping periods be of equal length*. Since the period length is arbitrary, this requirement provides no real drawbacks.

We can now express in matrix notation the drawdown at each location (A through E) over time (after 30, 60, and 90 days), due to pumping at the three wells during pumping periods 1, 2, and 3. First, we note that there are now nine pumping rates corresponding to the three well locations (first subscript) over three time periods (second subscript),

$Q_{1,1}, Q_{2,1}, Q_{3,1}$	$Q_{1,2}, Q_{2,2}, Q_{3,2}$	$Q_{1,3}, Q_{2,3}, Q_{3,3}$
pumping during period 1	pumping during period 2	pumping during period 3

Second, we have a larger system of drawdown-response equations,

$$\mathbf{RQ} = \mathbf{S} \tag{69}$$

where $\mathbf{R}$ = response matrix of drawdown-response coefficients showing the influence of a unit well on drawdowns;
$\mathbf{Q}$ = column vector of pumping rates, $Q_{location,time}$; and
$\mathbf{S}$ = vector of drawdowns at locations A through E at the end of 30, 60, and 90 days.

Finally, we must see how the influences on hydraulic head changes due to initial and boundary conditions can be represented using the linear system approach. To do this let us consider the total hydraulic head. This total head is the sum of several components as shown in the equation below:

Total Head = (head from boundary influences) – (change in head resulting from initial conditions) – (drawdown due to flow stresses).

We return to the original example where pumping occurs for only one 30-day period. However, now let us introduce the influence of boundary

Table 8. Transient Response Matrix Displayed and Explained

Observation Locations	Drawdown Influence Due to Unit Pumping Days 0–30			Drawdown Influence Due to Unit Pumping Days 30–60			Drawdown Influence Due to Unit Pumping Days 60–90		
	Well 1	Well 2	Well 3	Well 1	Well 2	Well 3	Well 1	Well 2	Well 3
A	0.1698	0.4177	0.2321						
B	0.2978	0.1848	0.3311						
C	0.1893	0.2865	0.3835		0			0	
D	0.2031	0.1885	0.5450						
E	0.1307	0.3673	0.3623						
A	0.0936	0.1141	0.1042	0.1698	0.4177	0.2321			
B	0.1088	0.0988	0.1108	0.2978	0.1848	0.3311			
C	0.0992	0.1085	0.1129	0.1893	0.2865	0.3835		0	
D	0.1010	0.0994	0.1160	0.2031	0.1885	0.5450			
E	0.0893	0.1124	0.1121	0.1307	0.3673	0.3623			
A	0.0592	0.0660	0.0625	0.0936	0.1141	0.1042	0.1698	0.4177	0.2321
B	0.0631	0.0609	0.0645	0.1088	0.0988	0.1108	0.2978	0.1848	0.3311
C	0.0603	0.0642	0.0651	0.0992	0.1085	0.129	0.1893	0.2865	0.3835
D	0.0610	0.0611	0.0660	0.1010	0.0994	0.1160	0.2031	0.1885	0.5450
E	0.0571	0.0655	0.0649	0.0893	0.1124	0.1121	0.1307	0.3673	0.3623
	Drawdown after 30 days due to pumping during days 0–30			Zeros (No drawdown after 30 days due to pumping during days 30–60)			Zeros (No drawdown after 30 days due to pumping during days 60–90)		
	Drawdown after 60 days due to pumping during days 0–30			Drawdown after 60 days due to pumping during days 30–60			Zeros (No drawdown after 60 days due to pumping during days 60–90)		
	Drawdown after 90 days due to pumping during days 0–30			Drawdown after 90 days due to pumping during days 30–60			Drawdown after 90 days due to pumping during days 60–90)		

conditions and initial conditions. Suppose the aquifer has boundaries such that a constant uniform gradient exists at steady state when there is no pumping. Also assume that previous pumping (prior to initiation of pumping from our three wells) left a remnant cone of depression as initial conditions. Heads will change over time as this remnant cone "fills in" due to natural recovery.

The influences of boundary conditions, initial conditions, and the three individual pumping wells are shown both in plan view and cross section in Figure 17. The cross sections are drawn through observation location B and are parallel to the natural direction of groundwater flow. The total head in the aquifer is the sum of the five influences shown in Figure 17. The total head at location B at 30 days is represented symbolically and with values from the figure as

$$H_{\text{total at B}} = H_{\text{boundary}} - \Delta H_{\text{initial}} - \sum_{j=1}^{3} r_{B,j}\, Q_j \tag{70}$$

$$78.8 = 86.3 - 3.4 - 4.1$$

where $H_{boundary}$ is the head at location B under the natural hydraulic gradient and $\Delta H_{initial}$ is the drawdown at location B due to the remnant cone that remains at 30 days. The last term is the drawdown at location B due to pumping that can be written as

$$\sum_{j=1}^{3} r_{B,j}\, Q_j = r_{B,1} Q_1 + r_{B,2} Q_2 + r_{B,3} Q_3$$

$$4.1018 = (0.2978)(4.0) + (0.1848)(5.0) + (0.3311)(6.0)$$

$$4.1018 = 1.1912 + 0.9240 + 1.9866$$

It should be noted that there are temporal influences on total head due to initial conditions. At 30 days the total head changed due to the last two terms in the Equation (70); these are recovery of the initial heads and drawdown induced by pumping. At 60 days these two influences will be different than at 30 days as the initial heads recover and pumping continues.

As discussed later, in the simulation-optimization approach we will separate the influences on drawdown into manageable and unmanageable components. The manageable component in the case above consists of the influences due to pumping, while the unmanageable component corresponds to "natural" changes due to initial conditions, boundary conditions, and preexisting pumping or recharge. This distinction is

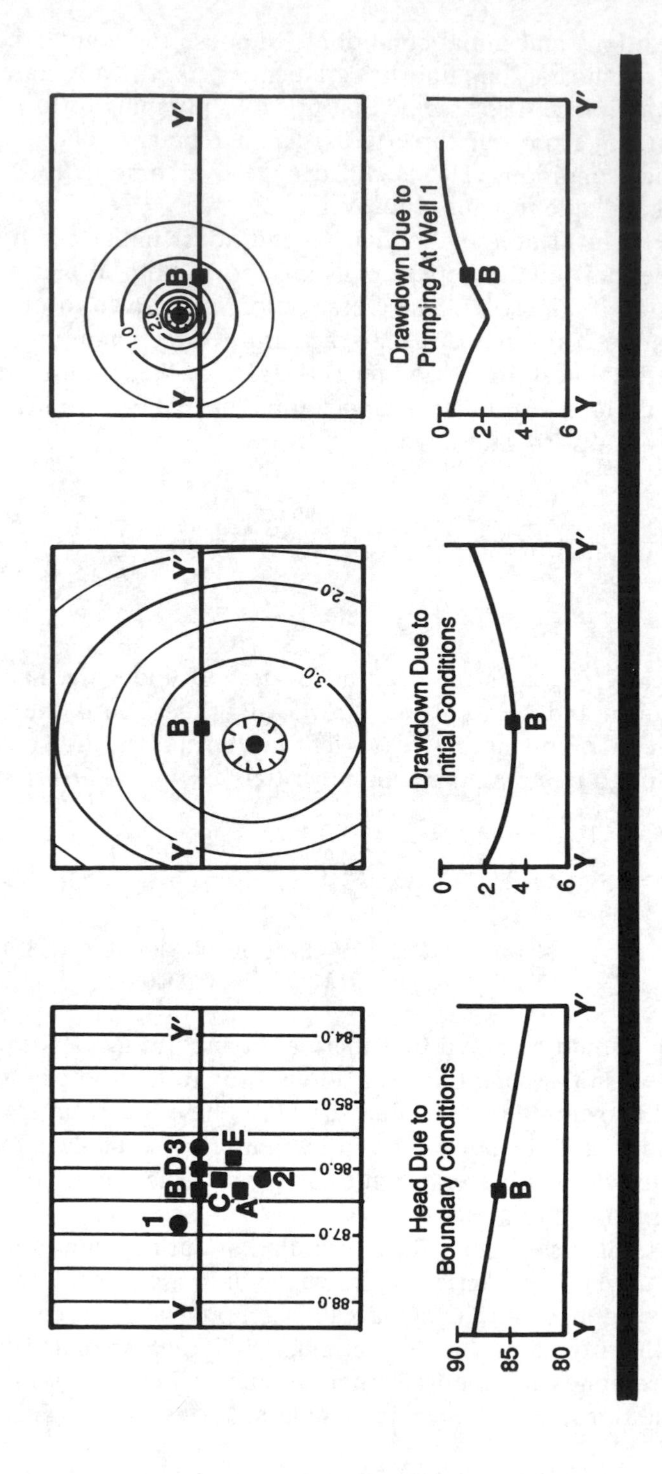
Head Due to
Boundary Conditions
Drawdown Due to
Initial Conditions
Drawdown Due to
Pumping At Well 1
Y
Y'
B
1
2
3
D
C
A
E
88.0
87.0
86.0
85.0
84.0
90
85
80
0
2
4
6
1.0
2.0
3.0

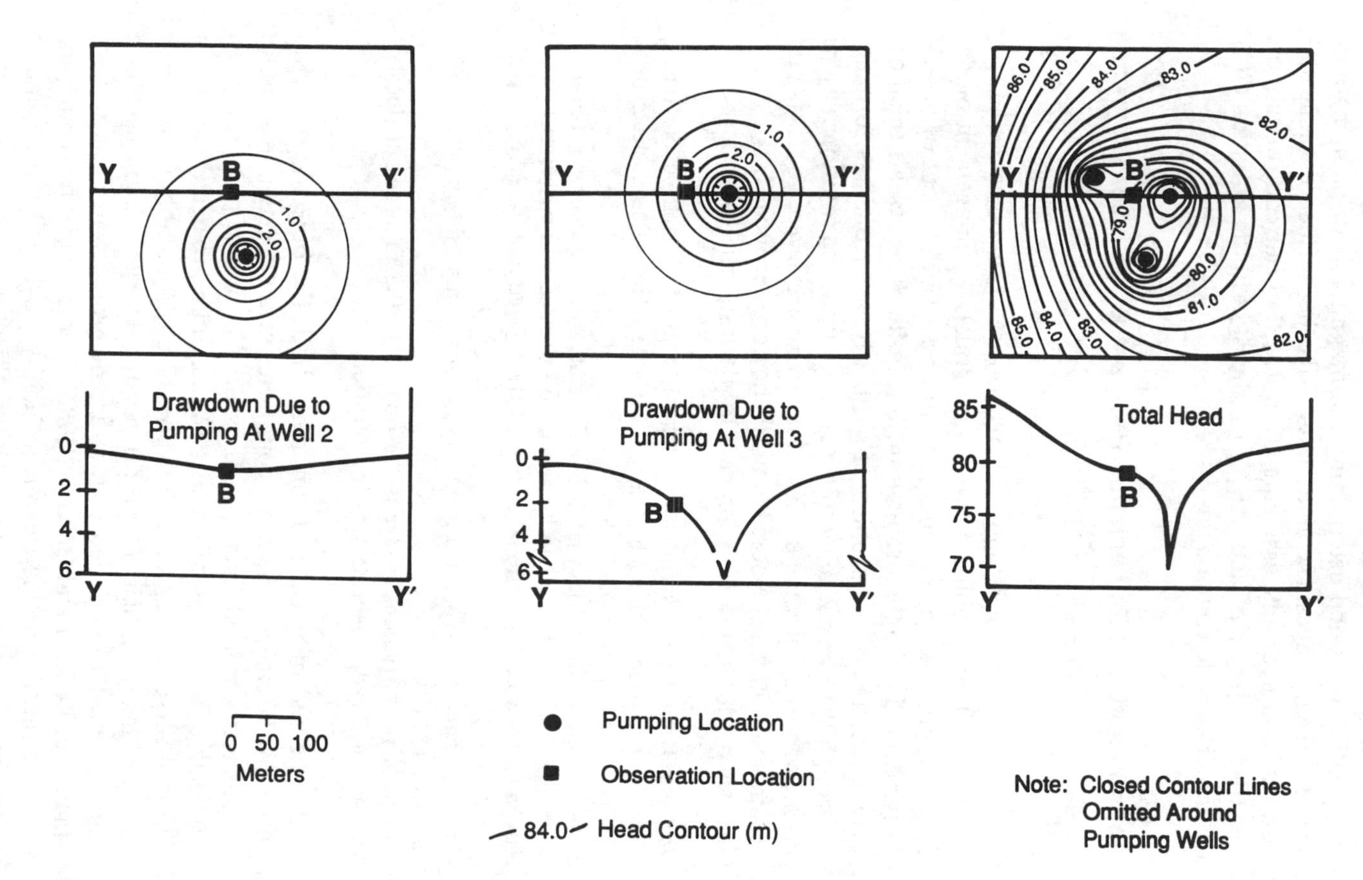

Figure 17. Water-level maps showing transient influences.

important because in the case of aquifer remediation we are interested only in locating wells and determining the magnitudes of pumping and recharge for the manageable components. All other influences on heads, drawdowns, and hydraulic gradients are not within our control and are lumped together [$H_{boundary}$ and $\Delta H_{initial}$ on the right in the Equation (70)] as the unmanageable component.

C. Formulation of a Remediation Problem as a Linear Program

1. Definitions

In most cases, optimization problems consist of an objective and an associated set of constraints. As with any model, a management model is only an approximation. It does not capture every element of a real-world system in its simple formulation, but rather it represents the most important features of the problem. Frequently, after one has formally gone through the exercise of problem formulation, the solution may be apparent, or it may become obvious that there are severe conflicts and no design solution will be possible until they are reconciled.

One begins the formulation process by defining the following:

Decision Variables—unknown quantities that can be managed or controlled. Associated with each decision variable is its value or level of activity (e.g., pumping rate).
State Variables—quantities that define the state of the system (e.g., head or concentration). They generally depend on the decision variables.
Auxiliary Variables—additional conveniently defined quantities (e.g., local velocities).
Objective Function—goal that is minimized or maximized (e.g., cost, total pumping, risk).
Constraints—restrictions that must be obeyed in the final design (local drawdowns, head, gradients, or concentrations).

In the method of linear programming, both the objective and constraints must be linear. The most common linear objective in groundwater management modelling is to minimize the sum of pumping rates. This objective function is known as an "economic surrogate" because it implies that there is a cost objective, but the cost function has not been or cannot be made explicit. This is often satisfactory if a significant cost of aquifer remediation is tied to the rate of groundwater removal for treatment (see Lefkoff and Gorelick, 1986). For example, total pumping may dictate the size of the treatment facility, cost of disposal, and to some degree the cost of pumping. Other objective functions related to cost and remediation time are discussed in Greenwald and Gorelick (1989).

Constraints may be of several types. The most straightforward constraint is a **simple bound** on an individual decision or state variable. For example, a particular pumping rate might be restricted,

$$Q_5 \leq 2.3 \text{ L/s}$$

or hydraulic head at a particular location might be limited,

$$H_{22} \leq 58 \text{ m}.$$

A second type of constraint is an **inequality** that is placed on groups of decision variables. For example, a limit might be placed on total pumping from several wells,

$$Q_1 + Q_2 + Q_3 \leq 5.8 \text{ L/s}$$

or the seepage velocity in the x-direction might be specified to be at least 10 times the seepage velocity in the y-direction,

$$V_x - 10\, V_y \geq 0.$$

A third type of constraint is an **equality**. For example, a specified pumping rate (supply) must be obtained from any of three wells,

$$Q_1 + Q_2 + Q_3 = 10 \text{ L/s}$$

or we might have a mass balance constraint that says the total rate of removal of water from two specific wells, Q_1 and Q_2, must equal the total rate of injection into three other wells, Q'_1, Q'_2, and Q'_3,

$$Q_1 + Q_2 - Q'_1 - Q'_2 - Q'_3 = 0.$$

With the above definitions in mind, we will now consider a problem of aquifer management that can be formulated using linear programming. It will take advantage of the notion of a response matrix. To recap, the linear simulation-optimization models of the type we will now discuss fall under the heading of **hydraulic management models** and the particular approach of including the simulation model in the constraints is known as the **response matrix method**.

2. Illustrative Problem

Figure 18 shows the problem displayed previously in Figure 14, but now there are only three remediation wells and the goal is to capture the contaminant plume before it crosses the site boundary just downgradient of the wells. In addition, the hydraulic gradient must be maintained to keep the plume from spreading laterally. The objective is to minimize the total pumping rate over the **planning horizon**, the total period during which the system is being managed. The constraints are of three types. The first are inequalities that ensure inward (reversed) hydraulic gradients are maintained toward the zone of contamination and toward the remediation wells. These are shown in Figure 18 as gradient constraints. The second are bounds that restrict the pumping rates themselves, including a bound that limits the total pumping to 8 L/s. Finally, we set up our problem so that all variables must be nonnegative. This is a simple bound.

We will consider one **management period** corresponding to the flow system at steady state. A management period or planning period is the time during which the decision variables must remain constant. In some cases we are interested in allowing the pumping rates at each well to change over time. For example, there is one set of rates from 0 to 90 days, another between 90 and 180 days, and another between 180 and 270 days. This can be done by adding additional decision variables corresponding to the pumping rates during additional management periods. For simplicity of explanation, we will only consider one management period in this example.

Things are made simpler by noting that an additional physical constraint already exists; the two outside wells will pump at identical rates. This is because, in our example, the system is symmetric and we would expect identical rates at these wells. This fact makes the problem easier to explain because it is reduced to only two decision variables, pumping rate at the central well and pumping rate at the two outside wells. In essence, because wells 1 and 3 pump at the same rate, we can replace the objective function containing three decision variables:

$$\text{Minimize } Z = Q_1 + Q_2 + Q_3$$

with an equivalent objective that contains two decision variables:

$$\text{Minimize } Z = 2Q_A + Q_b$$

where $2Q_A = Q_1 + Q_3$, and $Q_B = Q_2$.

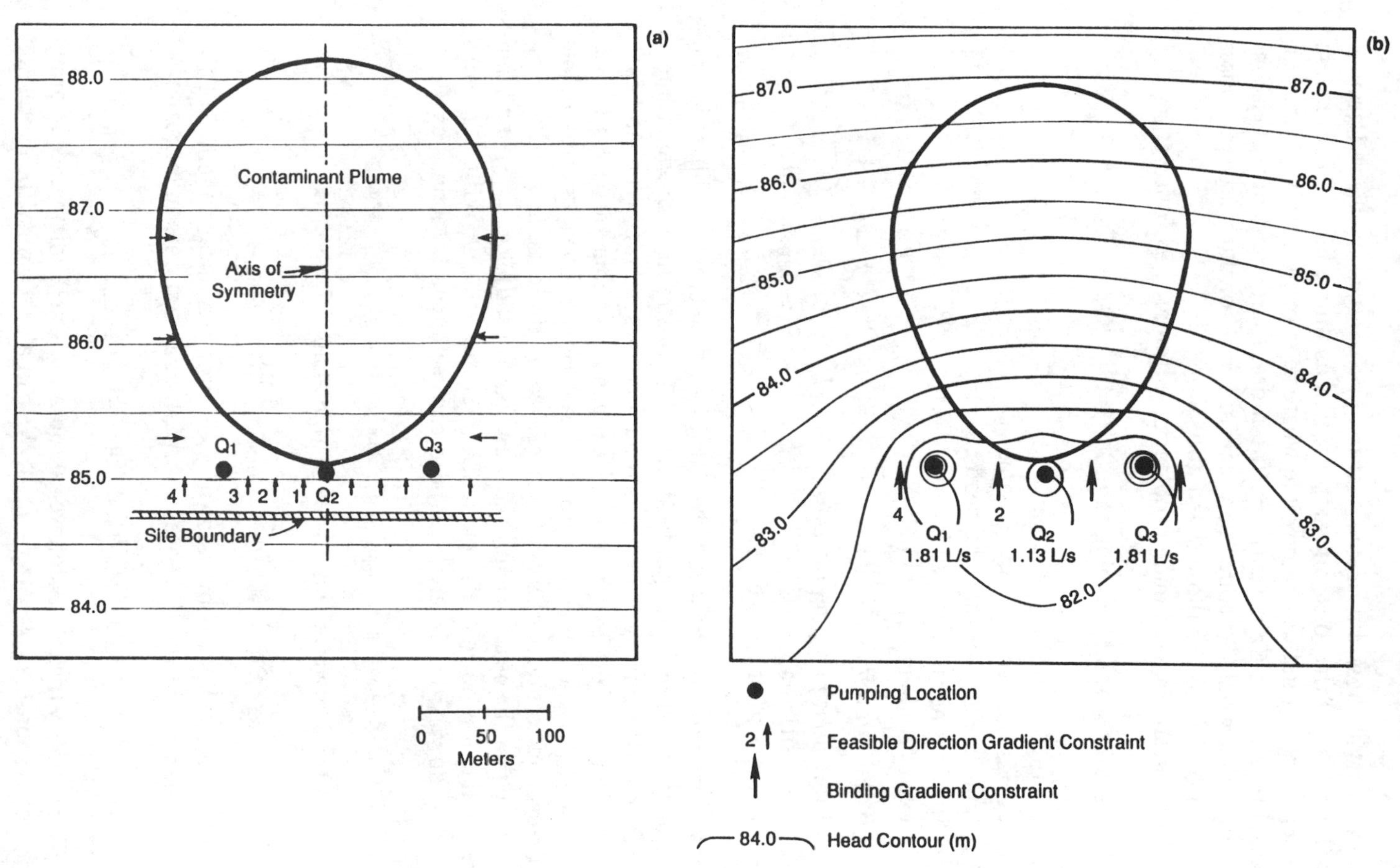

Figure 18. Typical contaminant plume with three remediation wells that can be used for plume capture.

As we will show, this means that the problem can be viewed graphically with all of the constraints represented as simple lines. Of course, in general, this would not be possible where tens or hundreds of decision variables and constraints exist. Furthermore, there are many other constraints that would be added in a real case, but for illustrative purposes we will only consider those noted above.

The Linear Programming problem is formulated as follows:

Minimize Z = total pumping rate	$2Q_A + Q_B$
Subject to	
Heads outside the contaminated zone are greater than the heads inside (one such constraint for each "gradient control pair")	$H_{out} \geq H_{in}$
Total pumping is limited	$2Q_A + Q_B \leq 8$ L/s
Individual pumping rates are restricted and are nonnegative	$0 \leq Q_A \leq 3.0$ $0 \leq Q_B \leq 3.0$

We will now use the ideas developed from the previous subsection on linearity to express the "gradient control" constraint above as a linear function of pumping. This constraint involves a "difference-in-head" limitation. The difference in head between two control points is H_{out} minus H_{in}. In a numerical groundwater flow model, **control points** correspond to nodes of a finite difference or finite element grid.

Note that the head is simply the unmanaged head, $H_{unmanaged}$, at a point minus the drawdown due to pumping, S, at the wells, Q. Written in terms of drawdowns, this difference-in-head constraint can be defined as

$$H_{out} - H_{in} = (H_{out_{unmanaged}} - S_{out}) - (H_{in_{unmanaged}} - S_{in}) \geq 0 \qquad (72)$$

moving the known quantities to the right-hand side we have

$$S_{in} - S_{out} \leq H_{in_{unmanaged}} - H_{out_{unmanaged}}. \qquad (73)$$

Now recall that the drawdown, S, is simply a linear function of the pumping rates at the three wells. A particular response matrix row is

$$\mathbf{r_{in}}^t \mathbf{Q} - \mathbf{r_{out}}^t \mathbf{Q} \leq (H_{in} - H_{out})_{unmanaged.}$$

To simplify the expression we can obtain a single "difference in drawdown" response, $\mathbf{r_{diff}}^t = (\mathbf{r_{in}}^t - \mathbf{r_{out}}^t)$. We can also simplify the expression for the "difference-in-unmanaged-head" as $H_{diff} = H_{in} - H_{out}$. So our constraint reduces to

$$\mathbf{r}_{\mathbf{diff}}^{\mathbf{t}}\, \mathbf{Q} \leq H_{diff}. \tag{74}$$

H_{diff} is calculated by the flow model using only the initial and boundary conditions.

Of importance here is that each "gradient control" constraint has now been converted into a simple linear function in which the "gradient control response" is proportional to the rate of pumping at the three wells. It is of the same form as a drawdown response constraint except now the response coefficients reflect the change in gradient. All problems involving hydraulic gradient control lead to constraints involving differences in head, such as that above.

A key concept is that of a **control pair**. The local gradient is defined across a **control pair** that is simply two neighboring observation points (finite difference or finite element nodes in a numerical flow model). Once a control pair is defined, the local direction and magnitude of flow can be constrained. In our case, each control pair corresponds to one point near the contaminant plume, H_{in}, and one point further from the plume, H_{out}. Such hydraulic gradient control constraints that use control pairs are shown in Figure 18. When constraints involve a control pair, the response coefficients reflect the change in gradient across the control pair due to pumping. This differs from constraints on hydraulic head in which the response coefficients represent the change in head at a control location.

3. Graphical Representation of Linear Programming Problem and Solution

We are interested in a **management solution** that consists of a set of values for the decision variables. Because the formulation only contains two decision variables, a graphical representation of the optimization problem is possible. Figure 19 is a graph with axes corresponding to values of pumping, Q_A and Q_B. Table 9 shows the algebraic form of the linear programming problem in Figure 18. Our graphical representation of the management solution is shown in Figure 19. The coefficients were obtained by solving Equation (66) to generate difference-in-head values and then substituting into Equation (74). The optimal pumping rates are indicated in Figure 18b.

Each constraint, when written as an equality, plots as a straight line on Figure 19. It defines a **boundary equation**. For example, consider the

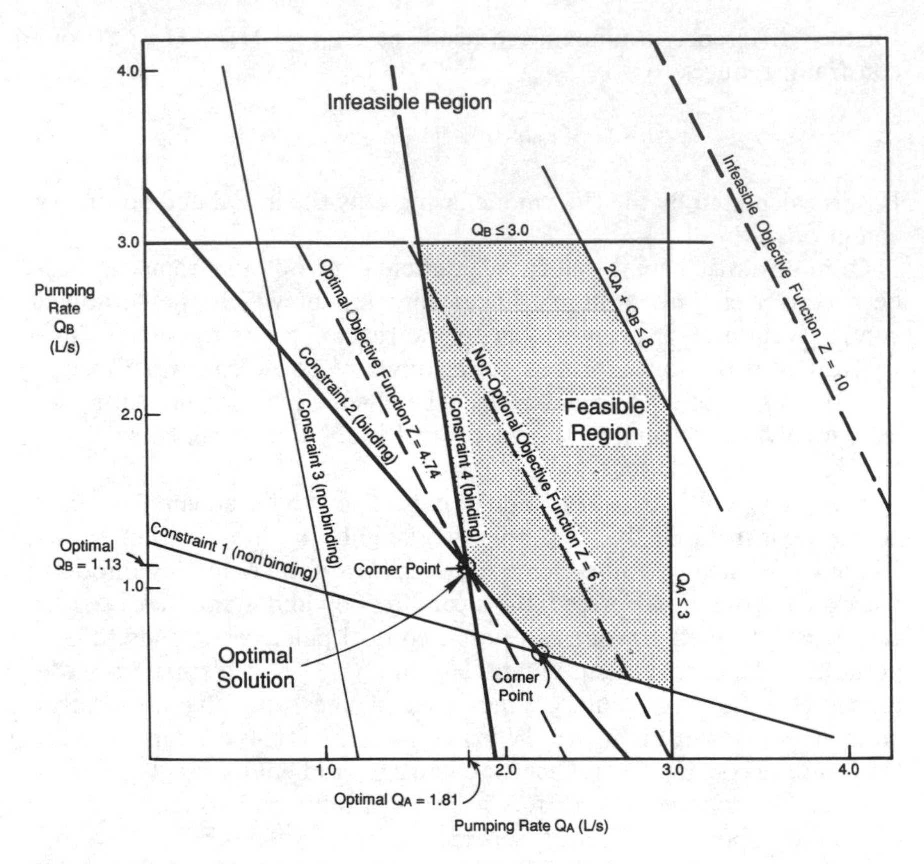

Figure 19. Graphical solution to two-variable remediation problem.

Table 9. Algebraic Formulation of Two-Variable Hydraulic Gradient Control Problem

Minimize Total Pumping,	$Z = (2.0)\ Q_A$	+	$(1.0)\ Q_B$		
Subject to					
Gradient Constraint 1:	$(-0.0236)\ Q_A$	+	$(-0.0786)\ Q_B$	$\leq$	-0.10
Gradient Constraint 2:	$(-0.0364)\ Q_A$	+	$(-0.0302)\ Q_B$	$\leq$	-0.10
Gradient Constraint 3:	$(-0.0840)\ Q_A$	+	$(-0.0159)\ Q_B$	$\leq$	-0.10
Gradient Constraint 4:	$(-0.0509)\ Q_A$	+	$(-0.0069)\ Q_B$	$\leq$	-0.10
Total Pumping Restriction:	$(2.0)\ Q_A$	+	$(1.0)\ Q_B$	$\leq$	8.0
Individual Pumping	$(1.0)\ Q_A$	+		$\leq$	3.0
Restrictions:			$(1.0)\ Q_B$	$\leq$	3.0
	$(1.0)\ Q_A$			$\geq$	0.0
			$(1.0)\ Q_B$	$\geq$	0.0

pumping restriction $Q_A \leq 3.0$. It is represented by the boundary equation, $Q_A = 3.0$, which is a simple vertical line, and the area to the left of the vertical line, which corresponds to $Q_A < 3.0$. The region on the figure defined by the vertical line and the area to its left represents feasible values for Q_A if the pumping restriction were the only one in the problem. The area to the right of that boundary equation represents pumping values, Q_A, which are greater than 3.0 and therefore violate the pumping restriction. There are eight other linear inequality constraints that make up the problem, and they are shown in Figure 19. Together these boundary equations define two regions, one "inside and including" the boundaries and another "outside." The inequalities in this case define a polygon surrounded by the boundary equations. This is known as the feasible region. The remaining area, outside the polygon is known as the infeasible region.

The solution to the optimization problem must lie within the feasible region, for it is only in this region that all of the constraints are satisfied. Also shown on the figure is the objective function in the form of a sequence of straight lines, which satisfies $Z = 2Q_A + Q_B$ for different values of Z. As the value of Z decreases, these lines approach the origin of the figure ($Q_A = 0$, $Q_B = 0$), corresponding to decreasing total pumping values. As the objective value decreases, the objective equation approaches a boundary of the feasible region.

Let us consider the status of the graphical solution for sequentially decreasing values of Z. First, for Z greater than 8.0 the solution is infeasible; the objective function lies outside of the feasible region. For values of Z less-than-or-equal to 8.0 the solution is feasible, indicating that there are combinations of pumping rates that satisfy all of the constraints. When Z is decreased further, the objective function reaches a corner point near the lower boundary of the feasible region. A **corner point** is the intersection of two boundary equations.

We now ask: When is the objective value, Z, at a minimum such that all the constraints are obeyed? It is clear that this minimum value is arrived at when the objective equation is in contact with the boundary of the feasible region. In fact, the **optimal solution will be at a corner point**. In this case, it is that corner point closest to the origin. Allowing the objective value to decline beyond this optimal corner point moves the objective function outside of the feasible region where the constraints are violated. At optimality some of the inequality constraints will be met as strict equalities and are called **binding constraints**. In our example, the binding constraints correspond to gradient control pairs 2 and 4 and their mirror images on the right side of the line of symmetry, as shown in Figure 18. At optimality, the remaining constraints do not dictate the solution as they are easily met. Although they were superfluous, this was not known prior to discovering the optimal solution. These are known as **nonbinding constraints.**

For our example problem, the optimal values of the decision variables are Q_A = 1.81 L/s and Q_B = 1.13 L/s. Together, these are the minimal pumping values that satisfy all of the constraints. The design pumping rates are Q_1 = 1.81, Q_2 = 1.13, and Q_3 = 1.81 L/s as shown in Figure 18b.

Some important facts about the solution of linear programming problems may be extracted from the graphical solution. First, the optimal solution (if one exists) will lie at a corner point of the feasible region. Second, if the objective function meets only one feasible corner point and a lower value of the objective function cannot be found without entering the infeasible region, then the solution is optimal. For **linear** programming problems, if an optimal solution exists, then the objective value is truly minimized (or maximized) and **global optimality** occurs; better solutions in the form of lower objective values cannot exist. This is not necessarily true for nonlinear problems where **locally optimal solutions** might exist that are not globally optimal. Third, if the objective function is parallel to a boundary of the feasible region (for example, if the total pumping limitation were the only nontrivial constraint), then two corner points will be optimal. The objective value is at a minimum in either case. If this occurred in our example it would tell us that more than one set of pumping rates could be found, each of which would minimize total pumping. The solution would **not be unique** but would be optimal. More importantly, if there is more than one optimal solution then there are infinitely many solutions. All points along the intersection of the objective equation and the feasible region are optimal. The above features are also common to more complicated multiple-variable linear programming solutions.

The solution to a linear programming problem must terminate in one of three ways:

1. The solution(s) may be **optimal** as discussed above; the value of the objective function is at a minimum or maximum.
2. There may be no possible **activity levels** of the decision variables that satisfy the constraints, in which case the solution terminates as **infeasible**. This is displayed in Figure 20a where a constraint is added that states that Q_A plus $2Q_B$ must be greater-than-or-equal-to 9 L/s. In that case, there is no feasible region.
3. Finally, the solution may be **unbounded**. This is displayed graphically

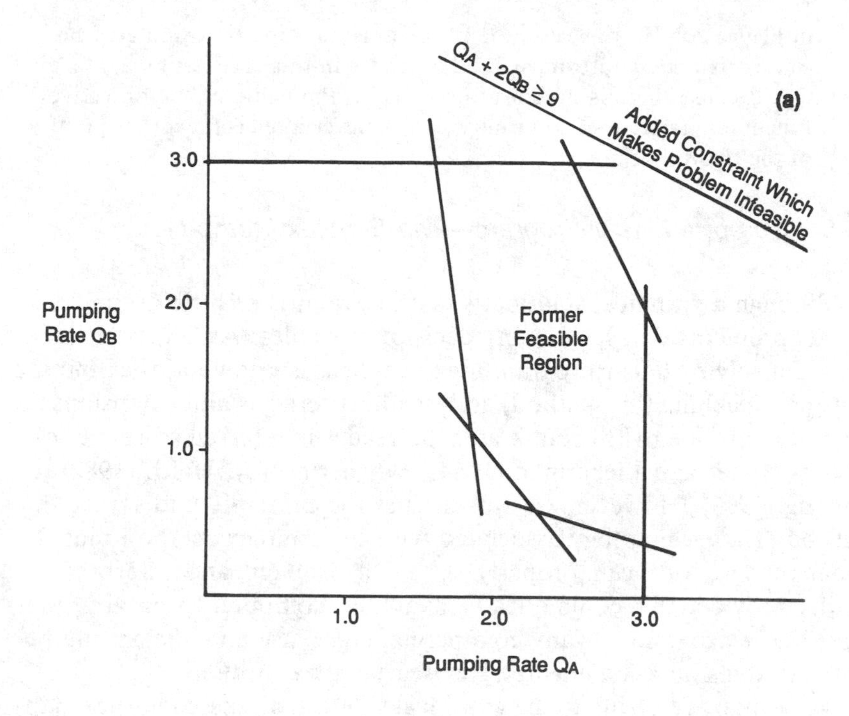

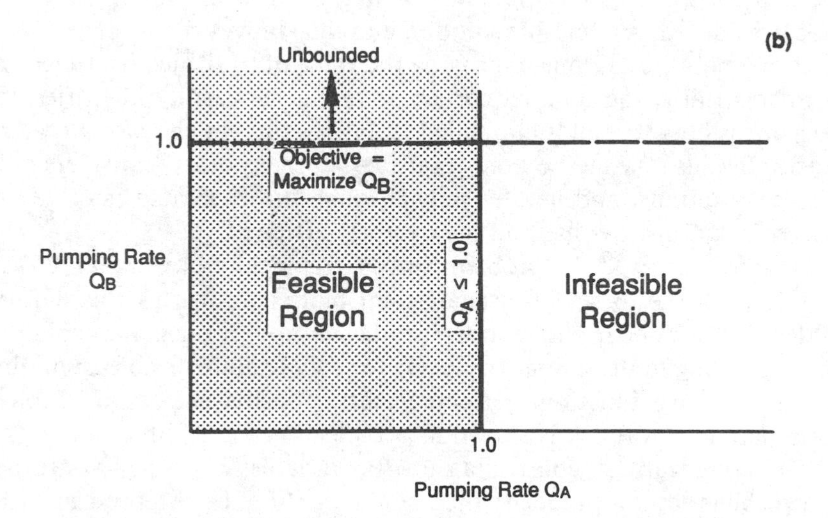

Figure 20. Graphical constraint sets (a) showing solution is infeasible and (b) unbounded solution.

in Figure 20b. In this case, the problem is improperly formulated because it is **underconstrained**. The objective in this case is to maximize Q_B. Because there is no constraint on Q_B, the value of the objective function may increase to infinity. It will never intersect a corner point of the feasible region.

4. Solution of a Linear Program—The Simplex Method

Although a graphical solution is instructive, it is of little use in "real-world" problems that have many decision variables. An algebraic technique for solving linear programming problems is known as the **Simplex Method**. Teaching this method, and its idiosyncracies and extensions, is beyond the scope of this book and the reader is referred to text books such as Hillier and Lieberman (1984), Wagner (1975), Haith (1982), or Dantzig (1963). However, we will discuss the principle underlying this method. The terminology associated with the Simplex Method must be understood so one can properly set up a problem and interpret the results provided by commercially available computer programs that solve linear programs. Many commercial codes use an efficient mathematical extension known as the **Revised Simplex Method**.

We begin by referring to the graphical solution discussed in the Chapter V.C.3. Recall that the optimal solution occurred at a corner point of the feasible region. Intuitively, one might expect an effective search for the optimum to involve inspection of the corner points. This is precisely the approach adopted in the Simplex Method. It is an iterative method that first identifies a **feasible solution** and then moves from corner point to corner point, each time improving the value of the objective function. Mathematically, the optimal corner point can be readily identified (as can infeasiblity or an unbounded solution) in which case iteration ceases. Because the method only considers corner points, considers only feasible solutions, and iterates only through those points that give improved solutions, the method is extremely efficient.

Now we return to the problem solved graphically in Chapter V.C.3. The first step is to set up a **coefficient matrix known as the simplex tableau**. The coefficient matrix for the above problem has columns corresponding to the unknowns Z, Q_A, and Q_B, and rows corresponding to the objective function and constraints. The mathematical problem formulation of Table 9 is shown as a tableau in Table 10.

The nonnegativity constraints on the variables Q_A and Q_B are not shown but they are implied. In fact, in the Simplex Method and the commercially available linear programming solvers, nonnegativity of variable values is the default.

Table 10. Tableau Form of Two-Variable Hydraulic Gradient Control Problem Shown in Figure 18

Constraint	Variable Names			Constraint		
Number	Z	Q_A	Q_B	Type	RHS	Explanation
0	1	−2	−1	=	0	Objective function
1		−0.0236	−0.0786	≤	−0.10	Gradient Constraint 1
2		−0.0364	−0.0302	≤	−0.10	Gradient Constraint 2
3		−0.0840	−0.0159	≤	−0.10	Gradient Constraint 3
4		−0.0509	−0.0069	≤	−0.10	Gradient Constraint 4
5		2.0	1.0	≤	8.0	Total Pumping Constraint
6		1.0		≤	3.0	Pumping Rate Restriction
7			1.0	≤	3.0	Pumping Rate Restriction

Although our purpose here is to discuss the solution to this linear programming problem, it is important to recall how groundwater simulation is incorporated into the management formulation. Simulation is accomplished using the response matrix that converts pumping rates into drawdown, hydraulic head, and hydraulic gradient responses. These simulation-based constraints that involve response coefficients are 1 through 4 in Table 10.

To apply the Simplex Method, the inequalities appearing as constraints must be converted into equalities. This is done by adding a variable to each inequality that represents the "slack" or play between the constraint written as an inequality and the equivalent constraint written as an equality. For example, one constraint says that the total pumping rate $2Q_A$ plus Q_B is limited to 8 L/s. The constraint is

$$2Q_A + Q_B \leq 8 \text{ L/s}, \tag{75}$$

which may be rewritten as

$$2Q_A + Q_B + \text{Slack}_1 = 8 \text{ L/s} \tag{76}$$

where Slack_1 is a **slack variable**. This is called the equality representation. As long as Slack_1 is non-negative, the original constraint and the augmented constraint (the one in equality form) are equivalent. Constraints of the greater-than-or-equal-to form can be converted to equality form by adding a **surplus variable** (negative slack).

The definition of slack variables is only provided here to give some idea of how the simplex procedure works and some knowledge of how to interpret the solution of a linear programming problem. The user of a commercial linear programming routine only enters the original problem formulation constraints and an objective; so in practice, one can ignore the issue of slack variables because they are added automatically by the solution routine. Their values will appear in the solution and they do have physical significance. If a particular slack variable is zero, this indicates that the constraint is binding; the constraint that contained that slack variable is one that dictates the design solution. If the slack activity is non-zero, this indicates two things: First, the constraint was satisfied but was nonbinding and did not matter in that particular design solution; it turned out to be unnecessary. Second, the value of the slack variable indicates how much more restrictive that constraint would have to be before it dictates the design solution.

The Simplex Method defines two types of variables that are exchanged one for the other during the optimization process, thereby selecting decision variables (wells and pumping rates) that will be active and those that are inactive in the optimal solution for remediation design. First are **non-basic variables**, each of which has the value of zero in the optimal solution. If, for example, a pumping rate (decision variable) is non-basic in an optimal solution, that well is inactive because by definition **non-basic variables are set to zero**. There are also **basic variables, whose values define the activity levels** of those variables in the optimal solution. If a pumping rate (decision variable) appears as a basic variable in an optimal solution, the well pumps at the optimal rate (activity level) of that basic variable. The final solution consists of both basic and non-basic variables.

The solution proceeds by identifying an initial **basic feasible solution**. This stage of the optimization procedure of the Revised Simplex Method is known as **Phase I**. Briefly, Phase I identifies an initial basic feasible solution. Finding an initial feasible solution is not difficult in most aquifer remediation problems of the type considered here. For example, in our case this could be done by setting the pumping rates at the remediation wells to their maximum limits. If the solution should prematurely terminate in Phase I, this indicates no feasible solution exists. Perhaps this is a consequence of inadequate well siting, not permitting enough potential well sites in the design formulation, or conflicting constraints on hydraulic gradient directions or magnitudes. In such a case, the infeasible constraint(s) will be indicated in the aborted solution and the formulation should be reexamined, reformulated, and the problem rerun.

Once an initial basic feasible solution is found then iteration proceeds by inspecting new basic feasible solutions during each iteration of the Simplex Method. This optimization stage of the Revised Simplex Method, during which feasibility is always maintained, is known as **Phase II**.

In Table 11 we show the solution at the three critical stages of the

Table 11. Stages of Linear Programming Solution for Example Problem

Variable	Status	Activity	Explanation
A) Starting Solution–Infeasible			
Z	Objective	0.0	
Q_A	Non-basic	0.0	Non-basic variables equal zero
Q_B	Non-basic	0.0	Non-basic variables equal zero
Slack 1	Basic	−0.10	Slack in gradient constraint 1
Slack 2	Basic	−0.10	Slack in gradient constraint 2
Slack 3	Basic	−0.10	Slack in gradient constraint 3
Slack 4	Basic	−0.10	Slack in gradient constraint 4
Slack 5	Basic	8.0	Slack in total pumping constraint
Slack 6	Basic	3.0	Slack in pumping limit on Q_A
Slack 7	Basic	3.0	Slack in pumping limit on Q_B
B) Phase I Solution–Initial Feasible Solution			
Z	Objective	8.0	
Q_A	Basic	2.5	Alternate feasible solutions exist
Q_B	Basic	3.0	
Slack 1	Basic	0.19497	Slack in gradient constraint 1
Slack 2	Basic	0.08159	Slack in gradient constraint 2
Slack 3	Basic	0.16002	Slack in gradient constraint 3
Slack 4	Basic	0.04787	Slack in gradient constraint 4
Slack 5	Non-basic	0.0	Slack in total pumping constraint
Slack 6	Basic	0.5	Slack in pumping limit on Q_A
Slack 7	Non-basic	0.0	Slack in pumping limit on Q_B
C) Phase II Solution–Optimal			
Z	Objective	4.75139	
Q_A	Basic	1.81302	Optimal value of Q_A
Q_B	Basic	1.12535	Optimal value of Q_B
Slack 1	Basic	0.03133	Slack in gradient constraint 1
Slack 2	Non-basic	0.0	Gradient constraint 2 is binding
Slack 3	Basic	0.07193	Slack in gradient constraint 3
Slack 4	Non-basic	0.0	Gradient constraint 4 is binding
Slack 5	Basic	3.24861	Slack in total pumping constraint
Slack 6	Basic	1.18698	Slack in pumping limit on Q_A
Slack 7	Basic	1.87465	Slack in pumping limit on Q_B

optimization procedure. First is the solution corresponding to the initial augmented (infeasible) problem, solution A in the table. Second in the table is solution B which is at the end of Phase I which identifies a feasible solution. Third in the table is C which is the solution at the end of Phase II which is optimal. Such solutions and information are based upon the results of an optimization routine, MINOS (Murtagh and Saunders, 1983).

Inspecting the optimal solution we note three important items. First, the minimal objective value is derived from the two variables, Q_A and Q_B, which became basic variables during the course of solution. Second, the slack activities, corresponding to gradient control constraints 2 and

4, are zero and those variables are non-basic. This means that those constraints are binding and are referred to as "tight constraints." As we saw in the graphical solution, these two constraints define a boundary of the feasible region. In fact, they define a corner point that dictates the minimum value of the objective. Third, the remaining slack variables are basic. Their values are measures of the ease with which the constraints are met. For example, consider the difference-in-head for the first control pair in the optimal solution. The augmented constraint was

Gradient Response Due to Pumping	+	Gradient Slack for Control Pair 1	=	The Manageable Gradient
$(-0.0236)Q_A + (-0.0786)Q_B$	+	$Slack_1$	=	-0.10
-0.13133	+	0.03133	=	-0.10

In the optimal solution the above constraint was met with room to spare, or slack, of 0.03133. The hydraulic gradient due to pumping, -0.13133, is known as the **row activity**. It is the value of the left-hand side of the constraint ignoring the slack activity. As another example of the interpretation of a constraint at optimality, consider the limitation on total pumping, which is nonbinding constraint number 5 in the solution,

$$(2.0)Q_A + (1.0)Q_B \leq 8.0.$$

Written in augmented form we have

$$(2.0)Q_A + (1.0)Q_B + Slack_5 = 8.0$$

and in the optimal solution we obtain

Total Pumping at Optimality	+	Slack Pumping Capacity	=	Upper Bound on Total of Pumping Rates
4.75139	+	3.24861	=	8.0

D. General Simulation — Management Modelling Using AQMAN

1. Introduction

This chapter addresses the simulation-management method for aquifer remediation design that involves larger scale and more complex problems than those discussed so far. However, we still assume that the flow system can be modelled using a two-dimensional representation and again we restrict the discussion to "on-the-shelf technology." Therefore, we will rely on existing simulation methods and optimization routines and concentrate on the linkage between the two. In approaching the remediation design problem it makes no sense to reinvent the wheel with regard to either optimization methods or simulation methods, because robust techniques and codes are available.

The key linkage in simulation-optimization modelling is the response matrix demonstrated in the previous chapter. In general, simulation of aquifer behavior requires a numerical model to take into account complicated aquifer geometry, heterogeneous transmissivity fields, and complex boundary conditions. For most problems a versatile and fairly unrestricted simulation model must be used to generate the response matrix.

Here we describe the response matrix generator AQMAN (Lefkoff and Gorelick, 1987). It employs the aquifer simulation model of Trescott et al. (1976), from here on referred to as the Trescott model, which is a fully documented, easy to use, and well-tested code. AQMAN (pronounced Ack-Man—short for AQuifer MANagement) is a FORTRAN-77 code that uses the Trescott model for two-dimensional aquifer simulation to generate aquifer responses. AQMAN controls the execution of the Trescott model, which is run repeatedly to construct the response matrix. It then sets up a special optimization data file that contains this response information and other elements of a management formulation. **This optimization data file, known as the MPS file (MPS stands for Mathematical Programming System), is an efficient representation of the input tableau for a linear programming problem**. It is in a standardized format, known as MPS format, so that commercially available linear programming solvers can recognize the objective and constraints and then provide a solution. The topics covered in this chapter summarize material found in the AQMAN documentation.

Chapter V.A.2 listed 10 steps that must be followed to develop and solve an aquifer management problem. Five of these steps must be completed before AQMAN can be used. First among them (see steps 1, 2, and 4) was the creation of an appropriate simulation model for the particular site of interest. It is assumed that the hydrologist has a good understanding of the aquifer and can use the Trescott model to represent its behavior. This is a prerequisite to using AQMAN. One should not attempt any optimization method without this level of understanding. The next two steps used specifically in aquifer management modelling are development of a menu of design considerations (step 3) and formulation of the remediation problem as a simulation-optimization problem

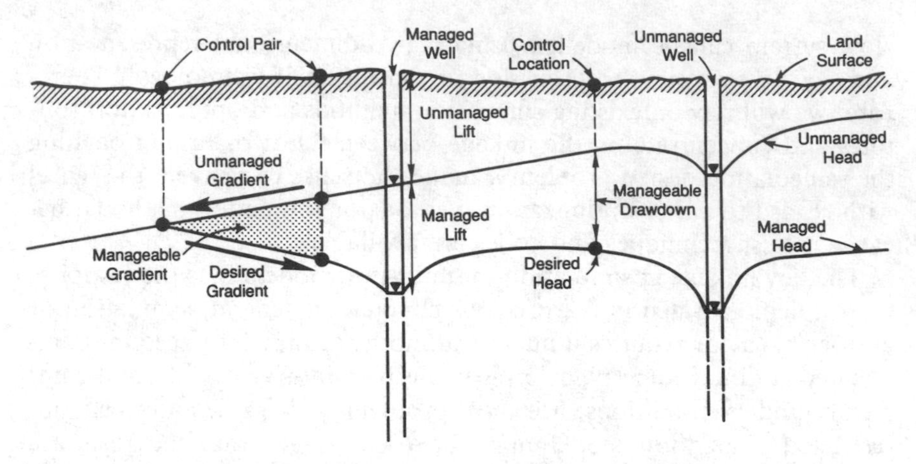

Figure 21. Cross section showing key terms used in response matrix method.

(step 5). These are independent engineering design steps that do not require a computer program of any sort. Once these first five steps are complete, AQMAN can be used.

Given a remediation problem that has been formulated as a linear programming problem (or quadratic programming problem as discussed later), AQMAN performs the task of generating the response matrix and constructing the input tableau data file. This file is then fed to a standard, independent, and available optimization routine that solves the optimization problem. With this framework in mind, we now provide a detailed discussion of the use of AQMAN.

2. Definitions Involving Stress and Response

In the response matrix approach it is convenient to define a series of terms, most of which help distinguish between activities and responses that are either part of the engineering design or those that are not. As noted earlier, this distinction enables us to effectively use the principle of linear superposition. The key terms are illustrated in the cross section shown in Figure 21 and are defined in the following paragraphs.

a. Definitions Involving Flow Stresses

A **managed well** is a location where pumping or injection may occur. Associated with each managed well are decision variables, representing the pumping rate during each management period at that location. The pumping or injection rate at a managed well is known as a **managed stress.**

An **unmanaged well** is a stress location that is not controlled by engineering design. For example, in a problem of aquifer remediation, a nearby irrigation well might pump at a predetermined rate and is not part of the remediation design but does influence it; its rate is not controllable as far as contaminant capture is concerned and thus it is an **unmanaged stress**. In this context, the term unmanaged "well" is somewhat euphemistic in that it can actually refer to the location of any fluid source or sink, such as a leaky river.

b. Definitions Involving Head and Drawdown Responses

Unmanaged heads are those that would exist in the absence of managed stresses. They are the "background" uncontrolled heads and may vary, over space and over time due to initial conditions, boundary conditions, and unmanaged stresses.

Manageable drawdown is the difference between the unmanaged head and the head limitation imposed at a **control location** in a design formulation. Manageable drawdown is the head decline we must create with our controlled pumping activities.

c. Definitions Involving Hydraulic Gradients and Velocities

Frequently in aquifer remediation design one is interested in controlling local hydraulic gradients and/or the direction and magnitude of seepage velocities. The following definitions apply.

A **control pair** consists of two locations between which a hydraulic gradient or seepage velocity must be restricted. As in our previous example, a response coefficient can be determined that reflects the effect of a managed stress on the "difference-in-drawdown" between the control pair locations.

The **manageable gradient** is defined for a particular **control pair** as the magnitude of the hydraulic gradient under unmanaged heads minus the magnitude of the gradient desired in the engineering design.

The **manageable velocity** is defined for a particular control pair as the seepage velocity under unmanaged heads minus the seepage velocity desired in the engineering design.

The latter two terms are analogous to that given for manageable drawdown, except they involve the notion of a control pair rather than a control location.

3. Objective Function

a. Linear Objective

AQMAN can handle simulation-management problems of two types, which differ only in the form of their objective function. The two types are linear programming problems and quadratic programming problems. Our example has been of the linear programming type, so it will be covered first. AQMAN assumes that one is interested in minimizing or maximizing the total of the pumping rates. The objective function is of the form:

Minimize or Maximize

$$Z = \sum_{n=1}^{N} \sum_{i=1}^{I} C_{i,n} Q_{i,n} \tag{77}$$

where N = total number of management periods;
I = total number of wells;
$C_{i,n}$ = coefficients for well i during period n, as explained below; and
$Q_{i,n}$ = pumping rates at well i during period n, $[L^3/T]$.

In AQMAN the coefficients in the objective, $C_{i,n}$, are assumed by default to be 1.0. This means the default objective function is to minimize or maximize the sum of the pumping rates over space and time. This can, of course, be changed by manually editing the MPS file. For example, in our small sample problem of the previous chapter we required the coefficient associated with Q_A, which represented the two outside wells, to have a value of 2.0. This was because symmetry dictated that Q_A was really one decision variable which we weight twice in the objective function to account for the fact that it represents pumping from two wells (recall $2Q_A = Q_1 + Q_3$). In such a case, the coefficient in the objective function for Q should be changed from 1.0 to 2.0 in the MPS file.

b. Quadratic Objective

AQMAN also allows one to specify an objective that minimizes the present value of the pumping costs, excluding fixed capital costs. This is a quadratic function of the pumping rates, hence the optimization problem will involve quadratic programming. The quadratic nature of the cost function is easily seen. Consider the cost of pumping at a single well. It is a product of the pumping rate and the energy required to lift the water to the land surface. As one increases pumping, the head in the well declines and the lift, which depends on pumping, increases. The pumping rate now enters multiplicatively into the cost, once because the cost is related to the rate of pumping and again because pumping creates greater lift.

The mathematics used to define this quadratic function involves the response matrix. The total pumping cost, Z', for a transient problem, is

$$Z' = \sum_{n=1}^{N} \sum_{i=1}^{I} C_{i,n} L_{i,n} Q_{i,n} \tag{78}$$

where Z' = total pumping cost [$];
$L_{i,n}$ = total lift at well i during period n, [L], and
$C_{i,n}$ = unit cost of pumping per unit lift at well i during period n, [$/(L^3/T)/L].

Note that the lift is caused by the joint influence of all pumping wells over all time plus the influence of the unmanaged heads. We can break up the lift into the **unmanaged lift**, L^*, which is the distance from the land surface to the unmanaged head, and the **managed lift**, S, which is the additional drawdown caused by pumping over all time and space. This can be expressed as

$$L_{i,n} = L^*_{i,n} + S_{i,n} \tag{79}$$

and

$$S_{i,n} = \sum_{k=1}^{n} \sum_{j=1}^{I} r_{i,j,(n-k)} Q_{j,k} \tag{80}$$

where each response coefficient, r, is the drawdown response at well i during period n due to pumping at well j during period k.

The total pumping cost is then minimized.

Minimize

$$Z' = \sum_{n=1}^{N} \sum_{i=1}^{I} \left[C_{i,n} L^*_{i,n} Q_{i,n} + \sum_{k=1}^{n} \sum_{j=1}^{I} C_{i,n} r_{i,j,(n-k)} Q_{j,k} Q_{i,n} \right] \tag{81}$$

The first terms are the linear components that account for the cost of pumping associated with the unmanaged lift. The second terms are quadratic (note the product of pumping rates, Q) and reflect the cost due to the managed lift. AQMAN uses the equation in this form to minimize total pumping cost. The unit cost coefficients are discounted using

$$C_{i,n} = (C^*_{i,n})/(1 + d)^m \tag{82}$$

where C^* = undiscounted unit cost,
d = monthly discount rate, and
m = the number of months.

This adjusts future costs to the present on a monthly basis. In order to handle this objective, quadratic programming is used. To solve an optimization problem involving a quadratic objective subject to linear constraints, a special subroutine is required that computes the value of the objective function. It also computes the gradient of the objective function, which is given by the partial derivative of the cost function with respect to a change in pumping. Both the objective function value, given any set of pumping rates, and the objective function gradients are needed in the quadratic programming optimization search method. This subroutine, called FUNOBJ, is included in the AQMAN documentation.

4. Constraints

Although we have defined the general types of constraints for a linear programming problem, the constraints appearing in aquifer remediation models represent a particular subset. Constraints may restrict the value of the decision variables for pumping or injection, may place limitations on a state variable such as hydraulic head, or may exist to simply define a useful quantity such as velocity or hydraulic gradient.

a. Constraints on Hydraulic Heads

Aquifer simulation describes changes in hydraulic head values due to managed and unmanaged stresses. This information is the key element in aquifer management modelling and the one for which AQMAN is most useful. It is also the foundation for other simulation-based constraints on velocities and hydraulic gradients. The bulk of the work done by AQMAN is to control the repeated executions of the flow simulation model to produce response coefficients and then transform them into the user's constraints in the MPS file. Consider an example similar to

Table 12. Definition of Unmanaged Head Values

Observation Location	$H_{unmanaged}$	= $H_{boundary}$	− $\Delta H_{initial}$
A	82.8	86.3	3.5
B	82.9	86.3	3.4
C	82.7	86.1	3.4
D	82.7	86.0	3.3
E	82.5	85.9	3.4

that presented in Chapter V.B.2 in which there are five control locations at which there are restrictions on simulated hydraulic heads. This management problem has three decision variables that correspond to pumping at Wells 1, 2, and 3. The constraints on the simulated hydraulic heads at the five control locations are

$H_A \leq 60$ Head A must not exceed 60 m
$H_B \leq 85$ Head B must not exceed 85 m
$H_C \geq 40$ Head C must equal or exceed 40 m
$H_D = 50$ Head D must equal 50 m
$H_E \leq 80$ Head E must not exceed 80 m.

These constraints can be represented using a response matrix by transforming them into equivalent constraints on drawdowns (changes in head). Recall that the total hydraulic head is simply the sum of manageable and unmanageable components. For example, given the above head constraints, AQMAN first computes the transient unmanaged heads that would exist in the absence of any managed stresses. The unmanaged heads are sums of the influences of the boundary and initial conditions:

$$H_{unmanaged} = H_{boundary} - \Delta H_{initial}$$

In the example in Chapter V.B.2, a constant uniform gradient exists at steady state, there is no pumping, and a remnant cone of depression exists due to previous pumping. For this case, the values of unmanaged head are shown in Table 12.

Next, AQMAN determines the manageable drawdown by subtracting the limitation on head from the unmanaged head. This is the drawdown that must be achieved by combined pumping from Wells 1, 2, and 3. The manageable drawdown is computed in Table 13.

Then the drawdown response matrix is formed, by running the simulation model once for each pumping well, to describe the linear conversion of pumping rates into drawdowns at locations A, B, C, D, and E. For this example, the drawdown response matrix at 30 days was given in

Table 13. Computation of Manageable Drawdowns

Unmanaged Site	Component	–	Requirement	=	Manageable Drawdown
A	82.8	–	60	=	22.8
B	82.9	–	85	=	–2.1
C	82.7	–	40	=	42.7
D	82.7	–	50	=	32.7
E	82.5	–	80	=	2.5

Table 6. Combining the response matrix values with the manageable drawdowns gives the expressions in Table 14.

This response matrix representation is equivalent to the head constraints shown above. Two important items should be noted in this example. First, the direction of the inequality changes when converting from head restrictions to drawdown restrictions; $\leq$ becomes $\geq$ and $\geq$ becomes $\leq$. AQMAN automatically keeps track of this switch so the user simply enters the true head constraint. Second, note that in this case the second constraint (corresponding to site B) will always be met because the unmanaged head is already within the requirement and remains so even if pumping rates are zero. The other four constraints will require some pumping to create the required drawdowns.

b. Constraints on Seepage Velocities and Hydraulic Gradients

Of particular interest in aquifer remediation are restrictions on flow directions and flow magnitudes. These constraints are generated in response matrix form using AQMAN. Suppose we require a local groundwater velocity component in the x-direction, V_x, from control location 1 to control location 2 to be at least 0.01 m/s or

Table 14. Response Matrix Representation of Head Constraints

		Drawdown Response Matrix		Manageable Drawdown	Interpretation of Constraints
(0.17) Q_1	+	(0.42) Q_2	+ (0.23) Q_3	$\geq$ 22.8	Drawdown at A is at least 22.8
(0.30) Q_1	+	(0.18) Q_2	+ (0.33) Q_3	$\geq$ –2.1	Drawdown at B is at least –2.1
(0.19) Q_1	+	(0.29) Q_2	+ (0.38) Q_3	$\leq$ 42.7	Drawdown at C is at most 42.7
(0.20) Q_1	+	(0.19) Q_2	+ (0.55) Q_3	= 32.7	Drawdown at D must equal 32.7
(0.13) Q_1	+	(0.37) Q_2	+ (0.36) Q_3	$\geq$ 2.5	Drawdown at E is at least 2.5

$$V_x \geq 0.01. \tag{83}$$

The velocity, defined over a control pair, is

$$V_x = (K/n_e \Delta L)(H1 - H2) \tag{84}$$

where K = hydraulic conductivity within the control pair,
n_e = the effective porosity in the vicinity of the control pair,
ΔL = the distance between the control pair locations (in the x-direction in this example), and
H1 and H2 = the hydraulic heads at the control pair locations.

Combining the two above equations, the velocity, defined in terms of the head difference, can be constrained as

$$(K/n_e \Delta L)(H1 - H2) \geq 0.01. \tag{85}$$

Again we define the total head as the difference between the unmanaged head and the drawdown,

$$H1 = H1_{unmanaged} - S1$$
$$H2 = H2_{unmanaged} - S2.$$

The difference in head can be related to the difference in drawdown as

$$H1 - H2 = (H1_{unmanaged} - S1) - (H2_{unmanaged} - S2) \tag{86}$$

or

$$H1 - H2 = -S1 + S2 + H1_{unmanaged} - H2_{unmanaged}. \tag{87}$$

Substituting the above right-hand side for H1 – H2 in the velocity definition, we obtain

$$(K/n_e \Delta L)(- S1 + S2 + H1_{unmanaged} - H2_{unmanaged}) \geq 0.01 \tag{88}$$

which, when we isolate drawdowns on the left side and all known quantities on the right, is equivalent to

$$S1 - S2 \leq H1_{unmanaged} - H2_{unmanaged} - 0.01(n_e \Delta L/K). \tag{89}$$

The final step is to note that the drawdowns are a linear function of the pumping rates given by a response matrix. For example, if we consider two pumping wells and replace the drawdown values S1 and S2 by their drawdown response to pumping, **RQ**, the above constraint becomes,

$$(r_{1,1}Q_1 + r_{1,2}Q_2) - (r_{2,1}Q_1 + r_{2,2}Q_2) \leq H1_{unmanaged} - H2_{unmanaged} - 0.01(n_e\Delta L/K) \quad (90)$$

where the response coefficients follow the form that $r_{1,2}$ is the drawdown response at site 1 due to pumping at Well 2.

As seen above, the local velocity constraint is written as a linear function of pumping rates. This is the form of the velocity constraint that is placed into the MPS file by AQMAN. If one wishes to simply constrain the hydraulic gradient instead of the velocity, the distance ΔL should be substituted for the factor $(n_e\Delta L/K)$. For example, if we simply wanted to constrain the gradient to be greater than 2 percent of the distance ΔL between the observation points that make up the control pair, we have

$$(H1 - H2)/\Delta L \geq 0.02. \quad (91)$$

c. Retarded Velocities

In groundwater contamination problems for which both advective flow and chemical retardation are being considered, one may wish to constrain the velocity of the contaminated groundwater. If so, one can easily incorporate a retardation factor. This **retardation factor, R_f,** is defined as the ratio of the velocity of a conservative tracer to that of the contaminant. Mathematically, it is a scaling factor on the velocity,

$$V_{unretarded} / V_{retarded} = R_f \quad (92)$$

or

$$V_{retarded} = V_{unretarded} / R_f = (K/n_e\Delta LR_f)(H1 - H2). \quad (93)$$

Thus, to convert a constraint on groundwater velocity into a constraint on retarded contaminant velocity, it is necessary to use the factor **$(n_e\Delta LR_f/K)$ instead of $(n_e\Delta L/K)$./**

d. Constraints on Pumping and Recharge

The simplest type of constraint is a **simple bound** on individual pumping or injection rates. Such constraints do not involve the response matrix but just the decision variables corresponding to pumping or injection. For example,

$$Q_1 \leq 20.$$

Other straightforward types of restrictions are the

1. **demand constraint**, which requires that total pumping during each management period meet some predetermined demand,

$$\sum_{i=1}^{I} Q_{i,n}, \geq 44.8 \text{ L/s} \qquad \text{for each management period n}$$

2. **capacity constraint**, which limits the total pumping or recharge rate during each management period,

$$\sum_{i=1}^{I} Q_{i,n}, \leq 18.2 \text{ L/s} \qquad \text{for each management period n}$$

3. **balance constraint**, which forces designed pumping totals (index i) to equal or exceed artificial recharge totals (index j) during each management period,

$$\sum_{i=1}^{I} Q_{i,n} - \sum_{j=1}^{J} Q_{j,n} \geq 0.0 \qquad \text{for each management period n}$$

or in the case of a **strict balance**,

$$\sum_{i=1}^{I} Q_{i,n} - \sum_{j=1}^{J} Q_{j,n} = 0.0 \qquad \text{for each management period n}$$

Because there are so many variations of these types of constraints, AQMAN does not automatically put them into the MPS file. They are easily entered into either the BOUNDS section or the COLUMNS section of the MPS file.

e. Definition Constraints on Head and Velocity

Often it is convenient to define a variable as a linear function of other variables, using a **definition constraint**. For example, one might wish to define a velocity deviation as the difference between a managed velocity and a predefined target velocity. Additional restrictions could then be placed on the velocity deviation or on sums of the deviations. AQMAN can automatically provide definitions of either head variables or velocity variables. For our example problem in Chapter V.2, the heads at location A, H_A, and location D, H_D, are defined according to the values of unmanaged head and drawdown response (see Tables 12 and 14) as

Head	=	Unmanaged Head	–	Drawdown Due to Pumping (From Unit Responses)
H_A	=	82.8		$-(0.17)\ Q_1 + (0.42)\ Q_2 + (0.23)\ Q_3$
H_D	=	82.7		$-(0.20)\ Q_1 + (0.19)\ Q_2 + (0.55)\ Q_3$

With H_A and H_D defined, one can have linear constraints that involve both of their values. For example, the average head in an area might be constrained to be less than some target value, or

$$(0.5)\ H_A + (0.5)\ H_D \leq 75.0.$$

One very important use of a velocity definition is to control vector sums of predefined velocity components. Such restrictions are useful because they help ensure that specified contaminated zones remain isolated or that uncontaminated water used for water supply only flows in specific directions. For example, consider the case presented in Figure 22. There the contaminated groundwater must not enter the restricted zone. One can define a feasible flow direction and an infeasible direction using constraints that involve both the magnitudes and directions of local groundwater velocities. This is accomplished using the idea of control pairs, which are now used to define velocity components in both the x- and y-directions.

Suppose we do not permit the contaminated groundwater to escape from the restricted zone. The constraint is

$$V_y/V_x \leq \text{Tan}\ \theta \tag{94}$$

or

$$(\text{Tan}\ \theta)\ (V_x) - V_y \geq 0. \tag{95}$$

Once the definitions of V_x and V_y are written [see Equation (84)], then one simply includes the latter constraint to restrict their ratio. This requires that each velocity component be defined as a separate variable.

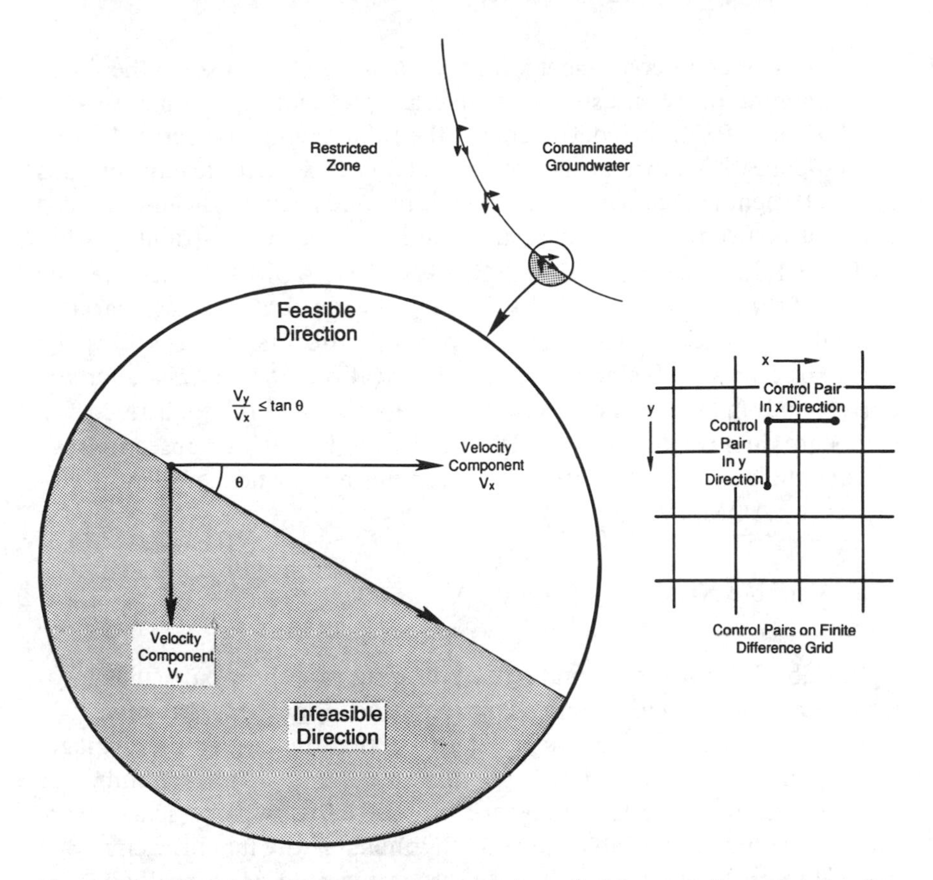

Figure 22. Use of local velocity components to control the magnitude and direction of groundwater flow.

5. Variable Naming and Row Labelling Convention Used by AQMAN

As shown in Chapter V.C.4, a linear programming formulation can be represented using a coefficient matrix known as a simplex tableau. Each coefficient in the tableau also appears within one of the original constraints or the objective. For example, recall the three-well gradient control problem in which the first gradient control constraint stated,

$$(-0.0236)\ Q_A + (-0.0786)\ Q_B \leq -0.10.$$

The first coefficient in this constraint is associated with the variable

Q_A and the second coefficient with the variable Q_B. To solve the linear programming problem using an available optimization routine, such as MINOS or MPSX, the coefficients in the tableau must be entered into a special data set known as an MPS file. The MPS file not only contains the coefficient associated with each variable and each constraint but also contains a specific name for each variable and each constraint.

The MPS file is generated by AQMAN. First AQMAN calculates the values of the tableau coefficients. Then it assigns names to each variable and labels to each constraint which provide information regarding the nature of each variable or constraint, the location index number, and the corresponding management period. These labels and the tableau coefficients are then written to the MPS file. In the following subsections we discuss the naming convention used for variables and constraints employed by AQMAN.

a. Variable Names

Variable names corresponding to pumping rates at a particular location during a particular management period appear as the letter "Q" followed by a five-digit number. The first two digits give the management period and the next three digits give the well index number or control pair number. Examples are presented below, but spaces were inserted within the variable name to "hyphenate" the label for clarity of presentation. When using AQMAN these spaces do not actually appear in the variable label.

Q 03 025 pumping during Management Period 3 at Well 25.

The number of digits in each "hyphenated field" indicates the size limitations that AQMAN places on the possible number of management periods and the number of pumping wells. For example, AQMAN allows up to 99 management periods with up to 999 wells in each period.

Gradient or velocity variables that are directly constrained (ones that do not appear in a definition constraint) begin with the letters **"DIF"** because they are based on a difference-in-head across a control pair. A variable corresponding to the constraint on velocity might be

DIF 18 007 velocity at end of Period 18 over Control Pair 7.

If state variables such as heads or auxiliary variables such as velocities are defined specially by a definition constraint, the letter prefix labels follow the convention:

"H"	begins a definition constraint for hydraulic head.
"G"	begins a definition constraint for hydraulic gradients or velocities (recall they only differ by a scaling factor in the constraint).

In the case of a head definition, the last four numbers indicate the control location, while for a velocity or gradient variable definition only the last three numbers give the control pair index. Examples of specially defined variables are

H 01 1258	head at end of Period 1 for Well 1258.
G 02 088	velocity at end of Period 2 for Gradient Control Pair 88.

b. Row Labels

Row labels, which include labels for the constraints and the objective, are of three types in AQMAN. First is

OBJ	the objective function row.

There is only one such row and time or space indices are not needed. The next two types of rows have a naming convention similar to those for the variable names. The label indicates the type of row, the management period of concern, and the location. Again spaces were inserted in the example names for clarity; they do not appear in the true names. A row that corresponds to a head (converted by AQMAN to drawdown) constraint begins with **"DR"** and is of the form:

DR 11 0331	drawdown at end of Period 11 at Control Location 331.

A row that corresponds to a gradient or velocity constraint begins with the letters "DIF" and is of the form:

DIF 11 002	gradient or velocity at the end of Period 11 for Control Pair 2.

6. Representing the Simplex Tableau Using the MPS File

AQMAN controls execution of the Trescott flow model to generate the response matrix. It then produces a simplex tableau in the form of a coefficient matrix, which represents the linear programming objective and constraints. This coefficient matrix is produced in MPS format that can be understood by any one of several available and robust linear programming problem solvers, such as MPSX of IBM, MPSIII of Ketron, or MINOS of Stanford University. This special form of the simplex tableau is known as the MPS file and is in MPS format. MPS is an acronym for Mathematical Programming System and is discussed in great detail in IBM Document No. H20–0476–2 and Murtagh and Saunders (1983).

The motivation for the format of this special MPS file lies in the fact that most linear programming problems result in a tableau that is sparse. That is, one which has a lot of zero elements. MPS format was designed to take advantage of this feature and only stores the non-zero elements in the coefficient matrix. Any element that does not appear in the coefficient matrix as a non-zero entry is presumed to have a value of zero. The structure and format of the MPS file is discussed in the following chapter.

a. Anatomy of the MPS File

Figure 23 shows the overall structure of the MPS file. It consists of seven sections, each defining a different part of the linear programming problem. All but the first and last sections contain data for the problem. The first and last are flags that define the beginning and end of the file. Although AQMAN automatically writes out these seven sections based on information provided to it, it is important that the user understand MPS format so that he or she can properly modify the MPS file. The brief discussion below does not give the formatting details but rather concentrates on the structural elements of the MPS file.

To make our discussion of the MPS file concrete, we return to the three-well gradient control example and show how the MPS file generated using AQMAN applies to this problem. However, here we ignore the fact that symmetry exists in the problem displayed previously in Figure 18. The example, called GRADCON3, shown in Figure 24, now contains eight gradient control constraints. The problem has three wells, Q_1, Q_2, and Q_3, as decision variables. In addition, our "unit pumping rates" will be in L/s or thousandths of m^3/s. The simplex tableau for the three-well gradient control problem is shown in Table 15.

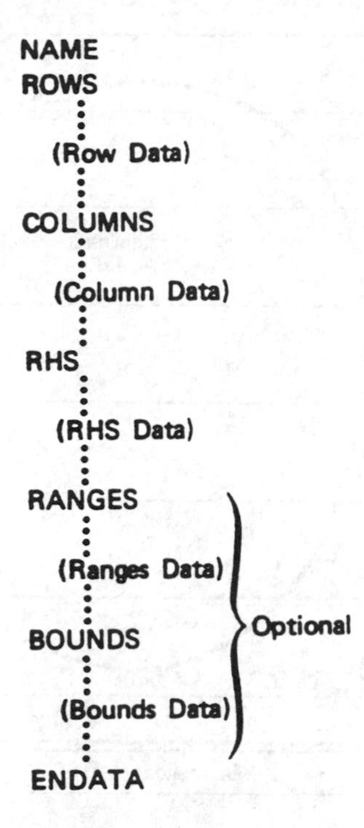

Figure 23. Overall structure of the MPS file for linear programming problems.

As seen below, AQMAN writes only the objective and the response matrix constraints 1 through 8 in the MPS file. Constraint 9 on total pumping and Constraints 10, 11, and 12, which are simple bounds on individual pumping rates, must be added by editing the file produced by AQMAN.

We now provide a description of the sections of the MPS file.

b. Name Section

The first line in the file is the NAME of the problem. It begins with the keyword NAME and is followed by a brief title, if any, for the problem. For example, if our problem was entitled GRADCON3, the NAME line would be

```
NAME        GRADCON3
```

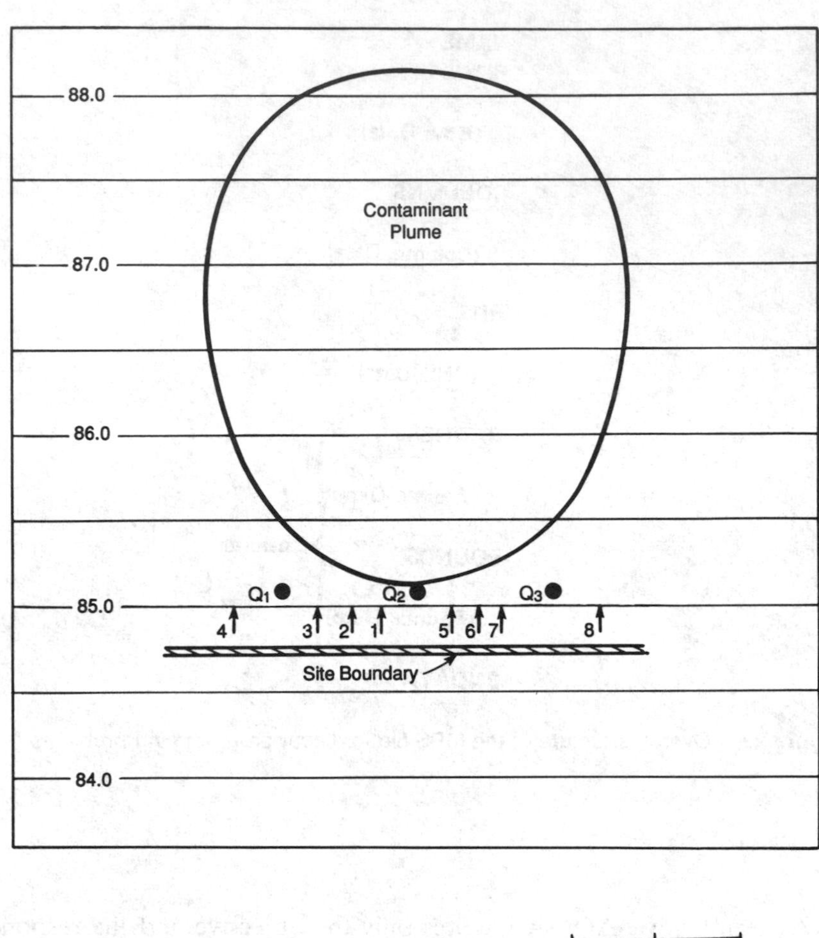

Figure 24. Three-well gradient control example with eight control pairs.

Table 15. Simplex Tableau for Three-well Gradient Control Example

Index	Variable Names			Constraint			
	Q_1	Q_2	Q_3	Type	RHS	Row Label	Explanation
Z	0.001	0.001	0.001	=	0	OBJ	Objective
1	−0.015876	−0.078634	−0.007687	≤	−0.10	DIF01001	Gradient Pair 1
2	−0.030226	−0.030173	−0.006204	≤	−0.10	DIF01002	Gradient Pair 2
3	−0.078713	−0.015849	−0.005280	≤	−0.10	DIF01003	Gradient Pair 3
4	−0.046852	−0.006888	−0.004029	≤	−0.10	DIF01004	Gradient Pair 4
5	−0.007687	−0.078634	−0.015876	≤	−0.10	DIF01005	Gradient Pair 5
6	−0.006204	−0.030173	−0.030226	≤	−0.10	DIF01006	Gradient Pair 6
7	−0.005280	−0.015850	−0.078713	≤	−0.10	DIF01007	Gradient Pair 7
8	−0.004029	−0.006889	−0.046852	≤	−0.10	DIF01008	Gradient Pair 8
9	1.0	1.0	1.0	≤	8.0	TOTALQ	Total Pumping
10	1.0			≤	3.0	BOUNDS[a]	Limit on Q_1
11		1.0		≤	3.0	BOUNDS[a]	Limit on Q_2
12			1.0	≤	3.0	BOUNDS[a]	Limit on Q_3

[a]These constraints are treated as bounds on individual variables as discussed in Chapter V.D.6.f.

c. ROWS Section

The second line in the MPS file begins the ROWS section and defines the name of each row and the type of each constraint. The section begins with the keyword ROWS and is followed by names that identify each row in the constraint set as well as the name of the objective. On each line of this section there is also a key which indicates the row type, that is, whether it is

KEY	ROW TYPE	EQUIVALENT MATHEMATICAL SYMBOL
G	greater than or equal to	≥
L	less than or equal to	≤
E	equal to	=
N	unconstrained (used for objective)	none

An example ROWS section with annotations is

```
ROWS
 L     DIF01001
 L     DIF01002
 L     DIF01003          (Note — these eight row names associated
 L     DIF01004          with the response matrix are
 L     DIF01005          produced by AQMAN)
 L     DIF01006
 L     DIF01007
 L     DIF01008          (Note — Constraint 9 on total pumping is not produced by AQMAN
 L     TOTALQ            and must be added to the ROWS section)
 N     OBJ               (Note — the Objective Row label is produced by AQMAN)
```

This section always will begin with the ROWS line, one line for each constraint, plus one "N" row for the objective row.

d. COLUMNS Section

This section, which begins with the label COLUMNS on its first line, provides the names of the variables in the linear programming problem. It also gives the coefficients associated with each variable in each row. Where coefficients are zero, these elements are left blank. This section is written in **column order**, where a column corresponds to a single variable. Column order means that the coefficients are written for one column at a time. Once all of the coefficients corresponding to a variable have been listed, then the next variable is considered, and the coefficients in each row for it are listed. Each line of the COLUMNS section allows coefficients from either one row or two rows of the tableau to be listed. The COLUMNS section corresponding to the three pumping variables, Q01001, Q01002, and Q01003, during pumping period 1 (the 01 part right after the Q label) will appear as

```
COLUMNS
     Q01001    OBJ          0.10000e-02
     Q01001    DIF01001    -0.15876e-01  DIF01002    -0.30226e-01
     Q01001    DIF01003    -0.78713e-01  DIF01004    -0.46852e-01
     Q01001    DIF01005    -0.76870e-02  DIF01006    -0.62035e-02
     Q01001    DIF01007    -0.52804e-02  DIF01008    -0.40293e-02
     Q01001    TOTALQ       1.0                                    (*line added)
     Q01002    OBJ          0.10000e-02
     Q01002    DIF01001    -0.78634e-01  DIF01002    -0.30173e-01
     Q01002    DIF01003    -0.15849e-01  DIF01004    -0.68888e-02
     Q01002    DIF01005    -0.78634e-01  DIF01006    -0.30173e-01
     Q01002    DIF01007    -0.15850e-01  DIF01008    -0.68889e-02
     Q01002    TOTALQ       1.0                                    (*line added)
     Q01003    OBJ          0.10000e-02
     Q01003    DIF01001    -0.76870e-02  DIF01002    -0.62035e-02
     Q01003    DIF01003    -0.52804e-02  DIF01004    -0.40293e-02
     Q01003    DIF01005    -0.15876e-01  DIF01006    -0.30226e-01
     Q01003    DIF01007    -0.78713e-01  DIF01008    -0.46852e-01
     Q01003    TOTALQ       1.0                                    (*line added)
```

Note that the response matrix information is written by AQMAN, but three coefficients are needed to define the restriction on total pumping and must be added to the MPS file as shown above.

e. RHS Section

The right-hand side of a linear programming problem gives the limits on drawdowns and gradients. The RHS section will be of the form:

```
RHS
     RHS       DIF01001    -0.10000e+00  DIF01002    -0.10000e+00
     RHS       DIF01003    -0.10000e+00  DIF01004    -0.10000e+00
     RHS       DIF01005    -0.10000e+00  DIF01006    -0.10000e+00
     RHS       DIF01007    -0.10000e+00  DIF01008    -0.10000e+00
     RHS       TOTALQ       8.0                                    (*line added)
```

Note that the right-hand side information is written by AQMAN for those constraints involving response matrix coefficients, but the upper limit on total pumping must be added to the MPS file as shown above.

f. BOUNDS Chapter and ENDATA Line

As an option it is possible to extend or restrict the default bounds on any decision variable, O $\leq$ Q $\leq$ ∞. The Bounds section possibilities are as follows:

LO	Lower bound	$L \leq Q$
UP	Upper bound	$Q \leq U$
FX	Fixed value	$Q = X$
FR	Free variable	$-\infty \leq Q \leq +\infty$
MI	Minus Infinity	$-\infty \leq Q$
PL	Plus Infinity	$Q \leq +\infty$

In our formulation we set upper bounds on the individual pumping rates. They are included as constraints in the MPS file shown in Table 16. These constraints are simple bounds on individual variables and are included in the BOUNDS section that would be

```
BOUNDS                                              (*entire section added)
 UP BOUND     Q01001      3.0000
 UP BOUND     Q01002      3.0000
 UP BOUND     Q01003      3.0000
ENDATA
```

The Bounds section must be added to the MPS file because AQMAN does not automatically produce it. The MPS file must end with the keyword ENDATA as shown above.

g. The Complete MPS File

The full MPS file, given by AQMAN along with the additional lines that account for constraints 9 through 12, is displayed in Table 16.

E. Generating Computer Solutions to the Optimization Problem

Given the MPS file, it is a simple matter to generate solutions to the optimization problem. MPS format is standard and, as mentioned previously, various available optimization programs expect it as input. In our examples we have used MINOS, which was developed at Stanford University and is available from that institution.

1. The Optimization Problem Specifications File

To use MINOS one must create the MPS file and one other brief file that describes the nature of the problem, such as whether it is a minimization or maximization problem. This additional information appears in a **specifications file** or **SPECS file**. For our linear programming problems, the SPECS file is shown in Table 17.

The SPECS file simply identifies

Table 16. MPS File for Three-Well Gradient Control Example

NAME	GRADCON3			
ROWS				
L DIF01001				
L DIF01002				
L DIF01003				
L DIF01004				
L DIF01005				
L DIF01006				
L DIF01007				
L DIF01008				
L TOTALQ				
N OBJ				
COLUMNS				
Q01001	QBJ	0.10000e−02		
Q01001	DIF01001	−0.15876e−01	DIF01002	−0.30226e−01
Q01001	DIF01003	−0.78713e−01	DIF01004	−0.46852e−01
Q01001	DIF01005	−0.76870e−02	DIF01006	−0.62035e−02
Q01001	DIF01007	−0.52804e−02	DIF01008	−0.40293e−02
Q01001	TOTALQ	1.0		
Q01002	OBJ	0.10000e−02		
Q01002	DIF01001	−0.78634e−01	DIF01002	−0.30173e−01
Q01002	DIF01003	−0.15849e−01	DIF01004	−0.68888e−02
Q01002	DIF01005	−0.78634e−01	DIF01006	−0.30173e−01
Q01002	DIF01007	−0.15850e−01	DIF01008	−0.68889e−02
Q01002	TOTALQ	1.0		
Q01003	OBJ	0.10000e−02		
Q01003	DIF01001	−0.76870e−02	DIF01002	−0.62035e−02
Q01003	DIF01003	−0.52804e−02	DIF01004	−0.40293e−02
Q01003	DIF01005	−0.15876e−01	DIF01006	−0.30226e−01
Q01003	DIF01007	−0.78713e−01	DIF01008	−0.46852e−01
Q01003	TOTALQ	1.0		
RHS				
RHS	DIF01001	−0.10000e+00	DOF01002	−0.10000e+00
RHS	DIF01003	−0.10000e+00	DOF01004	−0.10000e+00
RHS	DIF01005	−0.10000e+00	DOF01006	−0.10000e+00
RHS	DIF01007	−0.10000e+00	DOF01008	−0.10000e+00
RHS	TOTALQ	8.0		
BOUNDS				
UP BOUND	Q01001	3.0000		
UP BOUND	Q01002	3.0000		
UP BOUND	Q01003	3.0000		
ENDATA				

Table 17. Specs File for Three-Well Gradient Control Problem

```
BEGIN SPECS FOR GRADCON3
    MINIMIZE
    OBJECTIVE                 OBJ
    ROWS                      400
    COLUMNS                   600
    ELEMENTS                 4000
    BOUNDS                  BOUND
    ITERATIONS LIMIT          500
    MPS FILE                   10
    SOLUTION                  YES
    LOG FREQUENCY               1
END
```

1. The nature and size of the problem including the problem name; the type of objective; an upper limit on the number of rows, columns, and matrix elements; and the data-set name for the BOUNDS section (the second term on each line of the BOUNDS section). More than one bounds set can exist in the MPS file and MINOS will select the one named in the SPECS file. It is also wise to include a safety limit on the possible number of Simplex iterations.
2. The unit numbers assigned to the input file (the MPS file).
3. How much information is to be printed out during the course of the solution.

To obtain a solution one must simply run MINOS. It will read the SPECS file and the MPS file and then produce a SOLUTION file.

2. Discussion of Example Solution Given by MINOS

MINOS produces a SOLUTION file that shows three types of information. First is an echo of the SPECS file along with the key problem statistics. Second is a record of how the solution proceeded during each iteration. Third is the solution to the linear programming problem. We will concentrate on this third section of the SOLUTION file.

Table 18 shows the optimal solution to our example gradient control problem.

The information displayed in the solution begins with some preliminary information and then is divided into two sections. The relevant preliminary information is

1. The problem name;
2. The solution status as optimal, infeasible, or unbounded; and
3. The value of the objective function.

Table 18. Optimal Solution Using Minos and Aqman for Three-Well Gradient Control Problem

```
EXIT — OPTIMAL SOLUTION FOUND
No. of iterations                   8      Objective value              4.7512938554e-03
No. of degenerate steps             1      Percentage                              12.50
Norm of X                   1.334e+00      Norm of PI                          1.000e+00
Norm of X (unscaled)        2.915e+00      Norm of PI (unscaled)               1.000e+00
1
NAME        GRADCON3        Objective value      4.7512938554e-03
Status      OPTIMAL SOLN    Iteration    8       Superbasics    0
OBJECTIVE   OBJ    (MIN)
RHS         RHS
RANGES
BOUNDS      BOUND

SECTION  - ROWS
```

NUMBER	...ROW..	STATE	...ACTIVITY...	SLACK ACTIVITY	..LOWER LIMIT.	..UPPER LIMIT.	.DUAL ACTIVITY	..I
4	DIF01001	BS	−0.13121	0.03121	NONE	−0.10000	0.	1
5	DIF01002	BS	−0.10000	0.00000	NONE	−0.10000	0.	2
6	DIF01003	BS	−0.17012	0.07012	NONE	−0.10000	0.	3
7	DIF01004	UL	−0.10000	−0.	NONE	−0.10000	0.01741	4
8	DIF01005	BS	−0.13121	0.03121	NONE	−0.10000	0.	5
9	DIF01006	UL	−0.10000	−0.	NONE	−0.10000	0.02889	6
10	DIF01007	BS	−0.17012	0.07012	NONE	−0.10000	0.	7
11	DIF01008	UL	−0.10000	−0.	NONE	−0.10000	−0.00121	8
12	TOTALQ	BS	4.75129	3.24871	NONE	8.00000	0.	9
13	OBJ	BS	0.00475	−0.00475	NONE	NONE	−1.00000	10

1

SECTION 2 – COLUMNS

NUMBER	COLUMN.	STATE	...ACTIVITY...	.OBJ GRADIENT.	..LOWER LIMIT.	.UPPER LIMIT.	REDUCED GRADNT	M+J
1	Q01001	BS	1.81301	0.00100	0.	3.00000	−0.00000	11
2	Q01002	BS	1.12528	0.00100	0.	3.00000	−0.00000	12
3	Q01003	BS	1.81300	0.00100	0.	3.00000	−0.00000	13

ENDRUN

The first of the two solution sections gives information about the rows. For additional discussion about some of the terms used here, please refer to Chapter V.C.3. First is the row label followed by the state. The state indicates whether the constraint is at its upper limit (UL), is at its lower limit (LL), is equal to the right-hand side (EQ), or the constraint is not binding so the slack variable in that row is basic (BS). Although it did not occur in this example, it is possible to have cases in which alternate optimal solutions occur. This is indicated by the appearance of at least onc (A) in the row-state section. If the slack variable for a particular row is basic but the row is at one of its bounds, this is called a degenerate solution and is indicated by a (D). In the case of an infeasible solution, those rows that are infeasible are signaled with an (I).

Next is the row activity, which is the value of the left-hand side of the constraint when the slack activity is ignored. This is followed by the slack activity or the amount by which the constraint is met. A report of the upper and lower limits placed on each constraint follows. Finally, there is the **dual activity,** which is a sensitivity coefficient calculated during the course of optimization. The dual activity indicates the change in the objective value that would result from a unit change in the right-hand side of any particular constraint, provided that the solution remains feasible. Only a change in a binding constraint will influence the objective value, and the binding constraints are the only ones with nonzero dual activities. A nonbinding constraint is one that has no influence on the optimal solution. Therefore, its dual activity is defined as zero.

The second section of the solution output is the most important. It provides information about the columns that consist of decision and state variables. First is the column name and whether the variable is basic (BS) or non-basic and set to its lower limit (LL) or its upper limit (UL). Next are the results we are most concerned with: the activities or values of the variables. Here the optimal pumping rates in L/s are Q_1 = 1.81301, Q_2 = 1.12528, and Q_3 = 1.81301. Also reported is the objective gradient, which in a linear programming problem is simply the coefficient in the objective function for that variable. It corresponds to the unit pumping rate applied in AQMAN which, in this case, is 0.001 m^3/s (which equals 1 L/s). Finally are the upper and lower bounds placed on each variable. The remaining information may be important in nonlinear problems but is not of relevance here.

F. Experimenting With Alternative Remediation Design Strategies

When we are interested in exploring alternative remediation designs, repeated execution of a simulation-optimization model under various scenarios is a valuable tool. As discussed in the introduction to this chapter, the term "optimal solution" only has significance for a particular formulation. We can alter the formulation and the new optimal solution will represent a new alternative design. In this section we present a few examples of such alternative strategies and how each would be formulated using AQMAN. Solutions corresponding to the different strategies are also presented so that the optimal designs can be compared. Our graphical presentation provides some ideas for effectively viewing the large mass of numbers that come from an optimization model.

1. Hydraulic Gradient Control

In the example aquifer remediation problem used in this chapter, we introduced a simple strategy, hydraulic gradient control. For explanatory purposes we considered only three potential pumping wells. In fact, the hydraulic gradient control strategy can be expanded to include more wells. Furthermore, additional constraints can be added to our formulation, which selects only a specified "number of wells" from a large number of potential well sites. These constraints involve the introduction of integer variables where the value "one" indicates the presence of a well and "zero" the absence of a well. The technique used to solve such problems is an extension of linear programming known as **mixed-integer programming**. The term mixed-integer means that some decision variables have strictly integer values while other decision variables vary continuously, taking on any feasible value. In remediation design problems we are most interested in a subset of this technique called **zero-one mixed-integer programming**. Most commercially available optimization routines have a mixed-integer programming option. In the case of MINOS, a companion routine that is not part of MINOS and is known as MINT enables one to solve zero-one mixed-integer programming problems.

Table 19 presents a general hydraulic gradient control problem. The only conceptual difference between this formulation and the one presented in our earlier examples is the use of Constraints 2, 3, and 7, which involve integer variables. These are important because they force the solution to select only a specified number of pumping wells needed to

Table 19. Formulation of a General Hydraulic Gradient Control Problem

Minimize Z = Total pumping rate	$\sum_{i=1}^{n} Q_i$
Subject to	
1. The head inside the contaminated zone must be less than the head outside the zone (one such constraint for each "gradient control pair")	$H_{in} \leq H_{out}$
2. Define an integer variable, Y_i, for each well	$Q_i - mY_i \leq 0$
3. Restrict the number of active wells	$\sum_{i=1}^{n} Y_i = ACTWELLS$
4. Total pumping is limited	$\sum_{i=1}^{n} Q_i \leq QMAX_{total}$
5. Local head declines are limited	$H_k \geq HMIN_k$
6. Individual pumping rates are restricted	$0 \leq Q_i \leq QMAX_i$
7. The integer variables only have integer values	$Y_i = 0$ or 1

accomplish gradient control. Constraint 2 serves to "turn on" the integer counter at each location where a well is active. In that constraint, "m" is a number larger than any possible pumping rate. If pumping occurs at location i, Q_i will have a positive value and the constraint is automatically met by setting the well counter Y_i equal to 1. With the well counter established, Constraint 3 simply limits the number of active wells to some predetermined value. This limit on the number of wells can be experimented with in sequential solutions to the above problem. The solutions, each with a different number of wells, can then be compared. Figure 25 shows four solutions to the hydraulic gradient control problem in which the number of wells was restricted. There were 23 potential well sites and only a small number of wells were permitted to be active. By comparing the solutions, we see the reduced impact on the local flow field of using more and more wells to achieve gradient control. In addition, we see that the total pumping rate for the four-well case is only about one-fourth that for the one-well case.

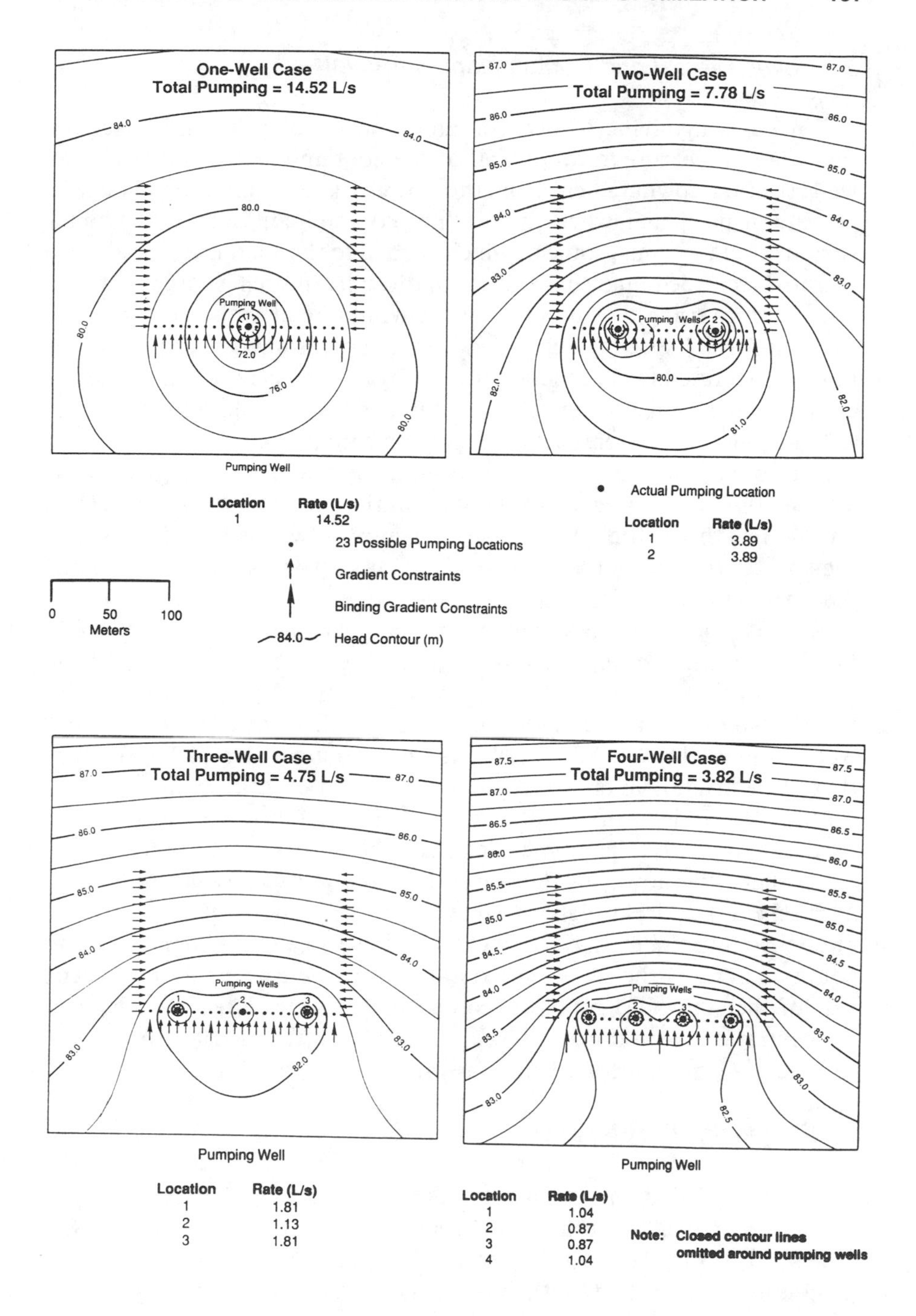

Figure 25. Hydraulic gradient control employing only a few wells.

2. Flow Reversal and Contaminant Removal

For cases involving sorbed contaminants, it may be an ineffective remediation measure to capture all of the contaminated groundwater at wells that are downgradient from the source of contamination. This will be true in particular when the original contaminant source contained various contaminants, some of which sorbed to the solid media and were therefore retarded, and some of which did not. In such a case, contaminant capture using downgradient wells would be effective in capturing the non-sorbed species, but would create a remediation scheme which must be operated for decades in order to wait for the retarded species to arrive. Under such circumstances, one may wish to explore a flow reversal and contaminant removal strategy. The contaminated groundwater is forced to migrate back toward the original source of contamination. In a sense, this strategy causes the contamination event to "run in reverse." By the time the sorbed contaminant is removed by flowing back toward the source, the non-sorbed contaminant has made its way back upgradient to the remediation wells.

Results from optimization employing a reversal and removal strategy are displayed in Figure 26. Pumping wells are placed on the upgradient side of a contaminated zone, and gradient control pairs surround the zone. Water is restricted to flow either upgradient toward the pumping wells or inward toward the flow axis of the plume. Except for the fact that there are now more gradient control pairs and the pumping wells are now upgradient of the plume, the formulation is identical to that used for the hydraulic gradient control case shown in Chapter V.F.1. The results are shown for the two-well, three-well, and four-well cases. This scheme requires roughly double the total pumping rate of hydraulic gradient control but may be more effective in rapidly removing the contaminants. If the source of contamination was near the upgradient side of the controlled zone, the contaminated water having the highest concentrations would be removed first. This was not true for the hydraulic gradient control scheme shown in Chapter V.F.1.

3. Screening Plus Removal

A strategy that can help reduce the volume of uncontaminated groundwater that is captured for treatment involves two sets of wells. One set serves to remove the regional flow, the uncontaminated natural groundwater flow that pushes the contaminated zone downgradient. These are called **screening wells** because they screen out the regional hydraulic gradient. Provided that the ambient groundwater is of good

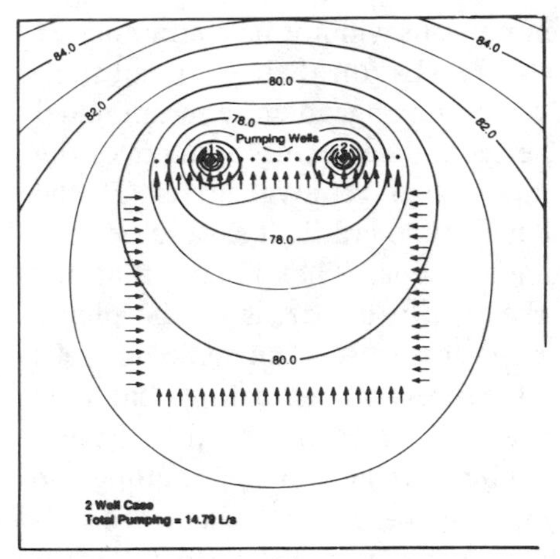

• Actual Pumping Location

Location	Rate (L/s)
1	7.395
2	7.395

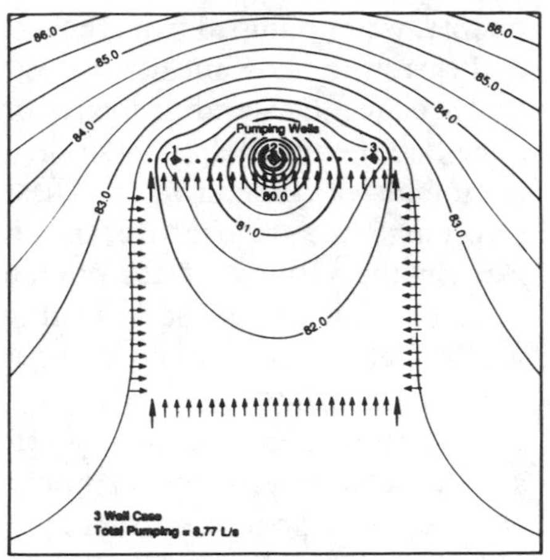

• Actual Pumping Location

Location	Rate (L/s)
1	1.35
2	6.07
3	1.35

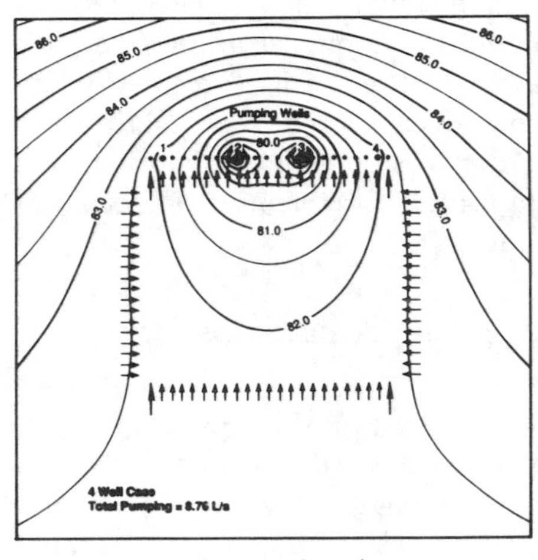

• Actual Pumping Location

Location	Rate (L/s)
1	0.66
2	3.72
3	3.72
4	0.66

• Possible Pumping Location

↑ Gradient Constraint

↑ Binding Gradient Constraint

-82- Head Contour (m)

0 50 100
Meters

Note: Closed contour lines omitted around pumping wells

Figure 26. Flow reversal and contaminant removal strategy.

quality, water pumped from the screening wells will not be contaminated and therefore need not be treated. These wells, on their own, substantially reduce the speed of migration of the downgradient contaminated zone and thereby help contain it. The second set of wells removes the contaminated groundwater. These are called **removal wells**. If the screening wells are effective, then the hydraulic gradient should be near zero in the vicinity of the contaminated zone. That means that the removal wells should be located in the **"hot spot"** areas of the plume where high concentrations of contamination exist. The removal wells will create rather symmetric cones of depression indicating that they collect contaminated groundwater from all directions. Such a system should be quite effective in stopping plume migration, containing the contaminated zone, and rapidly removing the contaminated groundwater. The formulation for the screening and removal strategy is shown in Table 20. The results are displayed in Figure 27.

To demonstrate this strategy we considered 5 potential removal wells and 23 potential screening wells. The number of active removal wells was restricted to three and the number of active screening wells was restricted to five. Results from optimization are shown in Figure 27. The screening wells are quite effective in capturing the regional flow, leaving a single removal well to pump less than 3 L/s to contain and remove the contaminated groundwater. As desired, a flow divide is created between the screening wells and the removal well. For comparison, the hydraulic gradient control strategy with one removal well was forced to pump more than 14 L/s when its role was to both capture the regional flow and the contaminated groundwater in order to stop plume migration. The use of screening wells may be quite useful in reducing the volume of water that must be treated after pumping.

4. Generalized Capture-Zone Design With Velocity Restrictions

The previous strategies rely on ensuring that local hydraulic gradients are maintained in certain directions. It is also possible to employ velocity restrictions in a remediation-design formulation using AQMAN. For example, we can consider a remediation system based on creating a specified capture zone using pumping and injection wells. In such a case, we specify the direction and magnitude of local groundwater velocities. This approach can be used to generalize the simple analytic capture zone analysis presented in Chapter IV. In fact, when compared with the analytic analysis, this approach is rather unrestricted in the following respects:

Table 20. Formulation of Screening Plus Removal Strategy

Minimize Z = Total pumping rate from the n-screening wells and the p-removal wells	$\sum_{j=1}^{n} Q_j + \sum_{k=1}^{p} Q_k$
Subject to	
1. The flow on the upgradient side of the plume must be reversed (one such constraint for each gradient control pair	$H_{down\text{-}grad} \geq H_{up\text{-}grad}$
2. Groundwater must not flow out of the) contaminated zone	$H_{out} \geq H_{in}$
3. Define an integer variable for each screening well	$Q_j - mY_j \leq 0$
4. Define an integer variable for each removal well	$Q_k - mY_k \leq 0$
5. Restrict the number of active screening wells	$\sum_{j=1}^{n} Y_j \leq SCRNWELLS$
6. Restrict the number of active removal wells	$\sum_{k=1}^{p} Y_k \leq RMVLWELLS$
7. Total pumping from the contaminated zone must be at least a specified rate	$\sum_{k=1}^{p} Q_k \geq QMIN_{total}$
8. Local head declines are limited	$H_L \geq HMIN_L$
9. Individual pumping rates are restricted	$0 \leq Q_i \leq QMAX_i$
10. The integer variables only have integer values	$Y_j = 0$ or 1 $Y_k = 0$ or 1

1. It is possible to analyze heterogeneous systems with complex boundaries and irregular recharge.
2. There is no restriction on the number of wells, and the optimal locations and number of wells are given as part of the solution.
3. There is no restriction that states that pumping rates at each well must equal one another.
4. Potential well placement is not restricted to a line.
5. Both pumping and injection wells may be included.
6. Additional constraints on well rates, local hydraulic heads, and velocities can be included in the design.
7. It is possible to treat steady-state or transient flow systems.
8. The design will be optimal with respect to the specified set of constraints and the objective of minimal pumping.

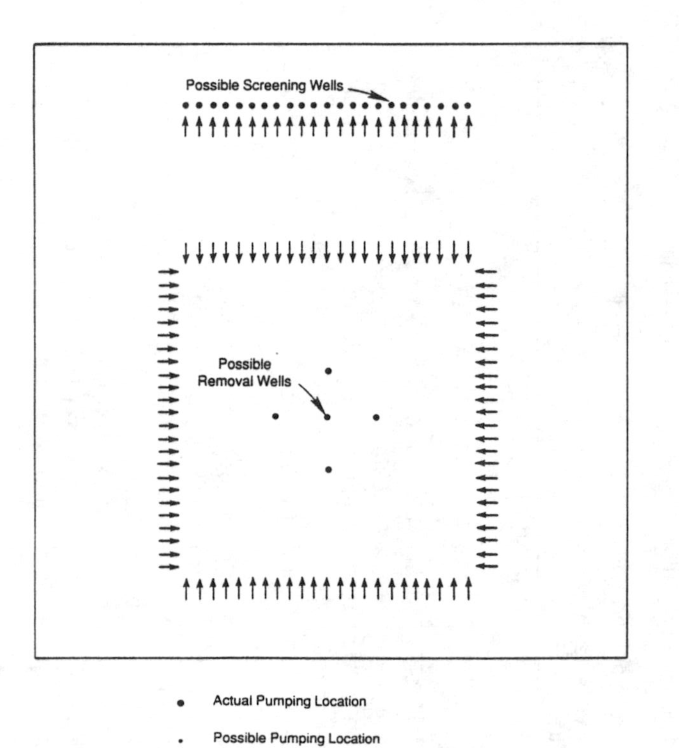

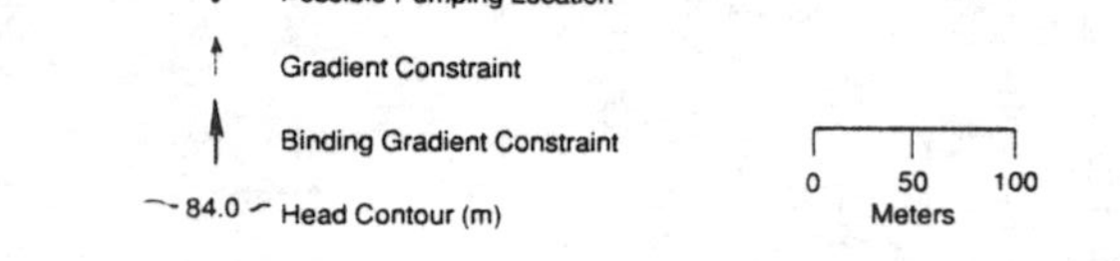

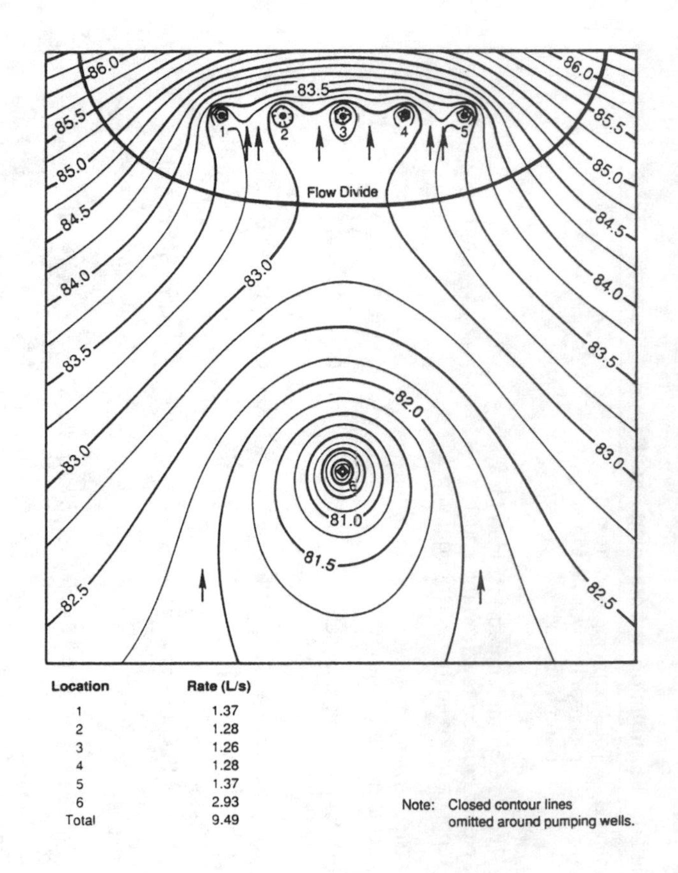

Location	Rate (L/s)
1	1.37
2	1.28
3	1.26
4	1.28
5	1.37
6	2.93
Total	9.49

Note: Closed contour lines omitted around pumping wells.

Figure 27. Results of screening plus removal.

To demonstrate capture curve design, consider Figure 28 and the formulation shown in Table 21. Here we include velocity constraints that define a feasible flow direction and an infeasible flow direction. Furthermore, we insist that a specified fraction of the water that is pumped and treated be reinjected upgradient of the contaminant plume to speed up the remedial action. Results are shown in Figure 28.

The solutions proved to be infeasible when 100 percent of pumped water was injected. This corresponds to complete recycling of the pumped and treated water. The injection mounds in that case were so large that the velocity constraints were violated. The same infeasibility was found when only one-half of the pumped water was injected. When only one-fourth of the pumped water was injected, all constraints were met, and an optimal pumping design was found. Four wells were required for plume capture and three injection wells were selected. The only injection well having a significant injection rate was located in the center of the line of potential wells. If integer constraints were employed in this problem, perhaps the number of wells could be reduced.

5. Transient Cases Involving Head Changes and Pumping Changes

Simulation-optimization using the response matrix approach is not limited to steady-state cases. The principle of linear superposition is valid for the case of transient flow. The four differences between a steady-state problem and a transient problem are

1. Initial hydraulic heads must be defined,
2. The storage coefficient or specific yield must be defined,
3. The length of management periods must be defined, and
4. The problem time frame must be determined.

AQMAN is set up to deal with transient cases. In such cases, the optimal solution will consist of a pumping schedule associated with each well rather than just a single pumping value as in the steady-state case. As an example, we now consider the three-well hydraulic gradient control case presented in Chapter V.F.1 (Figure 24), but treat the system as a transient one. We define the initial hydraulic heads to be the steady-state potentiometric surface in the absence of pumping. It is noteworthy that any initial hydraulic head surface could have been used. We define the length of each management period to be 30 days, and the problem time frame to be 24 management periods or 720 days. The hydraulic gradient control constraints must be met throughout the time frame of the problem.

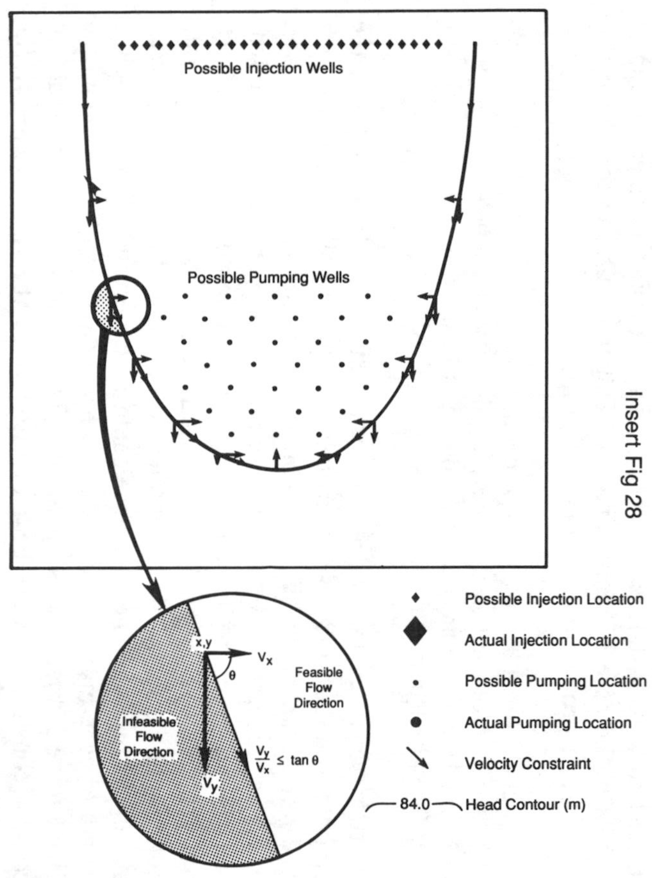

Insert Fig 28

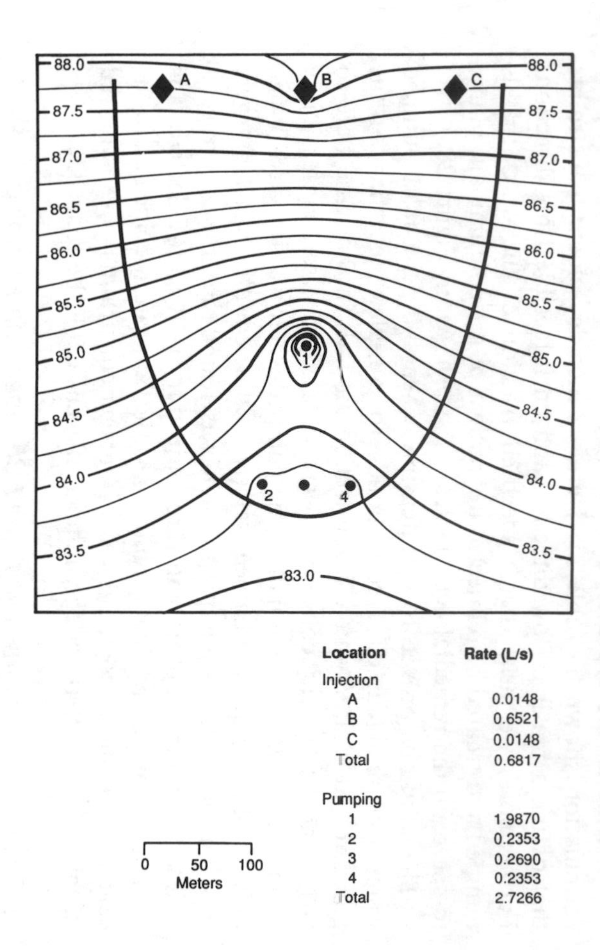

Location	Rate (L/s)
Injection	
A	0.0148
B	0.6521
C	0.0148
Total	0.6817
Pumping	
1	1.9870
2	0.2353
3	0.2690
4	0.2353
Total	2.7266

Figure 28. Capture-zone design strategy and solution.

Table 21. Formulation Involving Capture-Zone Design with Pumping, Treatment, and Recharge

Minimize Z = Total pumping rate	$\sum_{j=1}^{n} Q_j$
Subject to	
1. Define head differences in x- and y-directions using orthogonal control pairs, x,x′ and y,y′.	$(H_x - H_{x'}) = DIFF_x$ $(H_y - H_{y'}) = DIFF_y$
2. Define feasible and infeasible flow directions based upon the velocity components, where $V_x = (K/\Delta L_x n_e)$ $DIFF_x$, and $V_y = (K/\Delta L_y n_e)$ $DIFF_y$.	$(\mathrm{Tan}\ \theta)(V_x) - V_y \geq 0$
3. For n pumping wells and m injection wells, the discharge that is to be recycled after treatment is the total injection rate, where RECYCL is the fraction of pumped water that is recycled (RECYCL = 0.25).	$RECYCL \sum_{j=1}^{n} Q_j - \sum_{k=1}^{m} Q_k = 0$
4. Total pumping is limited	$\sum_{j=1}^{n} Q_j \leq QMAX_{total}$
5. Local head declines are limited	$H_k \geq HMIN_k$
6. Individual pumping rates are restricted	$0 \leq Q_j \leq QMAX_j$
7. The velocity components are unrestricted in sign	$-\infty \leq V_x \leq \infty$ $-\infty \leq V_y \leq \infty$

It is important to note that when using AQMAN for transient problems, each management period must be identical in length. This requirement was established because there is tremendous efficiency obtained when creating a transient response matrix with equal management periods as explained in Chapter V.B.2. If one wishes to have management periods of variable length, then it is best to select the smallest desired period as the management period length and additional periods as some multiple of that length.

The formulation is identical to that shown in Table 19 in Chapter V.F.1, however now there are 72 decision variables corresponding to the 30-day pumping rates at the three remediation wells for 24 periods. The results are displayed in Figure 29 in which the pumping schedule for each well is shown as a bar graph of pumping versus time. Two points are of interest here. First, the pumping rates are initially high because the gradient control must be quickly achieved by the end of 30 days. Second, the pumping rates decline rapidly to the optimal steady-state values obtained when only steady-state hydraulic gradient control was

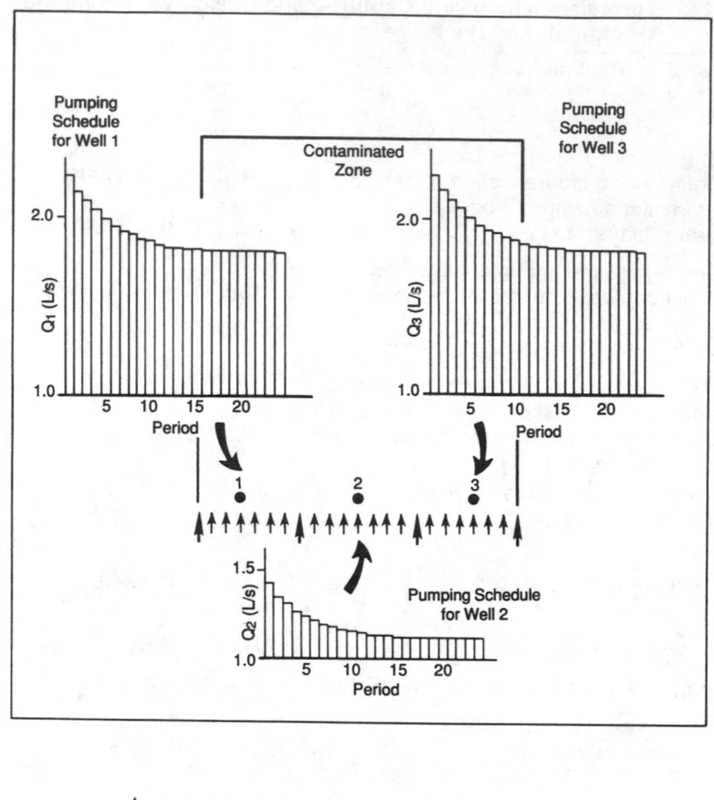

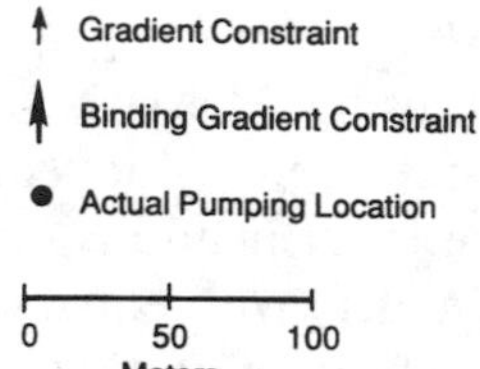

Figure 29. Transient case involving hydraulic gradient control.

considered (see Chapter V.F.1). The initial pumping rates exceed the steady-state values by about 25 percent at all three wells.

6. Approximating Unconfined Flow

When applying AQMAN to optimization problems involving unconfined aquifers, it is important to remember that AQMAN was written to solve the confined flow equation (Lefkoff and Gorelick, 1987). Successful application of AQMAN to unconfined flow problems requires the

Trescott code input data to be prepared in the form of a confined flow approximation of the unconfined flow system. This is accomplished by including the appropriate control parameters in the input file (Trescott et al., 1976) and replacing the hydraulic conductivity matrix with a transmissivity matrix. Nodal transmissivity values are obtained from simulated hydraulic heads and input values for hydraulic conductivity and aquifer bottom elevation.

For some cases involving unconfined aquifers, the drawdown will be a substantial fraction of the saturated thickness of the aquifer. In such situations, the solution to the groundwater flow equation for a confined aquifer is a poor approximation to the unconfined flow regime. The best alternative is to formulate the remediation design problem using nonlinear optimization techniques. Rigorous development of this approach is beyond the scope of this book, but an introduction to the issues is provided in Appendix B.

Fortunately, there is a simple, commonly used, iterative procedure using linear programming that can prove valuable for designing remediation schemes in unconfined aquifers. It is particularly useful when steady-state flow is considered but may be attempted for transient cases as well. The idea is to iterate between the results given by a linear programming solution, which gives the design pumping rates, and a stand-alone simulation model for unconfined flow to determine the saturated thickness. The procedure, which generally converges within three or four iterations, is

1. Select a remediation strategy, such as hydraulic gradient control.
2. Determine the saturated thickness map in the absence of remediation wells.
3. → Multiply the saturated thickness values by the hydraulic conductivities to arrive at a transmissivity map.
4. Assuming a fixed transmissivity map and a confined aquifer, run AQMAN to create an MPS file.
5. Solve the linear program to estimate design pumping rates.
6. See Stopping Rule. (Below)
7. Run a stand-alone simulation model under unconfined conditions using the pumping rates given in the linear programming solution. These results will give the saturated thickness map under pumping conditions.
8. ← Repeat the iterative sequence from Step 3.

Stopping Rule: Stop if the design pumping rates from any two successive linear programming solutions are within a specified tolerance of

Table 22. Hydraulic Gradient Control for an Unconfined Aquifer Showing Results from the Iterative Procedure

	Optimization Run	Pumping Rates in L/s			Total Rate	Percent Change
		Q_1	Q_2	Q_3		
	1	1.8130	1.1253	1.8130	4.7513	
						3.3
	2	1.7603	1.0743	1.7603	4.5949	
						0.1
converged→	3	1.7619	1.0769	1.7619	4.6007	
						0.01
	4	1.7619	1.0762	1.7620	4.6001	

perhaps 1 percent. For all practical purposes, these are then the optimal pumping rates.

This procedure was applied to the hydraulic gradient control example, presented in Chapter V.F.1, with three remediation wells. However, this time the aquifer was assumed to be unconfined with an initial saturated thickness of 50 m. The optimal pumping rates and the total pumping for sequential iterations are shown in Table 22. As seen, the pumping rates are only 3 percent less than those for the confined case, and the procedure converged after three iterations.

G. Embedding Method for Steady-State Hydraulic Management

1. Introduction

Another way in which flow simulation and linear optimization may be united into a single model is known as the **embedding method**. This method is very simple, but is limited to steady-state problems. In this method, the steady-state form of the flow equation is discretized using finite differences or finite elements, and this discretized set of linear equations serves as constraints in an aquifer management model. The simulation model is **embedded** directly into the linear programming formulation. This method was presented by Aguado and Remson (1974) and was applied to various hydraulic management problems by Aguado et al. (1977), Alley et al. (1976), Willis and Newman (1977), Schwartz (1976), Molz and Bell (1977), and Remson and Gorelick (1980). The key to the method is that because the flow equation is linear, its finite difference or finite element representation results in a system of linear equations. These discretized equations can be solved as equality constraints in a linear programming problem.

The benefits and limitations of the embedding method must be pointed out. This method is attractive because it provides the hydraulic heads at each finite-difference or finite-element node throughout the simulated aquifer. In addition, it is quite simple to allow a potential well to be located at each node throughout the system. In essence, the full simulated system is represented in the optimization model. On the other hand, it is rare in contamination remediation problems that additional constraints will exist on hydraulic heads throughout the entire simulated system; generally we are only interested in maintaining heads, drawdowns, or gradients at certain key locations. It will also be rare that wells can be placed anywhere in a simulated region; usually only a few potential well locations probably exist relative to the full domain of the problem. The size of the optimization problem using the embedding method is, therefore, much larger than when using the response matrix approach. Making the optimization problem compact is the virtue of the response matrix approach.

As discussed in Gorelick (1983), the embedding method is currently limited to steady-state problems. It is a difficult large-scale computational task to solve all of the finite-difference or finite-element equations simultaneously over time as well as space in a single optimization problem. For example, consider the case in which one wants to manage pumping for just 3 months so that an excavation area remains dewatered for the remainder of the year. Say that the aquifer is discretized into 1,000 nodes and the simulation requires 30 numerical time steps. The full simulation model that would have to be embedded into the constraint set of a linear programming formulation would involve 30,000 linear equations. Additional constraints reflecting dewatering and pumping restrictions would also exist and make the problem even larger. In addition to the cost of solving such a large problem, there can be numerical difficulties associated with solving them. The embedding method should therefore be limited to steady-state flow problems.

2. Description of the Method

If one is familiar with flow simulation methods and linear programming, then the embedding method is easily understood. To describe the embedding method let us again consider the problem presented earlier in this chapter involving hydraulic gradient control for contaminant plume capture. In this case, we will ignore the symmetry of the system and look at the full problem.

The formulation is shown in Table 23. The key point of interest is the set of equality constraints representing the groundwater flow equation. Each of the constraints is simply the steady-state flow equation written

Table 23. Formulation of Three-Well Gradient Control Problem Using Embedding

Mathematical Formulation	Explanation
Minimize $Z = Q_1 + Q_2 + Q_3$	Minimize total pumping
Subject to	
$\mathbf{AH} - \mathbf{DQ} = \mathbf{B}$	Groundwater flow equation
$H_{out} - H_{in} \geq$ for each gradient control pair	Gradient control limits
$Q_1 + Q_2 + Q_3 \leq Q_{max}$	Limit on Total Pumping
$0 \leq Q_1 \leq Q_1^*$	Limit on pumping at well 1
$0 \leq Q_2 \leq Q_2^*$	Limit on pumping at well 2
$0 \leq Q_3 \leq Q_3^*$	Limit on pumping at well 3

in finite-difference form for a particular node in the finite-difference network. Recall from Chapter II that numerical simulation of groundwater flow is accomplished by solving a system of algebraic equations in which the matrix **A** contains the finite-difference coefficients and the vector of unknown hydraulic heads is **H**. The matrix **D** is a diagonal matrix making the **DQ** the unknown sources and sinks term. Using simulation alone, this term is included, along with the boundary conditions, in the right-hand side vector **B**. However, in the case of a hydraulic management model, the sources and sinks for the manageable wells are unknown and must appear on the left-hand side of the equation. Here, the right-hand side vector only reflects the boundary conditions and not the sources and sinks.

In this problem we have one unknown state variable corresponding to head at each finite-difference node and one unknown decision variable for each manageable well. If the system is simulated with a finite-difference mesh containing 2,500 active nodes, there are 2,500 state variables. In this case there are only three potential pumping wells representing additional unknowns. The number of constraints is equal to the number of state variables, 2,500, plus the additional management constraints on pumping heads, and gradients. In this case there are 12 constraints on hydraulic gradients and 4 constraints on pumping rates giving a total of 2,516 constraints. Each constraint is linear as is the objective, so this constitutes a linear programming problem. When the linear programming problem is solved, the steady-state flow equation in finite-difference form is also solved because it is embedded in the constraint set. The solution yields the optimal pumping rates and the corresponding optimal head distribution at steady state. Let us compare the three main differences between this problem formulation and that using the response matrix approach.

First, in the response matrix approach the only variables are those corresponding to the potential pumping wells. Here the variables are both for pumping rates and hydraulic heads. In the embedding method, rather than 3 pumping variables, we have 2,500 head variables plus 3 pumping variables, for a total of 2,503.

Second, in the response matrix approach, the only constraints are those corresponding to gradient control and limitations on pumping. Simulation was implicitly involved through the linear relationship between pumping and hydraulic gradient response, **RQ**. In the embedding method, simulation is represented explicitly by solution of the discretized flow equation. This requires an additional 2,500 constraints.

Third, in the response matrix approach we do not know the full hydraulic head distribution from solution of the linear programming problem. Rather, the optimal pumping rates are given such that local hydraulic constraints are met. These pumping rates can then be used as values in a stand-alone simulation model to simulate the complete head distribution. In the embedding method, the head distribution is given as part of the linear programming solution. It is important to note that the head distribution from the stand-alone simulation model and that from the embedding approach will be identical.

The resulting optimal pumping rates using either the embedding approach or the response matrix method will be identical given the same set of hydraulic constraints and the same objective function.

References for Chapter V

Aguado, E., and I. Remson, "Groundwater Hydraulics in Aquifer Management," *Journal Hydraulics Division*, American Society of Civil Engineers, Vol. 100(HY), pp. 103–118, 1974.

Aquado, E., N. Sitar, and I. Remson, "Sensitivity Analysis in Aquifer Studies," *Water Resources Research*, Vol. 13, No. 4, pp. 733–737, 1977.

Ahlfeld, D. P., J. M. Mulvey, G. F. Pinder, and E. F.Wood, "Contaminated Groundwater Remediation Design Using Simulation, Optimization and Sensitivity Theory 1, Model Development," *Water Resources Research*, Vol. 24, No. 3, pp. 431–442, 1988.

Alley, W. M., E. Aquado, and I. Remson, "Aquifer Management Under Transient and Steady-State Conditions," *Water Resources Bulletin*, Vol. 12, No. 5, pp. 963–972, 1976.

Atwood, D. F., and S. M. Gorelick, "Hydraulic Gradient Control for Groundwater Contaminant Removal," *Journal of Hydrology*, Vol. 76, pp. 85–108, 1985.

Colarullo, S. J., M. Heidari, and T. Maddock, III, "Identification of an Optimal Ground-Water Strategy in a Contaminated Aquifer," *Water Resources Bulletin*, Vol. 20, No. 5, pp. 747-760, 1984.

Dantzig, G. B., *Linear Programming and Extensions*, Princeton University Press, 1963.

Deninger, R. A., "Systems Analysis of Water Supply Systems," *Water Resources Bulletin*, Vol. 6, No. 4, pp. 573-579, 1970.

Gorelick, S. M., "A Review of Distributed Parameter Groundwater Management Modelling Methods," *Water Resources Research*, Vol. 19, No. 2, pp. 305-319, 1983.

______, "Sensitivity Analysis of Optimal Groundwater Contaminant Capture Curves: Spatial Variability and Robust Solutions," *in Proceedings, NWWA Conference on Solving Groundwater Problems with Models*, pp. 133-146, National Water Well Association, Dublin, OH, 1987.

______, "Incorporating Assurance Into Groundwater Quality Management Models," *in Groundwater Flow and Quality Modelling*, edited by E. Custodio, A. Gurgui, and J. P. Lobo Ferreira, NATA ASI Series, Mathematical and Physical Sciences, Vol. 224, pp. 135-150, 1988.

Gorelick, S. M., and B. J. Wagner, "Evaluating Strategies for Groundwater Contaminant Plume Stabilization and Removal," *Selected Papers in the Hydrologic Sciences*, U.S. Geological Survey Water-Supply Paper Series 2290, pp. 81-89, 1986.

Gorelick, S. M., C. I. Voss, P. E. Gill, W. Murray, M. A. Saunders, and M. H. Wright, "Aquifer Reclamation Design: The Use of Contaminant Transport Simulation Coupled With Nonlinear Programming," *Water Resources Research*, Vol. 20, No. 4, pp. 415-427, 1984.

Greenwald, R. M., and S. M. Gorelick, "Particle Travel Times of Contaminants Incorporated Into a Planning Model for Groundwater Plume Capture," *Journal of Hydrology*, Vol. 107, pp. 73-98, 1989.

Haith, D. A., *Environmental Systems Optimization*, John Wiley and Sons, 1982.

Hillier, F. S., and G. J. Lieberman, *Operations Research*, Holden-Day Inc., 1984.

IBM, *Mathematical Programming System / 360, Version 2, Linear and Separable Programming Users Book*, Document No. H20-0476-2, IBM Corporation, New York, NY.

Lee, A. S., and J. S. Aronofsky, "A Linear Programming Model for Scheduling Crude Oil Production," *Journal of Petroleum Technology*, Vol. 213, pp. 51-54, 1958.

Lefkoff, L. J., and S. M. Gorelick, "Design and Cost Analysis of Rapid

Aquifer Restoration Systems Using Flow Simulation and Quadratic Programming," *Ground Water*, Vol. 24, No. 6, pp. 777–790, 1986.

______, *AQMAN: Linear and Quadratic Programming Matrix Generator Using Two-Dimensional Groundwater Flow Simulation for Aquifer Management Modelling*, U.S. Geological Survey Water-Resources Investigations Report 87–4061, 1987.

Maddock, T., III, "Algebraic Technological Function From a Simulation Model," *Water Resources Research*, Vol. 8, No. 1, pp. 129–134, 1972a.

______, *A Groundwater Planning Model: A Basis for a Data Collection Network*, Paper Presented at the International Symposium on Uncertainties in Hydrologic and Water Resource Systems, Int. Assoc. Sci. Hydrol., University of Arizona, Tuscon, AZ, 1972b.

Molz, F. J., and L. C. Bell, "Head Gradient Control in Aquifers Used for Fluid Storage," *Water Resources Research*, Vol. 13, No. 4, pp. 795–798, 1977.

Murtagh, B. A., and M. A. Saunders, *MINOS 5.1 User's Guide*, Technical Report Systems Optimization Laboratory 83–20R, 1983 (revised 1987).

Remson, I., and S. M. Gorelick, "Management Models Incorporating Groundwater Variables," in *Operations Research in Agriculture and Water Resources*, edited by D. Yaron and C. S. Tapiero, North-Holland, Amsterdam, 1980.

Rosenwald, G. W., and D. W. Green, "A Method for Determining the Optimum Location of Wells in a Reservoir Using Mixed-Integer Programming," *Society of Petroleum Engineering Journal*, Vol. 14, pp. 44–54, 1974.

Schwarz, J., " Linear Models for Groundwater Management," *Journal of Hydrology*, Vol. 28, pp. 377–392, 1976.

Trescott, P. C., G. F. Pinder, and S. P. Larson, *Finite-Difference Model for Aquifer Simulation in Two Dimensions With Results of Numerical Experiments*, U.S. Geological Survey Techniques of Water Resources Investigations, Book 7, Chapter C1, 1976.

Wagner, H. M., *Principles of Operations Research*, Prentice-Hall Inc., 1975.

Wagner, B. J., and S. M. Gorelick, "Optimal Groundwater Quality Management Under Parameter Uncertainty," *Water Resources Research*, Vol. 23, No. 7, pp. 1162–1174, 1987.

______, "Reliable Aquifer Remediation in the Presence of Spatially Variable Hydraulic Conductivity: From Data to Design," *Water Resources Research*, Vol. 25, No. 10, pp. 2211–2225, 1989.

Wattenbarger, R. A., "Maximizing Seasonal Withdrawals From Gas

Storage Reservoirs," *Journal of Petroleum Technology*, pp. 994–998, 1970.

Willis, R., and B. A. Newman, "Management Model for Groundwater Development," *Journal Water Resources Planning and Management*, Division of American Society of Civil Engineers, 103(WR1), pp. 159–171, 1977.

CHAPTER VI

Design Considerations for Complex Settings

The techniques presented in Chapters IV and V for the design of capture and containment remedial actions are very powerful tools, but they are not applicable to all groundwater contamination problems. The techniques described are limited by the assumptions that underlie the theoretical developments on which the methodologies are based. All of the material in Chapters IV and V is subject to the following limitations:

1. The methods apply to the remediation of contamination in the saturated zone. They do not apply to unsaturated-zone remediation.
2. The methods are limited to geologic formations that act as porous media with respect to groundwater flow. They are applicable to unconsolidated deposits, porous rock formations, or highly fractured nonporous rock that can be treated with an equivalent-porous-media formulation. They are not applicable to sparsely fractured rock that requires a discrete-fracture approach to groundwater flow.
3. The methods are most applicable to confined aquifers or unconfined aquifers in which the drawdowns created by the pumping well network are only a small percentage of the saturated thickness of the aquifer. The methods can also be applied cautiously to unconfined aquifers using the iterative procedure outlined in Chapter V.F.6
4. The contaminant plume that is to be remediated consists of dissolved contaminants. The methods are not applicable to immiscible liquids that exist as separate phases. However, the methods can be used to design capture and containment systems for the plume of dissolved constituents that emanate from an immiscible pool.
5. The dominant mechanism for the transport of dissolved contaminants into the pumping wells must be advection. The methods are useful as long as hydrodynamic dispersion does not cause contaminants to escape the capture zone. The influence of retardation can also be easily added to the analysis.
6. It is assumed that the contaminant concentrations are not so great as to alter the density of the groundwater. There is no density-driven convection. A uniform-density groundwater analysis is suitable.

In this chapter, we present a discussion of empirical design concepts that can be applied to remedial designs in settings that do not meet the limitations inherent in the techniques of Chapters IV and V. We will discuss the design considerations engendered by hydrogeologic complexities, contaminant complexities, and manmade complexities.

A. Hydrogeologic Complexities

The geologic and hydrologic processes that act to form and reshape the earth vary significantly from one location to the next and over time at a given location (Davis and DeWiest, 1966; Blatt et al., 1980). Because of these processes, most aquifers do not exist as a single horizontal stratum of uniform composition and thickness, and most do not have a single uniformly distributed source of recharge. Therefore, the drawdown cones and flowline patterns that result from operating extraction well fields in complex settings tend to be distorted and uneven. If such distortions occur at critical points in the zone of intended action, remediation may not be as effective as is required. For example, if an inward gradient is not maintained at all points around or under the plume, additional pumpage would then be required at certain locations. By examining some of the more common hydrogeologic complexities, it is possible to anticipate the modifications that must be made to design an effective pump-and-treat remediation in these more complex settings.

1. Unsaturated Zone

The unique processes that occur in the unsaturated zone are a common complicating factor for many remedial actions involving unconfined aquifers. The unsaturated zone can place a long-term limitation on the efficiency of a pump-and-treat remediation because of the potentially long residence time of fluids in the unsaturated zone. It may take a great deal of time for all drainable fluids to percolate down to the water table. The contaminants that are transported along with those fluids may be among the slowest to be removed by the remediation of the saturated zone.

To reduce the residence time of unsaturated-zone fluids, in situ soil washing or flooding can be used to create near-saturated flow conditions that will accelerate the flushing of those contaminants. The approach is effective because the hydraulic conductivity of any subsurface material is greatest at full saturation (Figure 30) and declines by orders of magnitude as the average moisture content in its pores is depleted (Freeze and Cherry, 1979; Hillel, 1980).

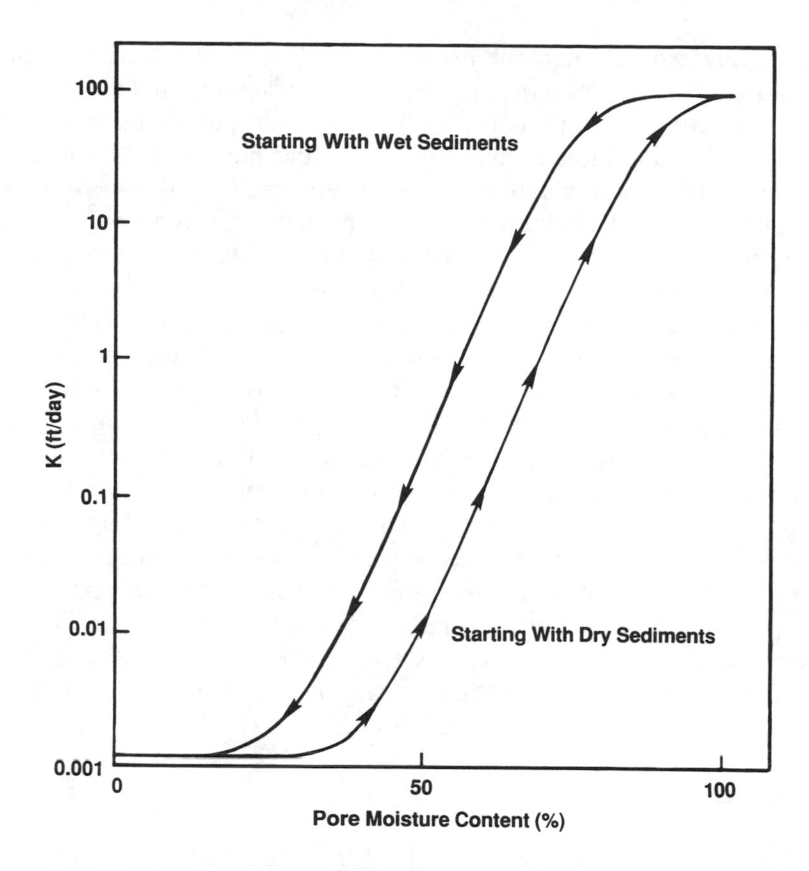

Figure 30. Effective hydraulic conductivities in unsaturated sediments.

Soil-venting or vacuum-extraction well fields are other means of removing contaminant residuals from the unsaturated zone so that their presence does not become the limiting factor for a pump-and-treat remedy of the underlying groundwater (Crow et al., 1987; Baehr et al., 1989; Thorstenson and Pollock, 1989). In the simplest version, small-diameter pipes with short, screened tips are driven to shallow depths and a vacuum is applied to draw soil vapor from the unsaturated zone. A pilot operation at a regulated site recovered more than 10,000 pounds of residual petroleum product, as vapor, at a rate of approximately 420 pounds (60 gallons) per day (Terra Vac, Inc., 1989).

A hypothetical example will illustrate the advantage provided by soil-venting as compared to natural flushing with pump-and-treat. If, instead of being removed as vapor, the residuals from a petroleum spill partition into percolating water and are then recovered by pumping the underlying groundwater, the extraction well field and treatment plant would have to be very large and expensive. Assuming that the percolating water contains 10 parts per million (ppm) of petroleum residuals, a minimum of 6 million gallons per day would have to be pumped and treated to remove residuals at the same rate as a 60-gallon-per-day soil-venting system. At a concentration of 1 ppm in the percolating water, 60 million gallons would have to be handled each day. It might not even be possible to sustain the necessary radii of influence for groundwater extraction wells capable of pumping 6 to 60 million gallons per day. In that case, the extraction wells would have to be operated at rates that produce only modest drawdowns, but they would then require much more time to remove the residuals than is needed using soil-vapor extraction. In cases where trapped residual contamination is known to remain above the water table, so that the unsaturated zone is likely to serve as a long-term contaminant source, it is wise to examine soil-vapor extraction as a possible adjunct remedial action to a pump-and-treat system designed to remediate contamination in the underlying saturated zone. Steam injection may further enhance removal of organic contaminants from the unsaturated zone (Hunt et al., 1988b). Alternatively, in-situ VOC removal using gas injection can be an effective alternative to pump-and-treat (Gvirtzman and Gorelick, 1991).

2. Water-Table Conditions

The design methods outlined in Chapters IV and V may not be directly applicable to unconfined aquifers under water-table conditions. The complexity that is introduced is that of delayed yield. When water tables are lowered by pumping, delayed-storage effects may come into play because of gravity drainage from formerly saturated sediments overlying the cone of depression. The finer grained sediments will release the drainable waters slowly, and the process may take weeks to come to equilibrium. Under these conditions, the equation of flow that is solved to calculate drawdowns in the aquifer due to pumping becomes nonlinear, and the assumption of linearity that underlies the conventional confined-aquifer analysis is no longer satisfied.

The complications introduced into the design of pump-and-treat remedial well fields by the presence of delayed-yield effects are twofold. First, the drawdowns produced by a pumping well at any point in the aquifer at a particular time after pumping commences will be smaller for a delayed-yield aquifer than for an instant-drainage aquifer (Figure 31), but the radius of influence will be larger. Well spacings and/or pumping rates calculated on the basis of the conventional linear equation for

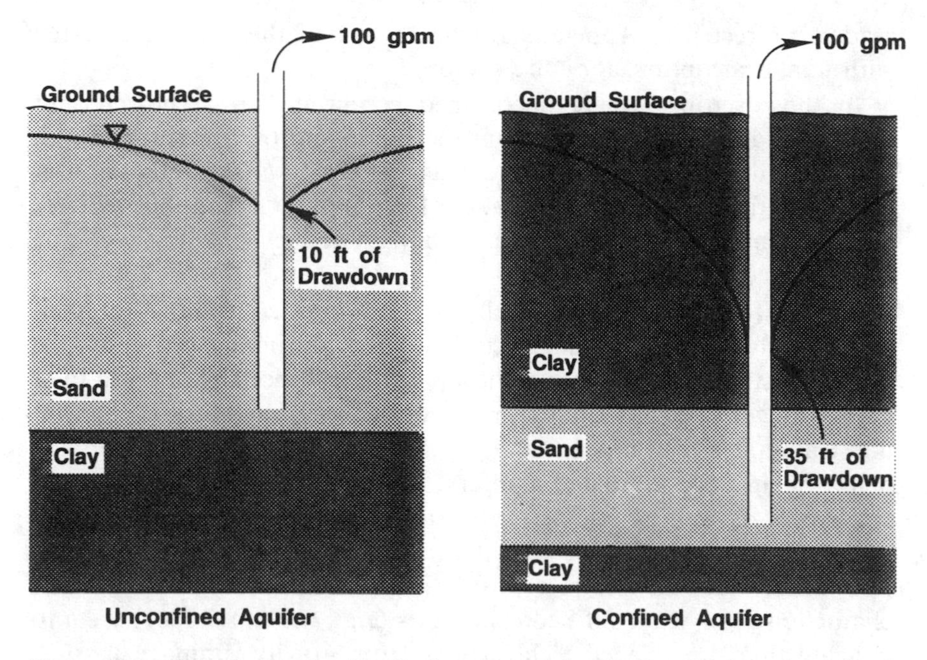

Figure 31. Drawdowns for identical pumpage from unconfined and confined aquifers with the same transmissivity.

confined aquifers may be significantly in error. They may not create the design gradients and capture-zone geometry that are required to achieve plume migration control. Second, the contaminants will be held in the cone of drainage longer than would be anticipated on the basis of conventional well hydraulics. The contaminant flux rates may be lower than expected, and the system would have to remain in operation for a longer time.

In those settings where the aquifer response to a pump test indicates that unconfined hydraulic conditions exist (e.g., storativity greater than 0.1), delayed storage effects may be significant. It is essential that drawdowns be observed for some time after equilibrium has apparently been reached during pumping tests and that proper estimates of transmissivity and specific yield be obtained using methods that take account of delayed yield (Freeze and Cherry, 1979; Walton, 1987; Kruseman and de Ridder, 1970). In the design of extraction-well networks for remediation, the concepts of Chapter IV can be used, but the capture zone must be determined using unconfined aquifer hydraulics. The optimization methods of Chapter V are applicable only as an approximation, and the reader is directed to Appendix B where some of the issues associated with nonlinear optimization are explored.

In those settings where unconfined conditions are indicated, but where the design drawdowns from the pump-and-treat extraction-well network will be only a small percentage of the total saturated aquifer thickness, the use of conventional confined-aquifer hydraulics will not lead to significant errors and the methods described previously can be used with confidence.

An aquifer that is not physically overlain by a confining stratum is traditionally defined as unconfined. If it has a confining stratum but the confining stratum is leaky, then the aquifer is traditionally referred to as being semiconfined. An aquifer need not be overlain by a single confining stratum (leaky or not) to exhibit semiconfined or confined hydraulic behavior. In unconsolidated aquifers, the aggregate effect of clay and silt lenses, as well as microstratification within sands and gravels, is often sufficient to produce confined or semiconfined responses. Field tests in alluvial and glacial deposits that produce storage coefficients characteristic of semiconfined conditions (e.g., 0.01 to 0.0001) seem to indicate that the delayed yield effects from gravity drainage in such situations are insignificant. Confined-aquifer hydraulics can be applied with reasonable confidence for well-field design in such cases.

3. Leaky Confined Aquifers

The flow rate of a well that pumps from a confined or semiconfined aquifer may exhibit an unsteady relationship to the drawdown produced by the pumping, even though the saturated thickness of the aquifer remains constant. For example, if the overlying confining stratum is leaky, it may allow significant recharge to the aquifer, thus affecting drawdown. Whether such hydraulic behavior is of concern in design studies depends on the geometry of the system and the magnitude of the leakage. Thick and areally extensive confining beds may require months for the establishment of steady-state release rates. The Hantush-Jacob equation and its associated families of type curves (Freeze and Cherry, 1979; Walton, 1987; Kruseman and de Ridder, 1970) may be used for the estimation of transmissivities and storativities from pump tests in such aquifers and for the calculation of drawdown patterns, gradients, and capture zones for remediation well-field design.

4. Low-Permeability Zones

Among the most complex hydrogeologic settings for pump-and-treat remediations are those that require removal of contaminants from low-permeability zones. Such zones may exist as relatively small isolated stringers or lenses in a higher permeability aquifer or as extensive strata that form thick aquitards separating major aquifers from one another. In unconsolidated hydrogeological environments, lenses and layers of clay and silt create low-permeability zones in sand and gravel aquifers. In sedimentary rock environments, extensive aquitards are often created by shale layers, massive sandstone formations, or unfractured evaporites such as salt and gypsum (Davis and DeWiest, 1966; Fetter, 1988). Sparsely fractured bedrock of igneous or metamorphic origin constitutes a low-permeability base to overlying shallow groundwater-flow systems.

In the course of a long-term groundwater contamination event, low-permeability zones may become contaminated by one of three mechanisms. First, many low-permeability zones provide enhanced potential for adsorption of contaminants from the groundwater onto the mineral surfaces of the aquitard materials (Mackay et al., 1986; Piwoni and Banerjee, 1989). Second, transport of contaminants into the low-permeability zones may occur by matrix diffusion (Freeze and Cherry, 1979; Feenstra et al., 1984; Keely et al., 1986). This transport mechanism involves the slow diffusion of contaminants into the matrix of the low-permeability materials on the basis of the concentration gradient that exists between the contaminated groundwater in the aquifer and the uncontaminated pore water in the aquitard material. Third, contaminant groundwater may flow into the low-permeability zones and may slowly flow out, thereby providing long term source of contamination.

Fine-grained sediments that possess low-permeability usually have orders-of-magnitude greater surface area per volume of material than coarse sediments (Corey, 1977; Bitton and Gerba, 1984). Much larger accumulations of contaminants may thus develop on the grain surfaces of low-permeability sediments by sorption, ion exchange, or other surface chemical processes (Enfield and Bengtsson, 1988; Bouchard et al., 1989), as compared with contaminant accumulations in a like volume of high-permeability sediments. Contaminant mass in excess of that required for surface chemical interactions is available for matrix diffusion into the pores of the low-permeability material. The higher the porosity of the low-permeability material, the greater is the storage available for diffusive flux.

The presence of contaminants in low-permeability zones has the potential to significantly complicate the design of a pump-and-treat remedial system (Satkin and Bedient, 1988; Osiensky et al., 1984; Keely, 1984). When an extraction-well network is installed, the more permeable zones of an aquifer system may be cleaned up relatively quickly, while the low-permeability zones retain their contaminants for a much longer time. The release mechanisms are the reverse of those that lead to the contamination of the low-permeability zones. These release mechanisms include desorption, slow outflow, and matrix diffusion toward the now cleansed aquifer waters. Contaminated low-permeability zones are troublesome not only because they are difficult to clean rapidly, but also because they may act as a continuing source of contamination for the more productive portions of the aquifer.

Let us first look at remediation issues associated with extensive low-permeability zones. The simplest approach to pump-and-treat remediation of contaminated formations that are uniformly low in permeability is one that accepts as a limitation the fact that the sustainable yield of each extraction well will be too low to allow for continuous operation of a submersible pump. The extraction well is set up to operate *on-demand*, by initially pumping the well dry with a small pump (e.g., 1 to 5 gallons per minute), and then allowing the well to recover to a preset level before the pump restarts.

Other approaches that have been considered at hazardous waste sites on fractured bedrock and in glacial till include the use of backhoes and tunnel-boring machines (such as those used to construct sewers and subways) to create horizontal galleries from which smaller laterals may be drilled and drained with submersibles or sump pumps. Large-diameter rock augers or other mining equipment might also be employed to sink shafts, from which smaller horizontal, vertical, and angled holes could be drilled to reach inward to the core of the rock like the roots of a tree for extensive dewatering. Fracturing of the formation by hydraulic or explosive means would also increase permeabilities and provide discharge faces for the diffusing fluids of the bulk matrix of the rock.

The more common, and more insidious, complication to pump-and-treat remediation is caused by the occurrence of low-permeability lenses of clay and silt in high-permeability sand-and-gravel aquifers. These lenses often have limited areal extents of only a few tens or hundreds of feet in any direction. Localized variations in the advective rate of groundwater flow in such settings (Anderson, 1979; Guven and Molz, 1986; Keely et al., 1986) result in rapid cleansing of the higher permeability sediments, which conduct virtually all of the flow (Figure 32), and much slower removal of the contaminants from the lower permeability sediments by matrix diffusion. The specific rate at which this diffusive

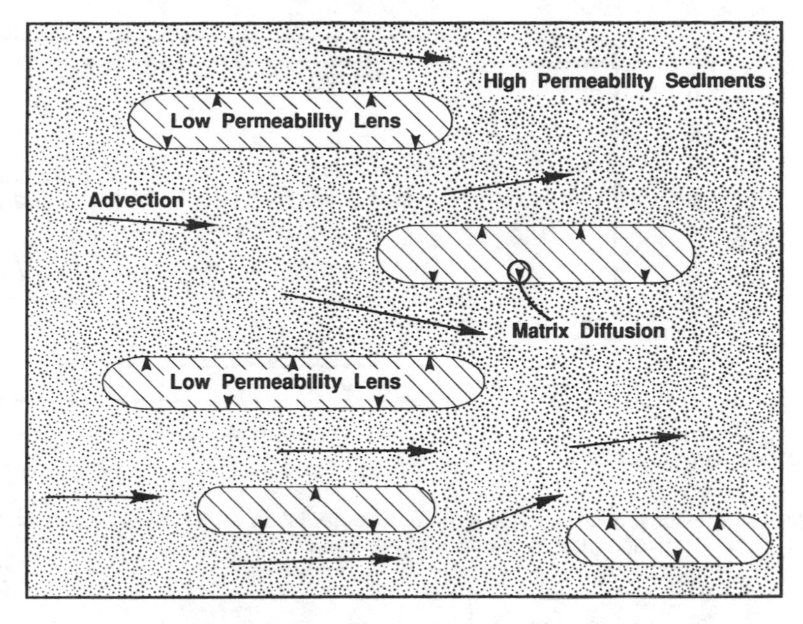

Figure 32. Permeability variations slow pump-and-treat recoveries.

release occurs is dependent on the difference in contaminant concentrations within and external to the low-permeability sediments. When the higher permeability sediments are cleaned up, the concentration gradient drawing contaminants from the lower permeability sediments is at its greatest. This gradient is exhausted only when the chemical concentrations are nearly equal everywhere. Alternatively, this process may be controlled by slow advection through the low permeability zones and not by diffusion.

The thicker a low-permeability stratum is, the more contaminant reserves it can hold, and the more likely it is that diffusion will dominate overall contaminant movement. Substantial occurrences of low-permeability materials in higher permeability formations may thus be a major limitation in determining the duration of pump-and-treat remediations there. In the worst situations, substantial occurrences of low-permeability materials will have become completely "saturated" with contaminants. The rate of diffusive transport of the contaminants from the core of each low-permeability microstratum to its contact surface with the enveloping high-permeability formation will dictate the minimum extra time needed for complete remediation. This extra time must be added to the nominal time required to remediate the high-permeability sediments alone. Diffusive transport times can be calcu-

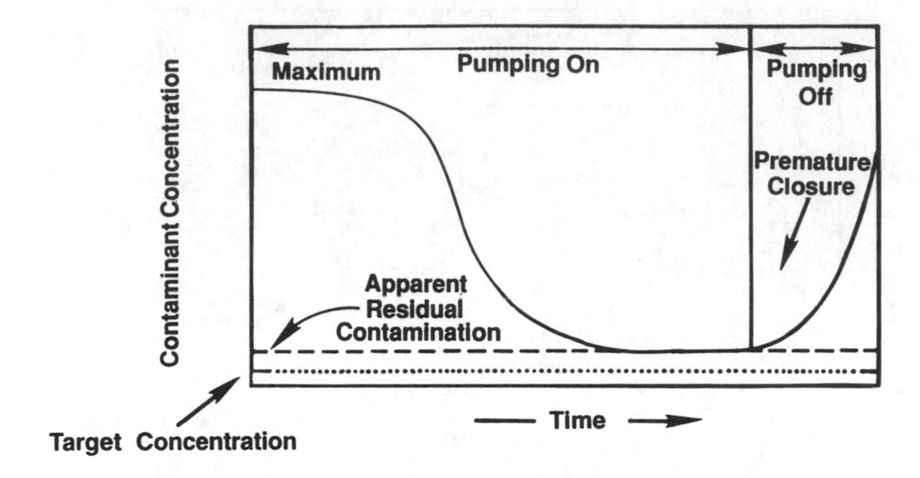

Figure 33. Conventional pump-and-treat remediation output with contaminant rebound after termination of pumping.

lated with the formulae provided by Freeze and Cherry (1979), Bear and Verruijt (1987), or de Marsily (1986).

If substantial occurrences of low-permeability materials are adequately characterized by well borings and tests, or by geophysical techniques, it may be possible to target these selectively for excavation, hydrofracturing, or other remedial methods. Alternately, if the heterogeneity of the setting is such that low-permeability materials are present as numerous occurrences of microstrata, each of very limited thickness, it is likely that they will defy physical removal or in situ treatment. In such cases, a conventional pump-and-treat remediation may not be the best solution.

Conventional pump-and-treat remediations are based on continuous operation of an extraction-injection well field. In these remedial actions, the level of contamination measured at monitoring wells may be greatly reduced in a moderate period of time, but low levels of contamination usually persist. In parallel, the contaminant load discharged by the extraction well field declines over time and gradually approaches a residual level in the latter stages (Figure 33). At that point, large volumes of water are being treated to remove small amounts of contaminants, and a site operator may be inclined to shut down the extraction wells under the impression that the remediation is "complete." However, if a significant amount of contamination is still present in the low-permeability zones

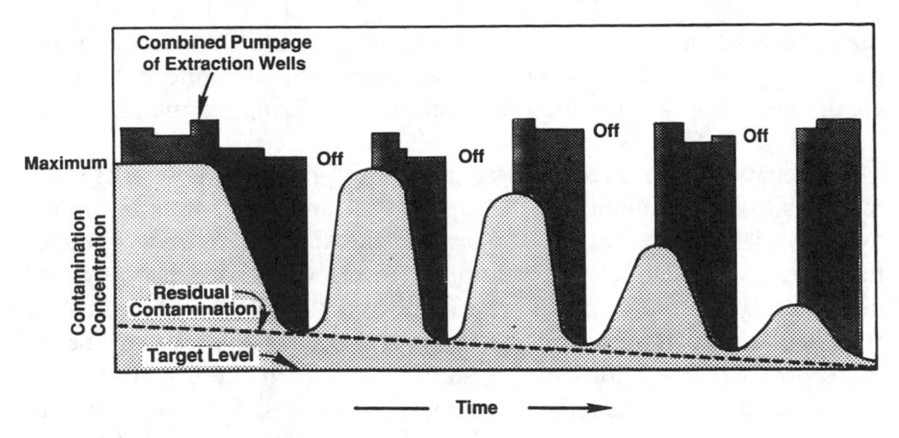

Figure 34. Pulsed pumping removal of residuals.

after cessation of pumping, matrix diffusion may cause concentrations in the high-permeability zones to increase (Figure 33).

To minimize the potential for this type of occurrence, alternative pumping strategies may be required. These may involve reconfiguring the remediation well field, adopting unconventional pumping schedules, or integrating pumping with other remediation technologies such as subsurface barrier walls.

An attractive alternative that can be applied to matrix diffusion remediation problems is pulsed pumping. Pulsed operation of hydraulic systems is the cycling of extraction or injection wells on and off in active and resting phases (Figure 34). The resting phase of a pulsed-pumping operation provides time for contaminants to diffuse out of low-permeability zones and into adjacent high-permeability zones. For sorbed contaminants and NAPL residuals, sufficient time can be allowed for equilibrium concentrations to be reached in local groundwater. Subsequent to each resting phase, the active phase of the cycle removes a minimum volume of contaminated groundwater, at the maximum possible concentrations, for the most efficient treatment.

By occasionally cycling only select wells, stagnation zones may be brought into active flow paths and remediated. This is important because the operation of any well field in an aquifer containing moving fluid results in the formation of stagnation zones downgradient of the extraction wells and upgradient of the injection wells. The stagnation zones are hydrodynamically isolated from the remainder of the aquifer, so mass transport into or out of the isolated water may occur only by diffusion. If remedial action wells are located within the bounds of a contaminant plume, such as for the removal of contaminant hot spots, the portion of the plume lying within their associated stagnation zones will not be effectively remediated. Changing the flowline pattern by changing pumping rates or well locations will bring stagnation zones into the active flow system.

Systematic methods for the design of pulsed-pumping networks using methods like those described in Chapters IV and V are now becoming available. Ultimately, such methods may allow for the optimal design of pumping-rate schedules and pump/rest cycle times for complex well networks, but at the present time a more operational approach must be followed. The customary approach is to design a remedial system for the more productive portions of the aquifer, and then monitor the water quality in the plume as remedial pumping proceeds. When the concentrations begin to fall, one waits for them to approach a relatively constant low value, and then pumping is stopped. If the concentrations rise again to an unacceptable level, pumping is reinstituted. The goal should be to maximize the ratio of contaminant mass removed per volume of groundwater discharged.

Pulsed-pumping systems may lead to lower pumped volumes than conventional systems and, hence, lower treatment costs. However, pulsed pumping incurs certain additional costs and concerns that must also be considered when comparisons are made with continuous pumping for site-specific applications. During the resting phase of pulsed-pumping cycles, peripheral gradient control may be needed to ensure adequate hydrodynamic control of the plume. In an ideal situation, peripheral gradient control would be unnecessary. Such might be the case where there are no active wells, major streams, or other significant hydraulic stresses nearby to influence the contaminant plume while the remedial action well field is in the resting phase. In these cases, the plume would migrate only a few feet during the tens to hundreds of hours that the system was at rest and that movement would be rapidly recovered by the much higher flow velocities back toward the extraction wells during the active phase. When there are significant hydraulic stresses nearby, however, plume movement during the resting phase may be unacceptable. Irrigation or water-supply pumpage, for example, might cause plume movement on the order of several tens of feet per day. It might then be impossible to recover the lost portion of the plume when the active phase of the pulsed-pumping cycle commences. In such cases, peripheral gradient control during the resting phase would be essential. If adequate storage capacity is available, it may be possible to provide gradient control in the resting phase by injection of treated waters downgradient of the remediation well field. Regardless of the mechanics of the compensating actions, their capital and operating expenses must be added to those of the primary remediation well field to determine the complete cost.

5. Fractured-Rock Aquifers

In some geologic settings, most of the groundwater flow occurs through fractures in low-permeability rock formations (Feenstra et al., 1984; Spayed, 1985; Barker et al., 1988). The flow in the fractures often responds quickly to rainfall events and other fluid inputs. The flow through the bulk matrix of the rock is, however, extremely slow; so slow that contaminant movement in the matrix by molecular diffusion may be quicker than movement by advection. If there is a high degree of fracturing, such that the material has moderate to high permeability, it is likely that its hydraulic behavior will be similar to that of unfractured, porous media (de Marsily, 1986; Mercer and Faust, 1981). Darcy's law, the Theis equation, and other porous media formulae may be applied with reasonable confidence in such cases for the design of remediation well fields.

Where there are only a few fractures that dominate flow through a stratum, the relationship of flow rate to drawdown may be nonlinear and nearly unpredictable. Careful delineation of the dominant fractures is important in such cases. One may expect narrow troughs of drawdown to develop and persist along the fractures during pumping (Davison and Guvenasen, 1985). Equations have been developed that allow prediction of drawdowns for situations where a single horizontal fracture or a single vertical fracture passes through the pumped wellbore, but there are no generically useful equations for multiborehole or multifracture situations.

The usual design approach in situations where a few irregular fractures dominate the groundwater flow system is one of trying to position each extraction well such that it intersects as many of the fractures as possible. In these cases, it is necessary to conduct several pilot tests of extraction wells to develop site-specific relationships of flow rate to drawdown. Such empirical relationships may show the effects of changes in the fracture width during pumping. Drawdown is likely to exhibit alternating acceleration and deceleration because of nonuniform drainage along the fracture face and the siphoning effects of irregular fractures. In those settings where the flow-rate drawdown relationship at one extraction well differs markedly from that of another close by, the remediation well-field design strategy must default to well-by-well installation and testing. After each extraction well is installed, it is tested singly and then in concert with all existing extraction wells to determine the short-term effectiveness in providing the desired gradient control of the contaminant plume; then the next extraction well is drilled in the location where additional control is needed.

6. Karst Aquifers

Two flow regimes occur in karst settings that rarely occur in other geologic settings (White, 1988). Both are a consequence of the fact that karst terrain is characterized by a solution-cavity structure. In one flow regime, *pipe flow*, fluids completely fill the solution cavities and channels; fluid movement may be described by the Hagen-Poiselle equation for nonturbulent pipe flow. In the other regime, *open-channel flow*, fluid movement occurs as subterranean streams through modest-to-large solution cavities and caverns; Manning's formula for turbulent open-channel flow may be applicable. Because of these complexities, the advection and dispersion of dissolved contaminants through the bulk of a karst aquifer may not be adequately describable by Darcy's law and other porous media concepts.

Dye tracers have been used to study fluid movement in karst aquifers, but such studies have yet to yield quantitative relationships that can be transferred from the study site to other sites. Qualitatively, it has been shown that karst streams and springs control the distribution of most of the flow through karst terrain, and dye tracer studies can be helpful in delineating these flow systems. Unfortunately, the preferred flow paths may change, depending on the magnitude of recharge or discharge events. So, while it may seem sensible to target locations for extraction wells on the basis of intercepting karst streams, one should not expect a consistent response; the drawdown and available flow rate may vary substantially over short periods of time.

Just as noted for those fractured rock settings where the flow-rate drawdown relationship at one extraction well differs markedly from that of another close by, the remediation well-field design strategy for karst aquifers must focus on well-by-well installation and testing. Each extraction well should be installed and tested singly and then in concert with all existing extraction wells to determine their short-term effectiveness in providing the desired gradient control of the contaminant plume, prior to proceeding to the next extraction well. The operation of well fields in karst terrain is sometimes accompanied by collapse of dewatered underground caverns and subsequent damage at ground level in the form of sinkholes. For safety and property protection, it is essential to review all available public records and published studies of the area.

If potentially hazardous conditions are suspected, geological engineering studies should be undertaken to support the remediation design.

B. Contaminant Complexities

In an ideal situation for capture and containment remediation, the physical, chemical, and biological properties of the plume would be similar to that of pure water. The plume would have the same density as freshwater; the contaminants it carries would be present only in dissolved form, and they would be in chemical equilibrium at all times. Many plumes approximate this character at their downgradient boundaries, but few exhibit such characteristics close to the source of contamination. Near the source, a plume is likely to have a density that differs from that of freshwater, and pure solvents and other free-product liquids may be present.

The design of capture and containment remediation must take into account the effect of these conditions on the geometry of the plume that is to be remediated. There are two issues: whether the contaminant is miscible or immiscible with water, and whether the contaminant is lighter or denser than water. Figure 35 shows the four possibilities: (1) light, immiscible plumes; (2) dense, immiscible plumes; (3) light, miscible plumes; and (4) dense, miscible plumes. Gasoline and its components (benzene, toluene, and xylene) constitute the most common light immiscible contaminants. Trichlorethylene, perchloroethylene, and other chlorinated solvents are the most common dense immiscible contaminants. Methanol and ethylene glycol provide examples of light and dense miscible plumes, respectively.

The concept of miscibility is discussed in Chapter II.E.1. Miscible fluids are infinitely soluble in water. Immiscible fluids have limited solubility in water and do not "mix" with it. At their boundaries, miscible fluids develop mixing zones with water, whereas immiscible fluids do not develop extensive mixing zones with water. Rather, immiscible fluids form "sharp" interfaces with water. Miscible fluids provide a highly soluble source for the development of a plume of dissolved contaminants that migrate in the downgradient direction. Immiscible fluids, despite their low solubility, can also act as a source of dissolved contaminants.

We will first discuss the implications of light and dense miscible plumes, and then the more important case of light and dense immiscible plumes (LNAPLs and DNAPLs).

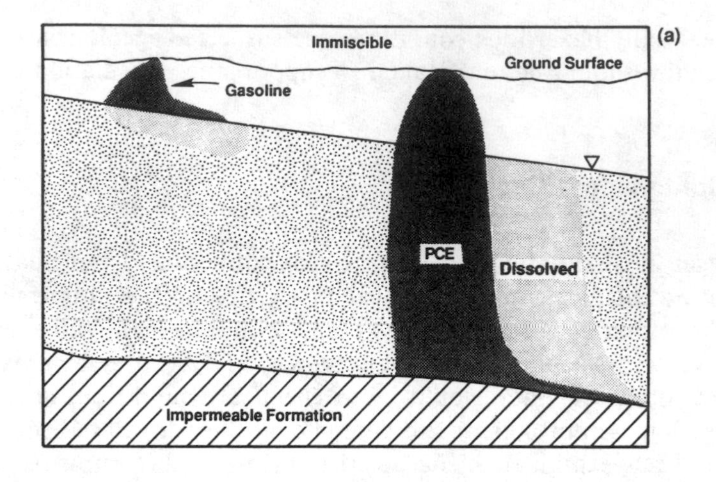

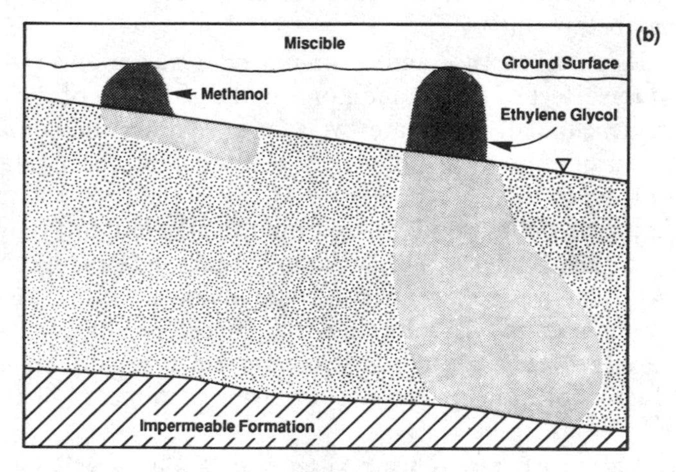

Figure 35. The four common types of contaminant plumes found in groundwater: (a) immiscible liquids such a gasoline (light) or PCE (dense); and (b) miscible liquids such as methanol (light) and ethylene glycol (dense).

1. Miscible Plumes

Acetone is an example of a liquid contaminant that has a low density relative to water ($\rho = 0.791$ g/cm^3) and is completely soluble in water. A plume of water contaminated with acetone will be characterized by completely dissolved contaminant and will exhibit a density less than the density of uncontaminated water.

Because of their low density and infinite solubility, contaminants such as acetone display a range of behaviors when encountering groundwater. If a large release of a low-density miscible contaminant reaches the water table as a large mass of liquid, the low density of the liquid will prevent the immediate mixing of the contaminant into the groundwater. Rather, the contaminant liquid will tend to pool on the water-table surface with mixing occurring at the interface between the pool of contaminant and the groundwater. This behavior has implications for emergency response remedial actions that may be used to recover pure product if such actions are implemented before a low-density miscible liquid is able to mix completely with groundwater.

After a spill of acetone or a similar liquid has completely dissolved in groundwater, the contaminant plume will have a density between that of the pure contaminant and that of groundwater. Large concentrations of contaminant in the dissolved plume will correspond to lower plume densities and a greater propensity for the plume to remain at the top of the aquifer as it is advected along with the groundwater flow. In such a case where a concentrated plume has migrated a limited distance from the source area, remedial action wells would commonly need to be screened only through the upper portion of the aquifer that is known to be contaminated.

After a low-density miscible plume has migrated a great distance from the source area, the effects of hydrodynamic dispersion will have diluted the plume, resulting in a density near that of freshwater. Additionally, the process of transverse dispersion will have contributed to the vertical spreading of the plume. At great distances from the source area, the plume will extend from the top of the aquifer to a depth greater than that observed near the source. Depending on the migration distance and the thickness of the aquifer, contamination may extend over the entire vertical thickness of the aquifer. It is clear, then, that a pump-and-treat remedial action would now involve drilling deeper wells, installing longer screened intervals, and pumping greater volumes of contaminated water.

Ethylene glycol is an example of a contaminant that has a high density ($\rho = 1.113 \text{ g/cm}^3$) relative to water and is completely soluble in water. An ethylene glycol plume will therefore be composed of completely dissolved contaminant. At high concentrations, the contaminated water will have a higher density than the surrounding freshwater.

In this sense, ethylene glycol acts very much like a saline brine. The dynamic behavior of brines and brackish waters has been studied extensively in groundwater reservoirs as a result of the problems associated with upconing of saline waters from shale beds into productive sandstone aquifers, geothermal investigations, and studies of saltwater intrusion into coastal aquifers. It is not unusual for very large mixing zones to form where a brine is in contact with freshwater (Jorgensen et al., 1982; Mercer and Faust, 1981; Mercado, 1985; Novak and Eckstein, 1988; Poole et al., 1989).

For most situations involving dense miscible contamination in contact with groundwater, the greater the difference in densities, the shorter the time required for formation of a large mixing zone. If the rate of release from the near-surface source is substantial, the contaminant solution may reach the water table in relatively undiluted form and high volume and, thus, begin vertical migration downward in the aquifer as a sinking plume. If the dense miscible plume encounters and begins to follow a preferential flow channel, the mixing that takes place quickly moves contaminants into surrounding uncontaminated material and produces a smoothed front.

As time proceeds, the mixing zone that is continuously generated depletes the miscible contaminant plume of much of its mass and thereby reduces its overall density and propensity to continue downward migration. This may be one reason why many landfill leachate plumes and brine pit plumes tend to stabilize prior to reaching the bottom of an aquifer.

Remediation of dense miscible plumes is complicated by the fact that freshwater will respond more quickly to the same hydraulic stress than will brines or other dense aqueous solutions. An extraction well that is screened across a formation containing both fresh water and brine pumps a disproportionately greater amount of freshwater than brine. The removal of a dense miscible plume can be done most effectively, therefore, by pumping only from the core of the plume. Selective vertical positioning of partial well screens in a few extraction wells will facilitate contaminant removal. Eventually, the dense portion of the plume will be removed, leaving only a low-concentration plume with a density that is similar to that of uncontaminated groundwater. The remediation well field can then be expanded for full-plume capture, with the assurance that all portions of the plume will respond as a constant-density solution.

Many miscible plumes contain contaminants that react with the groundwater, the formation, other contaminants, or biota in the subsurface. Reactions may transform the contaminant into a different chemical species, such as the biological transformation of mercury into methylmercury (Tinsley, 1979; Bitton and Gerba, 1984), in which case the toxicity and mobility of the contaminant (and the plume) may change substantially. It is important to understand the possible transformation reactions for a given situation so that the well-field design and monitoring-network design can incorporate contingencies for possible changes in plume dynamics and so that the treatment design can incorporate contingencies for changes in the chemical character of the treatment plant influent.

In many situations, ion-exchange and adsorption reactions take place and alter the transport rate of a miscible contaminant plume. Toxic metals are common constituents of miscible plumes and are subject to both mechanisms. Ion exchange occurs over a broad range of pH, but it is limited by the number of available exchange sites (high in clays, low in sands) and the presence of competing ions. Ion exchange is a fairly rapid process, and the assumption that chemical equilibrium is nearly instantaneous is likely to be accurate in many natural situations (Stumm and Morgan, 1981).

Adsorption and desorption processes have been widely studied in recent years (Goltz and Roberts, 1986, 1988; Woodburn et al., 1986; Curtis et al., 1986; Bouchard et al., 1988; Lee et al., 1988; Mackay et al., 1986; Valocchi, 1988; Bahr, 1989; Brusseau et al., 1989; Nkedi-Kizza et al., 1989; Piwoni and Banerjee, 1989). It is clear from these studies that desorption is not always a rapid process; hundreds of hours may be required before chemical equilibrium is established. This fact may be of little consequence prior to remediation when the migration velocity of a contaminant plume is relatively slow. When a pump-and-treat remediation is implemented, however, groundwater flow velocities will be significantly increased and chemical equilibrium may not be achieved or maintained. As shown in Figure 36, if sufficient contact time is not allowed due to high groundwater velocities, the affected water may be advected away from the adsorbed contaminant residuals before achieving equilibrium contaminant concentrations. Contaminants will then be released to the groundwater at concentrations that are much lower than would occur at full chemical equilibrium. This means that large volumes of slightly contaminated water are being produced for treatment over a long period of time. It would be much more efficient and economical to pump smaller volumes of highly contaminated water to remove the same mass of contaminants. Pulsed pumping (Chapter VI.A.4) may be employed to achieve the latter goal.

2. *Immiscible Plumes*

Immiscible organic liquids constitute a unique group of contaminants commonly encountered at contaminated industrial sites. These liquids are often broken into two categories: (1) fuel-related hydrocarbons and (2) chlorinated solvents. The former are typically immiscible liquids with

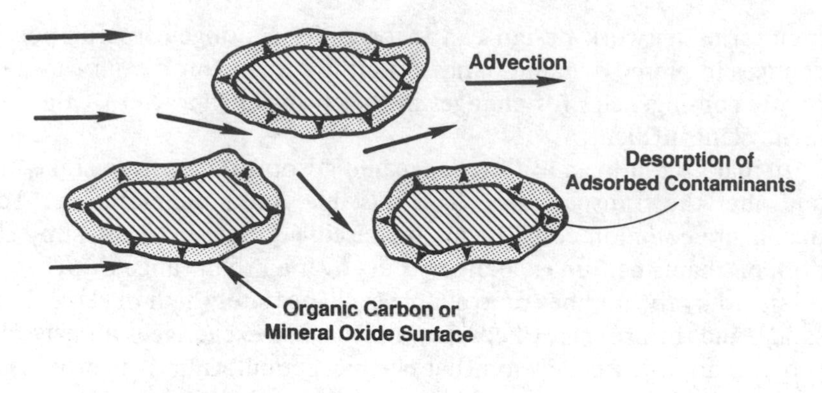

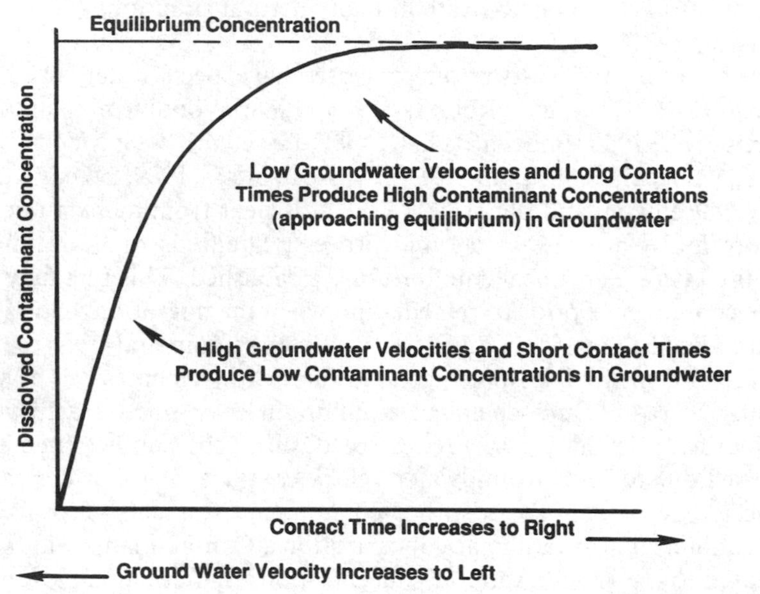

Figure 36. Relationship between groundwater velocity induced by a pump-and-treat system and concentration of dissolved contaminants that desorb from solid grains.

low density (LNAPLs) and float on water. The latter are typically immiscible liquids with high density (DNAPLs) and sink in water. As shown on Figure 37, the floaters are generally found at or near the water table and because of their bulk, may cause a local depression in the water table. The sinkers tend to move downward under the influence of gravity. When they encounter bedrock or some other aquitard formation at the base of the aquifer, they will spread out laterally. Their movement is controlled both by gravity and by the advective ground-

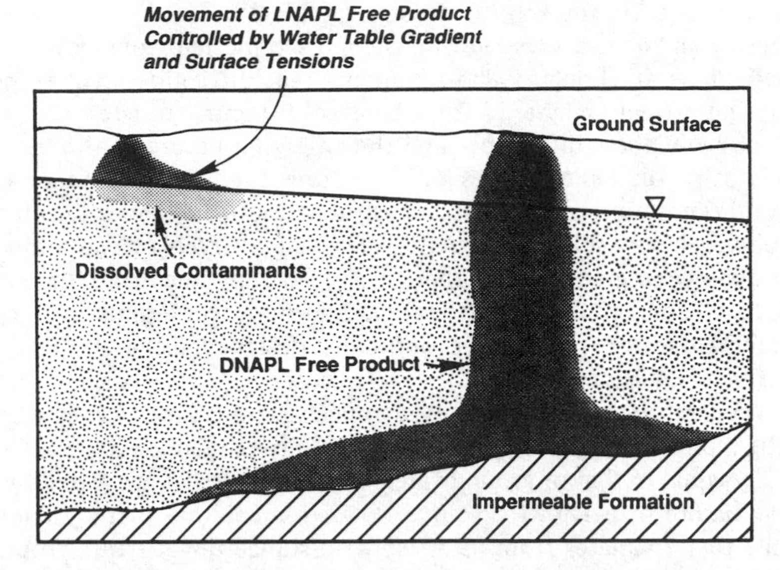

Figure 37. Plume density and miscibility dictate the hydraulic behavior of immiscible liquid spills that encounter groundwater.

water velocity. It is possiblc in some cases for the movement of a DNAPL plume to oppose the direction of groundwater flow. An example of such a case is when gravity-controlled flow moves a DNAPL down-dip on a bedrock aquifer bottom in a setting where groundwater is flowing in the up-dip direction. The mechanics of immiscible plume migration have been heavily studied in recent years (Abriola and Pinder, 1985; Faust, 1985; Hinchee and Reisinger, 1987; Hossain and Corapcioglu, 1988; Kueper and Frind, 1988; Pinder and Abriola, 1986). Cohen and Mercer (1993) discuss many of the special aspects of DNAPL behavior in the context of identifying and characterizing DNAPL contamination problems.

The remediation of LNAPL and DNAPL plumes in source regions is an active research area that is giving rise to some emerging technology (Stover, 1989; Testa and Paczkowski, 1989; Hunt et al., 1988a, 1988b). In principle, the optimization techniques outlined in Chapter V could be used in the design of skimming wells for LNAPLs and extraction wells for DNAPLs. In practice, however, there are many problems associated with the application of formalized optimization techniques to multiphase flow problems. In the first place, it is almost never possible to accurately know where DNAPL and LNAPL pools are located, or if some have been located, whether there are others. In addition, the

NAPL pools are unlikely to possess an ideal geometry; they exhibit fingering and have associated with them disconnected globules and ganglia. Hunt et al. (1988a, 1988b) summarize the difficulties in attempting to design extraction systems for DNAPL-contaminated sites.

In view of these difficulties and the emerging nature of the technology, further discussion of design procedures for NAPL contamination beyond the notion of containment is outside of our scope. Recall that despite their low solubility, NAPLs release low levels of contaminants that dissolve in groundwater; these dissolved constituents are the major source of concern. As noted earlier, hydraulic containment of these dissolved constituents may be optimally designed using the methods described in Chapter V.

In general, because the dissolved concentrations are low, the density of the plume of dissolved constituents that arises from NAPL sources is usually equal to that of groundwater. No unique issues are raised by the dense nature of a DNAPL source in the remediation of the dissolved plume that emanates from it. At some distance downstream from the source, the pump-and-treat design procedures described in our earlier chapters are appropriate. Closer to the source, however, there are issues that may affect remedial design.

Contaminants are released into the groundwater from a pool or globule of free product according to the principles of liquid-liquid partitioning (Gschwend and Wu, 1985; Stumm and Morgan, 1981). Generally, the lower the solubility of the NAPL in water, the slower the dissolution rate. In some cases, the dissolution of an immiscible fluid may take a great deal of time to produce its maximum concentration. This means that the duration of exposure of fresh groundwater to immiscible fluids during pump-and-treat remediation may be too short to allow maximum-dissolved contaminant concentrations to develop (Figure 38). This has been shown to occur in laboratory column experiments with extruded cores from a shallow aquifer that was contaminated with jet fuel. The concentrations of hydrocarbon components reached solubility limits when flows through the cores approximated natural conditions but decreased sharply as the flows were increased to remediation rates.

The mechanism is analogous to the limitation posed by slow chemical kinetics for some sorbing contaminants (Figure 36). The pumping-induced limitation on liquid-liquid partitioning will generate large volumes of mildly contaminated water under continuous operation of the remediation well field. Again, pulsed pumping may be an appropriate alternative for removing the plume of dissolved constituents, if it proves inefficient to pump and treat water having low contaminant concentrations.

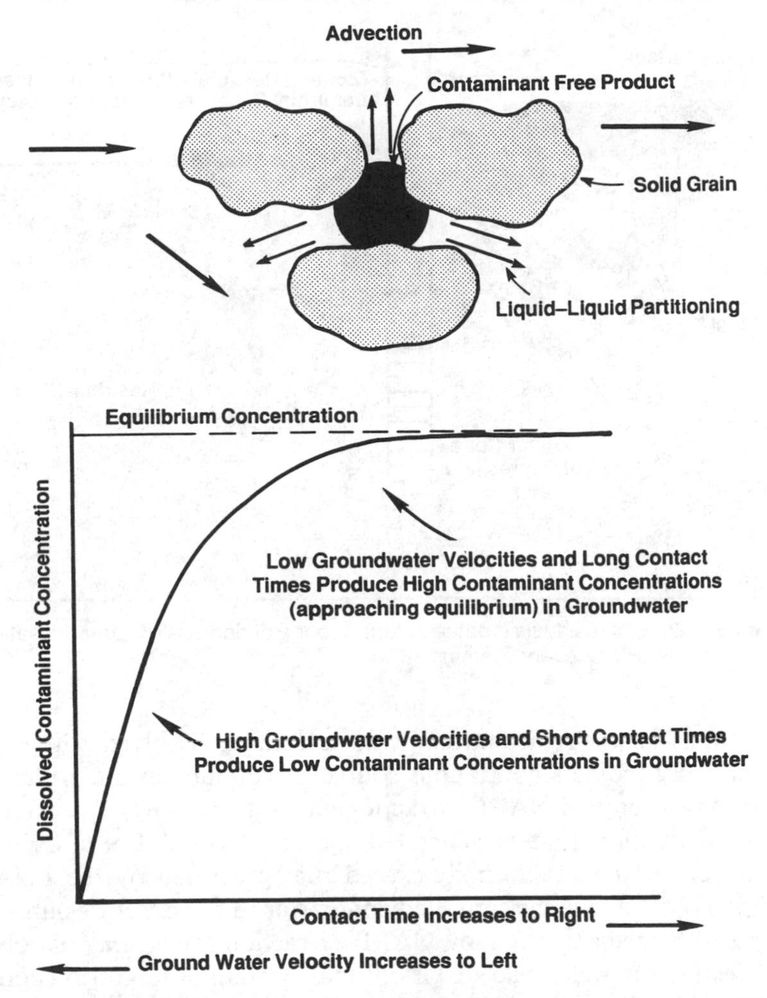

Figure 38. Relationship between groundwater velocity induced by pump-and-treat system and partitioning of contaminants between NAPL and groundwater.

It has never been possible to completely remove subsurface NAPL pools by extraction (Hunt et al., 1988a; Stover, 1989; Testa and Paczkowski, 1989). Both bouyant and dense immiscible fluids leave residual portions trapped in pore spaces by capillary tension. For example, consider an extraction well utilized to control local gradients such that free product (mobile LNAPL) flows into its cone of depression. After pumping ceases, the former cone of depression will contain trapped residual LNAPL below the water table (Figure 39). This has

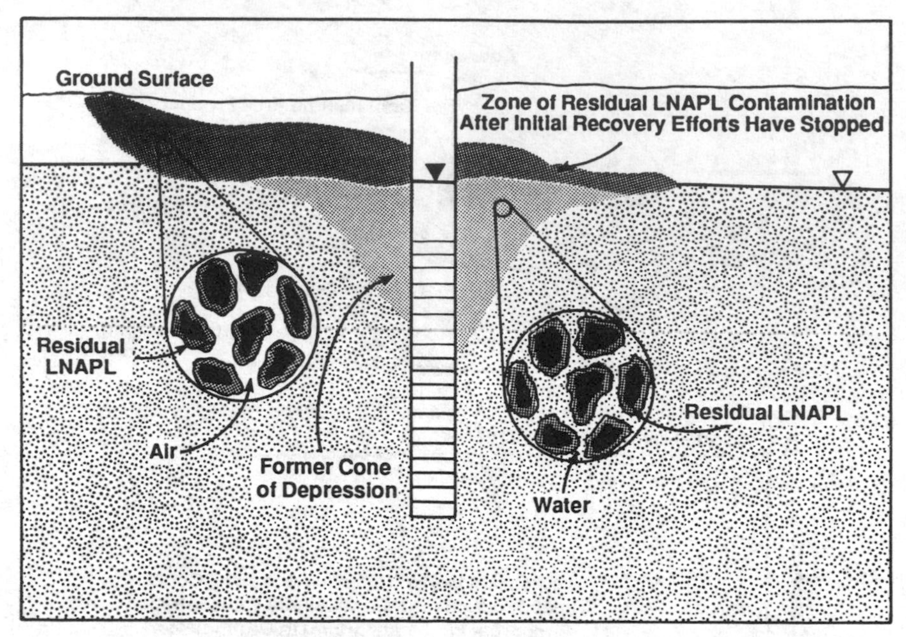

Figure 39. Zone of residuals created in former cone of depression after cessation of LNAPL recovery system.

implications for the remediation of the associated dissolved plume. The residual will become a continuous source of contamination, which will persist long after the LNAPL extraction well is turned off. The extent of the contamination that is generated by the residual LNAPL in the former cone of depression may exceed that generated by the LNAPL resting in place above the saturated zone prior to the onset of pumping. Integrated remedial design of NAPL-extraction wells and dissolved-plume extraction wells is needed if optimal cleanup or migration control is to be achieved at minimum cost. Future remedial efforts may require surfactants to 1) reduce interfacial tension, thereby removing ganglia and/or 2) increase the solubility of the non-aqueous phase, so it can be removed with the groundwater.

3. Effects of Facilitated Transport

The design methodologies for capture and containment remedial extraction-well networks outlined in Chapters IV and V do not take into account the mechanisms of facilitated contaminant transport described in Chapter II.F.4. The principal mechanisms identified there were co-solvation (Enfield, 1985; Nkedi-Kizza et al., 1985, 1987; Staples and

Geiselmann, 1988) and particle transport (Enfield and Bengtsson, 1988; Bengtsson et al., 1988). Both have the ability to increase the migration rates of contaminant plumes.

Cosolvation effects are generally limited to areas close to contamination sources where pure solvents and high dissolved concentrations of organic contaminants are often found. Cosolvation is of most concern for contaminants that are normally expected to be largely immobile, such as polychlorinated biphenyls. Little special consideration need be given to cosolvation during design of a remediation well field, as the potential concerns should be addressed by considerations associated with immiscible NAPL plumes (Chapter VI.B.2).

The transport of contaminants adsorbed on colloidal particles should not have a significant effect on remediation design. Pumping removes colloidal particles from the subsurface because they are suspended in the groundwater. Pumping can also remove sand and other settleable particles and these may also conduct contaminants that have sorbed to their surfaces, but the distances of transport of such particles are typically very short (i.e., tens of feet). Pumping of suspended and settleable particles can complicate the hydraulic operation of a pump-and-treat remediation because wells may clog. Treatment plants may also suffer operationally. Excessive loads of colloidal particles may reduce the effectiveness of various treatment processes such as air stripping and carbon filtration and may also encourage biofouling.

C. Man-made Complexities

Among the manmade complexities that can affect the design of pump-and-treat remediations are property access, local water-supply pumping, and institutional controls.

Difficulties in property access can affect the feasible suite of potential well locations in the design and optimization of a remedial extraction-well network. Most groundwater contamination sites occur in an urbanized industrial environment. The presence of roads, parks, and buildings; the complexity of land ownership; and the possible lack of cooperation from owners and/or municipal officials can all frustrate the remedial design process. In the first place, few landowners are anxious to have a portion of their property used for remediation unless they see a direct benefit or have a direct responsibility for the contamination. In addition, the presence of buried fuel tanks or underground utility lines may preclude drilling in certain areas. Lack of property access can create major difficulties for pump-and-treat remediations because the extraction wells generally have to be joined by a common piping network that will conduct the combined fluid outputs to the treatment plant.

Manmade complexity can also be introduced into the remedial design process if there are large water-supply wells nearby or if there is competing remedial pumping in the area. Seasonally operated pumps and on-demand pumping may occur locally, and these hydraulic impacts must be accounted for. In some cases, the rates and schedules of local pumping are unknown. If a pump-and-treat remedy is designed without knowledge of the presence of local well extractions, it may be unable to overcome the external hydraulic stresses and the system would then fail to operate as designed.

Institutional controls may also affect design. Drilling restrictions may be in effect for water conservation purposes. Drawdown limitations may be in effect for the same reasons or to minimize upconing of saline water or limit ground subsidence. Water rights may limit allowable pumping in fully adjudicated aquifers. Water-reuse requirements may be in effect to reverse the impact of excessive withdrawal rates. Fortunately, many institutional controls are negotiable, particularly if the potential impacts of operation of a pump-and-treat remediation can be shown to be controllable or of a temporary nature or if the treated water can be diverted for use in a municipal or industrial water-distribution system.

References for Chapter VI

Abriola, L. M., and G. F. Pinder, "A Multiphase Approach to the Modeling of Porous Media Contamination by Organic Compounds," *Water Resources Research*, Vol. 21, No.1, pp. 11–18, 1985.

Anderson, M. P., "Using Models to Simulate the Movement of Contaminants Through Groundwater Flow Systems," *Critical Reviews in Environmental Controls*, Vol. 9, No. 2, pp. 97–156, 1979.

Bahr, J., "Analysis of Nonequilibrium Desorption of Volatile Organics during Field Test of Aquifer Decontamination," *Journal of Contaminant Hydrology*, Vol. 4, No. 3, pp. 205–222, 1989.

Baehr, A. L., G. E. Hoag, and M. C. Marley, "Removing Volatile Contaminants from the Unsaturated Zone by Inducing Advective Air-Phase Transport," *Journal of Contaminant Hydrology*, Vol. 4, No. 1, pp. 1–26, 1989.

Barker, J. F., J. E. Barbash, and M. LaBonte, "Groundwater Contamination at a Landfill Sited on Fractured Carbonate and Shale," *Journal of Contaminant Hydrology*, Vol. 3, No. 1, pp. 1–25, 1988.

Bear, J., and A. Verruijt, *Modeling Groundwater Flow and Pollution*, D. Reidel Publishing Company, Boston, MA, 1987.

Bengtsson, G., C. G. Enfield, and R. Lindquist, "Macromolecules Facilitate Transport of Trace Organics," *Science of the Total Environment*, Vol. 67, pp. 159–169, 1988.

Bitton, G., and C. P. Gerba (editors), *Groundwater Pollution Microbiology*, Wiley Interscience/John Wiley & Sons Inc., New York, NY, 1984.

Blatt, H., G. V. Middleton, and R. C. Murray, *Origin of Sedimentary Rocks*, 2nd Ed., Prentice-Hall Inc., Englewood Cliffs, NJ, 1980.

Bouchard, D. C., C. G. Engield, and M. D. Piwoni, "Transport Processes Involving Organic Chemicals," in *Reactions and Movement of Organic Chemicals in Soils,* Soil Science Society of America, Special Publications Press, 1989.

Bouchard, D. C., A. L. Wood, M. L. Campbell, P. Nkedi-Kizza, and P. S. C. Rao, "Sorption Nonequilibrium during Solute Transport," *Journal of Contaminant Hydrology*, Vol. 2, No. 3, pp. 209–223, 1988.

Brusseau, M. L., P. S. C. Rao, R. E. Jessup, and J. M. Davidson, "Flow Interruption: A Method for Investigating Sorption Nonequilibrium," *Journal of Contaminant Hydrology*, Vol. 4, No. 3, pp. 223–240, 1989.

Cohen, R. M., and J. W. Mercer, DNAPL Site Evaluation, EPA/600/R-93/022, prepared for the Robert S. Kerr Laboratory, U.S. Environmental Protection Agency, Ada, Oklahoma, 1993.

Corey, A. T., *Mechanics of Heterogeneous Fluids in Porous Media*, Water Resources Publications, Fort Collins, CO, 1977.

Crow, W. L., E. P. Anderson, and E. M. Minugh, "Subsurface Venting of Vapors Emanating from Hydrocarbon Product on Ground Water," *Ground Water Monitoring Review*, Vol. 7, No. 1, pp. 51–57, 1987.

Curtis, G. P., P. V. Roberts, and Martin Reinhard, "A Natural Gradient Experiment on Solute Transport in a Sand Aquifer: 4. Sorption of Organic Solutes and its Influence on Mobility," *Water Resources Research*, Vol. 22, No. 13, pp. 2059–2067, 1986.

Davis, S. N., and R. J. M. DeWeist, *Hydrogeology*, John Wiley & Sons Inc., New York, NY, 1966.

Davison, C. C., and V. Guvenasen, *Hydrogeological Characterization, Modeling, and Monitoring of the Site of Canada's Underground Research Laboratory*, Atomic Energy of Canada Ltd., Report AECL-8676, 1985.

de Marsily, G., *Quantitative Hydrogeology: Groundwater Hydrology for Engineers*, Academic Press Inc.; Orlando, FL, 1986.

Enfield, C. G., "Chemical Transport Facilitated by Multiphase Transport Flow Systems," *Water Science and Technology*, Vol. 17, No. 9, pp. 1–12, 1985.

Enfield, C. G., and G. Bengtsson, "Macromolecular Transport of Hydrophobic Contaminants in Aqueous Environments," *Ground Water*, Vol. 26, No. 1, pp. 64–70, 1988.

Faust, C. R., "Transport of Immiscible Fluids Within and Below the Unsaturated Zone: A Numerical Model," *Water Resources Research*, Vol. 21, No. 4, pp. 587–596, 1985.

Feenstra, S., J. A. Cherry, E. A. Sudicky, and Z. Haq, "Matrix Diffusion Effects on Contaminant Migration from an Injection Well in Fractured Sandstone," *Ground Water*, Vol. 22, No. 3, pp. 307–316, 1984.

Fetter, C. W., *Applied Hydrogeology*, 2nd Ed., Merrill Publishing Company; Columbus, OH, 1988.

Freeze, R. A., and J. A. Cherry, *Groundwater*, Prentice Hall, Inc.; Edgewood Cliffs, NJ, 1979.

Goltz, M. N., and P. V. Roberts, "Interpreting Organic Solute Transport Data from a Field Experiment Using Physical Nonequilibrium Models" *Journal of Contaminant Hydrology*, Vol. 1, No. 1/2, pp. 77–94, 1986.

______ "Simulations of Physical Nonequilibrium Solute Transport Models: Application to a Large-Scale Field Experiment," *Journal of Contaminant Hydrology*, Vol. 3, No. 1, pp. 37–64, 1988.

Gschwend, P. W., and M. D. Reynolds, "Monodisperse Ferrous Colloids in an Anoxic Groundwater Plume," *Journal of Contaminant Hydrology*, Vol. 2, No. 4, pp. 309–327, 1987.

Guven, O., and F. J. Molz, "Deterministic and Stochastic Analyses of Dispersion in an Unbounded Stratified Porous Medium," *Water Resources Research*, Vol. 22, No. 11, pp. 1565–1574, 1986.

Gvirtzman, H. and S.M. Gorelick, "The Concept of In-situ Vapor Stripping for Removing VOCs from Groundwater," *Transport in Porous Media,* Vol. 8, 71–92, 1992.

Hillel, D., *Fundamentals of Soil Physics*, Academic Press; New York, NY, 1980.

Hinchee, R., and H. J. Reisinger. "A Practical Application of Multiphase Transport Theory to Ground Water Contamination Problems," *Ground Water Monitoring Review*, Vol. 7, No. 1, pp. 84–92, 1987.

Hossain, M. A., and M. Y. Corapcioglu, "Modifying the USGS Solute

Transport Computer Model to Predict High-Density Hydrocarbon Migration," *Ground Water*, Vol. 26, No. 6, pp. 717–723, 1988.

Hunt, J. R., N. Sitar, and K. S. Udell., "Nonaqueous Phase Liquid Transport and Cleanup: 1. Analysis of Mechanisms," *Water Resources Research*, Vol. 24, No. 8, pp. 1247–1258, 1988a.

______, "Nonaqueous Phase Liquid Transport and Cleanup: 2. Experimental Studies," *Water Resources Research*, Vol. 24, No. 8, pp. 1259–1269, 1988b.

Jorgensen, D. G., T. Gogel, and D. C. Signor, "Determination of Flow in Aquifers Containing Variable-Density Water," *Ground Water Monitoring Review*, Vol. 2, No. 2, pp. 40–45, 1982.

Keely, J. F., "Optimizing Pumping Strategies for Contaminant Studies and Remedial Actions," *Ground Water Monitoring Review*, Vol. 4, No. 3, pp. 63–74, 1984.

Keely, J. F., M. D. Piwoni, and J. T. Wilson, "Evolving Concepts of Subsurface Contaminant Transport," *Journal Water Pollution Control Federation*, Vol. 58, No. 5, pp. 349–357, 1986.

Kruseman, G. P., and N. A. de Ridder, *Analysis and Evaluation of Pumping Test Data, International Institute of Land Reclamation and Improvement*, Bulletin 11, Wageningen, The Netherlands, 1970.

Kueper, B. H., and E. O. Frind, "An Overview of Immiscible Fingering in Porous Media," *Journal of Contaminant Hydrology*, Vol. 2, No. 2, pp. 95–110, 1988.

Lee, L. S., P. S. C. Rao, M. L. Brusseau, and R. A. Ogwada, "Nonequilibrium Sorption of Organic Contaminants during Flow through Columns of Aquifer Materials," *Environmental Toxicology and Chemistry*, Vol. 7, No. 10, pp. 779–793, 1988.

Mackay, D. M., W. P. Ball, and M. G. Durant, "Variability of Aquifer Sorption Properties in a Field Experiment on Groundwater Transport of Organic Solutes: Methods and Preliminary Results," *Journal of Contaminant Hydrology*, Vol. 1, No. 1/2, pp. 119–132, 1986.

Mercado, Abraham, "The Use of Hydrogeochemical Patterns in Carbonate Sand and Sandstone Aquifers to Identify Intrusion and Flushing of Saline Water," *Ground Water*, Vol. 23, No. 5, pp. 635–645, 1985.

Mercer, J. W., and C. R. Faust, *Ground-Water Modeling*, National Water Well Association, Dublin, OH, 1981.

Nkedi-Kizza, P., M. L. Brusseau, P. S. C. Rao, and A. G. Hornsby, "Nonequilibrium Sorption during Displacement of Hydrophobic Organic Chemicals and ^{45}Ca through Soil Columns with Aqueous and Mixed Solvents," *Environmental Science and Technology*, Vol. 23, No. 7, pp. 814–820, 1989.

Nkedi-Kizza, P., P. S. C. Rao, and A. G. Hornsby, "Influence of Organic Cosolvents on Sorption of Hydrophobic Organic Chemicals by Soils," *Environmental Science and Technology*, Vol. 19, No. 10, pp. 975–979, 1985.

______, "Influence of Organic Cosolvents on Leaching of Hydrophobic Organic Chemicals through Soils," *Environmental Science and Technology*, Vol. 21, No. 11, pp. 1107–1111, 1987.

Novak, S. A., and Y. Eckstein, "Hydrogeochemical Characterization of Brines and Identification of Brine Contamination in Aquifers," *Ground Water*, Vol. 26, No. 3, pp. 317–324, 1988.

Osiensky, J. L., G. V. Winter, and R. E. Williams, "Monitoring and Mathematical Modeling of Contaminated Ground-Water Plumes in Fluvial Environments," *Ground Water*, Vol. 22, No. 3, pp. 298–306, 1984.

Pinder, G. F., and L. M. Abriola, "On the Simulation of Nonaqueous Phase Organic Compounds in the Subsurface," *Water Resources Research*, Vol. 22, No. 9, pp. 109S–119S, 1986.

Piwoni, M. D., and P. Banerjee, "Sorption of Volatile Organic Solvents from Aqueous Solution onto Subsurface Solids," *Journal of Contaminant Hydrology*, Vol. 4, No. 2, pp. 163–179, 1989.

Poole, V. L., K. Cartwright, and D. Leap, "Use of Geophysical Logs to Estimate Water Quality of Basal Pennsylvanian Sandstones, Southwestern Illinois," *Ground Water*, Vol. 27, No. 5, pp. 682–688, 1989.

Satkin, R. L., and P. B. Bedient, "Effectiveness of Various Aquifer Restoration Schemes Under Variable Hydrogeologic Conditions," *Ground Water*, Vol. 26, No. 4, pp. 488–499, 1988.

Spayed, S. E., "Movement of Volatile Organics Through a Fractured Rock Aquifer," *Ground Water*, Vol. 23, No. 4, pp. 496–502, 1985.

Staples, C. A., and S. J. Geiselmann, "Cosolvent Influences on Organic Solute Retardation Factors," *Ground Water*, Vol. 26, No. 2, pp. 192–198, 1988.

Stover, Enos, "Co-produced Ground Water Treatment and Disposal Options During Hydrocarbon Recovery Operations," *Ground Water Monitoring Review*, Vol. 9, No. 1, pp. 75–82, 1989.

Stumm, W., and J. J. Morgan, *Aquatic Chemistry: An Introduction Emphasizing Chemical Equilibria in Natural Waters*, 2nd Ed., Wiley Interscience/John Wiley & Sons, Inc., New York, NY, 1981.

Testa, S., and M. Paczkowski, "Volume Determination and Recoverability of Free Hydrocarbon," *Ground Water Monitoring Review*, Vol. 9, No. 1, pp. 120–128, 1989.

Terra Vac, Inc., Personal Communication to J. F. Keely, 1989.

Thorstenson, D. C., and D. W. Pollack, "Gas Transport in Unsaturated

Porous Media: The Adequacy of Fick's Law," *Reviews of Geophysics*, Vol. 27, No. 1, pp. 61–78, 1989.

Tinsley, I. J., *Chemical Concepts in Pollutant Behavior*, John Wiley & Sons Inc., New York, NY, 1979.

Valocchi, A. J., "Theoretical Analysis of Deviations from Local Equilibrium during Sorbing Solute Transport through Idealized Stratified Aquifers," *Journal of Contamination Hydrology*, Vol. 2, No. 3, pp. 191–208, 1988.

Walton, W. C., *Groundwater Pumping Tests: Design & Analysis*, Lewis Publishers, Inc.; Chelsea, MI, 1987.

White, W. B., *Geomorphology and Hydrology of Karst Terrains*, Oxford University Press, New York, NY, 1988.

Woodburn, K. B., P. S. C. Rao, M. Fukui, and P. Nkedi-Kizza, "Solvophobic Approach for Predicting Sorption of Hydrophobic Organic Chemicals on Synthetic Sorbents and Soils," *Journal of Contaminant Hydrology*, Vol. 1, No. 1/2, pp. 227–241, 1986.

CHAPTER VII

Performance Evaluation

A. Performance Evaluation Strategies

Performance evaluations may be conducted at a hazardous waste site for a variety of reasons. The two most common reasons are (1) to satisfy the compliance requirements of negotiated or court settlements and (2) to assess whether design modifications are required as part of the operational management plan. The strategy that one adopts for conducting performance evaluations must focus on the kind of data that will be collected and how those data will be presented for interpretations and decisions. It is best to start building a strategy, therefore, by examining the data collection studies performed previously at a site.

Let us consider data collection activities which follow the Superfund model. As described in Chapter I.B and illustrated in Figures 1 and 2, a Superfund site investigation proceeds through several stages. These are

- Preliminary assessment: existing information on contaminants, sources, pathways, and receptors is collected.
- Site inspection: preliminary field data are collected.
- Remedial investigation: detailed field data are collected and analyses are performed to fully characterize contaminants, sources, pathways, receptors, and risks.
- Feasibility study: alternative remedial actions are evaluated.

Strategies for performance evaluations should, therefore, view data gathering activities during the remediation as an extension of the data gathering conducted during the site characterization. For the sake of continuity and the ability to conduct meaningful comparisons at future dates, the performance evaluation program should (1) extend/continue the collection of key chemical and hydrodynamic data and (2) fill in data gaps from previous site characterization efforts, to the full extent that these are needed, for both compliance monitoring and operational management purposes.

Traditionally, the key controls on the form and quality of the technical data obtained during remediations have been the compliance criteria that are selected and the compliance point locations at which those criteria are to be applied. This traditional collection of samples for chemical analysis and measurement of water-level elevations has not been adequate for detailed performance evaluations. There is a need for ongoing improvement of the site characterization data base. This may be accomplished economically by a combination of periodic and opportunistic testing. Periodic testing should include such activities as biennial collection of aquifer-sediment samples from new boreholes for estimation of contaminant residuals by chemical extract analysis. Opportunistic testing should include such activities as geological logging and pump testing of each new well that is drilled for refinement of the information available on flow paths and velocities under various pumping conditions.

The collection of aquifer-sediment samples for identification and quantification of residual contamination is an example of a technical activity that has become routine to preoperational site characterization studies but which is rarely done during operation of a remediation. The latter is unfortunate because a measure of the magnitude of the reduction of contaminant residuals that is attained during a given time period can provide one of the few direct and uncompromising means of evaluating how well the remediation is performing. Together with operational data, the information from analysis of the sediments can be used to estimate the future rate of reduction of the contaminant residuals by the current well-field configuration and pumping schedule.

The continued hydrogeological characterization of the site during remediation can be of great help in assessing the need for remedial design modifications. It can provide the basis for understanding what the optimal well-field configuration and pumping schedule should be for the next forecast period and for projecting the system performance during that period. As the amount of available hydrogeologic data increases, the potential for successful modification of the existing remedial system also increases. One of the best reasons to actively manage the remediation in this way is the economics of the situation; pumping the minimum volume at the highest concentrations and in the minimum time frame is the most direct means of minimizing operational costs in a pump-and-treat system.

Note, however, that even if a remediation is operating successfully, occasional reevaluation of its performance is likely to be necessary. This is so because technological advances may raise the threshold of "optimum" performance (for example, an improvement in treatment efficiencies) and because estimates of the contaminant masses in reserve, their spatial distributions, and their effective rates of depletion are likely to change over time. As a consequence, there is a need to continuously collect routine monitoring and operational data during remediation. Operational data must be obtained regularly to achieve and maintain the most efficient and effective remediation. Certain data are critical, including (1) the flow rates of extraction and injection wells; (2) water-level elevations in all wells and piezometers; and (3) the concentrations of contaminants in monitoring wells, extraction and injection wells, and treatment-plant effluent.

In the following subsections we will look first at issues associated with the location of monitoring points and then at issues associated with the development of evaluation criteria at those locations. Throughout the discussions we will continue to distinguish between *compliance monitoring* (to assess regulatory compliance) and *performance monitoring* (to assess design performance).

1. Locations for Compliance Monitoring and Performance Monitoring

The importance of careful water quality monitoring near the downgradient boundary of a contaminant plume is widely recognized by hydrogeologists. It is this downgradient monitoring that allows potential receptors to take action in the event of a threatening plume advance. Monitoring within the plume is equally important, as it allows the effectiveness and efficiency of the remediation to be evaluated. In this context, effectiveness can be defined as the extent to which contaminant levels are reduced, and efficiency can be defined as the ratio of contaminant mass removed to water mass removed. This inner-plume data facilitate management of the remediation well field for best performance. For example, by reducing the flow rates of extraction wells that pump from relatively clean zones and increasing the flow rates of extraction wells that pump from zones that are highly contaminated, the efficiency of the system can be increased. By contrast, the exclusive use of monitoring points downgradient of the plume boundary does not improve one's understanding of plume behavior during remediation. Exclusive use of downgradient monitoring points only serves to indicate downstream plume advancements when contaminants are detected at the downgradient monitoring locations.

There are many kinds of monitoring points in use today. All of these, and others not traditionally employed, are useful for evaluation of regulatory compliance or design performance. In the following paragraphs we will discuss background monitoring points, public water-supply wells

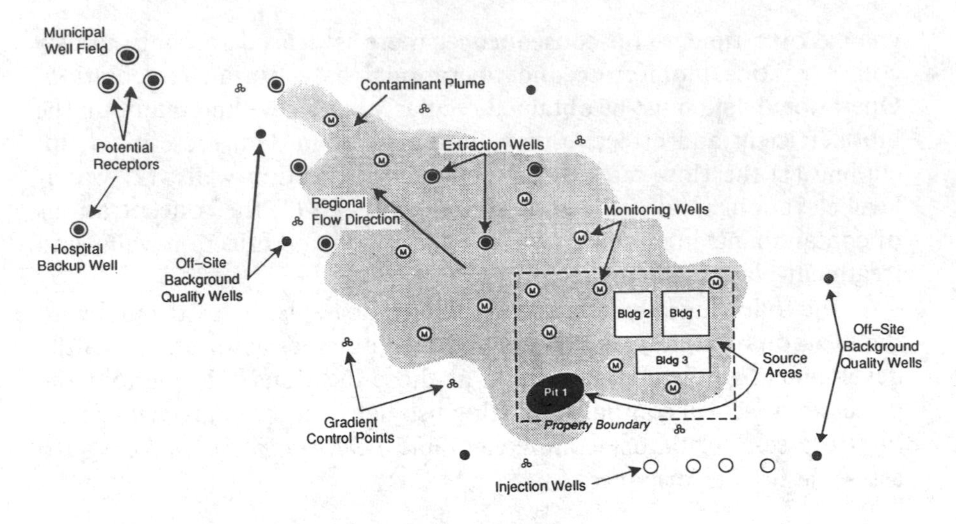

Figure 40. Compliance monitoring point locations.

as monitoring points, gradient-control monitoring points, and within-plume monitoring points.

Natural water-quality or *background* sampling locations are the most widely used compliance monitoring points. They are usually positioned a short distance upgradient or downgradient of the plume, although cross-gradient locations may also be established (Figure 40). Upgradient monitoring points are usually installed early in the site-investigation stage for the purpose of determining the quality of uncontaminated groundwater. Downgradient monitoring points are installed after the boundaries of the plume have been defined and prior to the implementation of a remedial system. Their purpose is to provide a warning of plume advance either before or after a remedial system has been placed in operation. The location of a downstream background water-quality well is chosen so that

1. It is neither in the plume nor in adjacent areas that may be affected by the remediation,
2. It is in an uncontaminated portion of the aquifer through which the plume would migrate if the remediation failed, and
3. Its location minimizes the possibility of detecting other potential sources of contamination (i.e., it is relevant to the target site only).

Data gathered at a background-compliance monitoring point located downgradient of the plume can be used to identify out-of-control conditions when a portion of the plume escapes the remedial action. For an anthropogenic contaminant, the water-quality standard specified for this kind of compliance monitoring point might be as low as the detection limit. For a naturally occurring contaminant, it might be defined in terms of a particular increase over the natural water-quality concentration. The latter is usually established with water-quality data from wells located upgradient of the source.

Data gathered from a background-compliance monitoring point located upgradient or cross-gradient of the plume of contamination may also serve to indicate out-of-control conditions, should the remediation generate unanticipated effects. The latter might occur, for instance, if injection wells, used to return treated water to the aquifer at the upgradient edge of the plume, simultaneously have the unintended effect of sweeping some of the contaminant plume outward. The upgradient and cross-gradient locations of background-compliance monitoring points may also be useful in providing a continuous check on the quality of the regional flow system, thus providing early warning of any new contaminants being transported into the remediation area.

Public water-supply wells located downgradient of a plume are a second kind of beyond-plume compliance monitoring points. The locations of these points are predetermined and fixed. These wells have been drilled in locations that are suitable for water supply development; they were not originally designed to serve as monitoring wells. The purpose of sampling these wells is to determine whether water-quality standards are met for the particular use of the water withdrawn. The standards typically specified for this kind of compliance point are those found in the National Primary Drinking Water Regulations. These regulations specify enforceable maximum contaminant levels (MCLs) for a large number of organic and inorganic contaminants. Water quality standards for public water-supply wells can also be based on maximum contaminant level goals (MCLGs), National Secondary Drinking Water Regulations, or a requirement to maintain existing water quality. Appendix D provides a brief summary of water-quality regulations that play a role in groundwater remediation problems.

The *gradient-control monitoring point* is a third kind of beyond-plume monitoring point. It is established specifically to determine the directions and magnitudes of horizontal and vertical hydraulic gradients. This type of monitoring is required for design performance evaluation and is being used with increasing frequency as part of regulatory-compliance monitoring networks. In this case, the monitoring "point" comprises a cluster of small-diameter piezometers that have very short, screened intervals. It is usually located just outside the perimeter of the plume. Hydraulic-head measurements are obtained from wells that have comparable screened intervals and are used in preparing hydraulic-head contour maps from which the directions and magnitudes of local horizontal hydraulic gradients can be determined. An evaluation of the vertical gradients can be made in similar fashion by comparison of water-level measurements from screened intervals located at different depths. Additionally, water-level measurements from different depths may be used to determine vertical variations in horizontal gradients. In a thick aquifer, an improperly designed remediation could result in a failure to control plume migration in a particular vertical segment of the aquifer. Careful evaluation of vertically distributed hydraulic-head data will allow one to identify and correct this type of problem.

Gradient-control monitoring networks are best established prior to remedial design. They can be used to determine the pre-existing pattern of hydraulic gradients at the site, which is a requirement of the remedial design process outlined in Chapters IV and V. After the remedial operation begins, they are used for performance evaluation.

Less often utilized than any of the foregoing are *within-plume monitoring points*. These are monitoring wells located within the perimeter of the plume. Many of these may have been installed during the site investigation phase, prior to the remediation, but others may be added subsequent to implementation of the remediation. They are needed to monitor the progress of the remediation within the plume. These can be subdivided into *on-site plume monitoring points* that are located within the property boundary of the facility that contains the source of the contaminant plume and *off-site plume monitoring points* that are located beyond the facility boundary but within the boundary of the plume.

The water-quality standards applied to off-site plume monitoring points used for regulatory compliance may require lower concentration limits than those imposed at on-site monitoring points. This is due to the tendency of regulatory agencies to impose stricter water-quality standards in the zone between the facility property line and the uncontaminated portions of the aquifer. These standards may be specified as risk-based alternate concentration limits (ACLs). A risk-based ACL is the maximum contaminant concentration that can be allowed without excess public or environmental health risk and is commonly determined during a health risk assessment. Because risk-based ACLs should account for unique or site-specific factors that may increase the risk posed by a contaminant, these ACLs may be lower than MCLs.

Because of their proximity to the source of contamination and the technical infeasibility of complete removal of the source at many sites, the water-quality standards for *on-site compliance monitoring points* range from local background levels to technology-driven ACLs. Technology-driven ACLs are commonly defined as the minimum concentration that can be obtained using the current accepted treatment technology.

For performance-assessment purposes, monitoring-well data will have little meaning without data from the extraction and injection wells. In particular, chemical concentration levels, flow rates, and water-level elevations are essential to active management of the remediation well field. These data are used with data from the various monitoring points to evaluate gradient control, water-quality trends, and estimated contaminant residuals. On the basis of these evaluations, appropriate system modifications are made to ensure regulatory compliance and maximum operational efficiency.

2. Evaluation Criteria for Compliance Monitoring and Performance Monitoring

It is convenient to identify four categories of compliance monitoring and performance monitoring criteria: chemical, hydrodynamic, treatment efficiency, and administrative.

First consider *chemical monitoring criteria*. These come into play in the form of water-quality standards for regulatory compliance. While several different Federal statutes impose limitations on chemical concentrations in freshwater, the EPA's 1985 National Contingency Plan (NCP) revision and the Superfund Amendments and Reauthorization Act of 1986 (SARA) were the first to provide the integrating concept of ARARs: the *applicable, relevant, or appropriate regulations* that derive from any and all Federal, regional, State, and local statutes. It is thus possible for *chemical criteria for regulatory compliance monitoring* to be based on a combination of

1. Maximum contaminant levels that have been promulgated for drinking water supplies under the Safe Drinking Water Act (SDWA) of 1974, as amended by the SDWA Amendments of 1986;
2. Alternate concentration limits that are risk-based or technology-based levels that are site-specific to groundwater Corrective Action Plans under the Resource Conservation and Recovery Act (RCRA) of 1976, as amended by the Hazardous Waste Amendments Act of 1984;
3. State Water Quality Standards that derive from the Clean Water Act of 1972 and other statutes that protect aquatic wildlife and their habitat; and
4. County, municipal, or regional restrictions on the quality of water that

may be discharged to local lands or other requirements for the return of pumped water in water-short areas.

Furthermore, the actual numerical values of the chemical standards may also be affected by detection limits and natural water quality. It may be, for example, that the acceptable concentration level for a given contaminant in health terms is actually much lower than can be physically determined with routine instrumentation. Or it may be that the natural levels of certain contaminants in certain settings are greater than the pertinent ARAR.

Hydrodynamic monitoring criteria serve to confirm nonchemical performance such as

1. Prevention of infiltration through the unsaturated zone,
2. Maintenance of an inward hydraulic gradient at the boundary of a plume of groundwater contamination, and
3. Provision of minimum flows in streams or wetlands.

Of these, the requirement that an inward hydraulic gradient be maintained at the periphery of a contaminant plume undergoing pump-and-treat remediation is the most common. It is imposed to ensure that no portion of the plume is free to migrate away from the zone of remedial action. Evaluations of the hydraulic gradients exerted by remediation well fields require many measurements. To assess this performance adequately, the hydraulic gradients must be measured accurately in three dimensions. The design of an array of piezometers for this purpose is not as simple as one might first imagine. Many points are needed to define the convoluted piezometric surface or water-table configuration that develops between adjacent pumping or injection wells. There are flow divides in the horizontal plane near active wells, and stagnation points and zones can form downgradient as the wells pump against the natural flow. In the vertical dimension there may be flow divides, too, because the hydraulic influence of any well may control the flow in only a portion of the full depth of the aquifer.

The performance standards that are applied to the treatment-plant portion of a pump-and-treat remediation are most often phrased in terms of *treatment-efficiency criteria*. Although simple comparisons of total organic chemical loading in the influent and the effluent of the treatment train give some idea as to overall treatment efficiency, it is important to make a more detailed examination of the influent and effluent streams. In some cases, individual compounds present limitations to the treatment process because of effluent quality standards for discharging to a local stream (e.g., requirements of a National Pollutant

Discharge Elimination System [NPDES] permit). In other cases, after some time of pumping VOC-laden water for treatment solely by air-stripping, nonvolatiles may begin to arrive in the influent and require the addition of carbon filtration to the treatment train. In still other cases, there is a need for examination of key locations within the treatment train to troubleshoot the effects of certain contaminants on treatment units. Iron and calcium, for example, may precipitate in an air stripper and dramatically decrease throughput.

Administrative monitoring criteria are codified governmental rules and regulations. They may include

1. Regulations for fire safety, electrical hazards, etc.;
2. Access-limiting administrative orders, such as drilling bans;
3. Proof of maintenance of site security; and
4. Reporting requirements, such as frequency and character of reports of compliance and operational monitoring.

Combinations of chemical, hydrodynamic, treatment-efficiency, and administrative monitoring criteria are generally stipulated at compliance monitoring points. The exact combination for a specific compliance monitoring point depends on its location relative to the source of contamination and the potential receptors.

3. Strategies for Selection of Locations and Evaluation Criteria

Each kind of monitoring point has a specific and distinct role to play in evaluating the progress of a remediation. In choosing chemical monitoring criteria, it is essential to recognize the interdependency of the criteria at different compliance monitoring-point locations. For example, one cannot justify liberal ACLs on site and have realistic expectations of meeting more stringent ACLs off site. Similarly, one cannot expect background-compliance monitoring points to remain free of contamination if the off-site plume ACL is chosen inappropriately.

Regulatory agencies must decide if the evaluations of compliance monitoring data should incorporate allowances for statistical variations in the reported values. If so, then what cutoff should be used (e.g., the average value plus two standard deviations)? Should evaluations consider each compliance monitoring point independently for violations or use an average of all compliance monitoring points? Finally, what method should be used to indicate that cleanup has been achieved? The zero-slope method, for example, requires one to demonstrate that contaminant concentrations have stabilized at their lowest levels before a remediation can be terminated. Stabilization is commonly considered complete when a plot of contaminant concentration versus time yields a flat or zero-slope line. Monitoring-network designers should be aware of regulatory policies when considering these matters.

The monitoring-point locations and criteria should be selected initially on the basis of a detailed site characterization, from which transport pathways prior to remediation are identifiable and from which the probable transport pathways during remediation may be predicted. In general, the impacts of remediation should be forecast prior to selecting monitoring-point locations and criteria. This will allow the designer to consider the effects of system operation on flow and transport pathways when making location and criteria selections. Common methods of forecasting remediation impacts are model studies and pilot studies.

Specifically, monitoring-point locations and evaluation criteria should be chosen with regard to transport pathways prior to and during the operation of the remedial system. Since the chemical composition and flow pattern in the zone of remedial action may change substantially as the remediation progresses, the compliance monitoring-point locations and criteria may need to change too.

Assuming that a remediation well field is operated with a constant system configuration and pumping schedule, it will still generate flow velocities that vary in space throughout the zone of remedial action. In part, this is due to variability in subsurface properties (e.g., permeability). Spatial variation in velocities also arises from the radial flow patterns imposed on the local flow system through the operation of one or more remedial wells (Figure 41).

Consider, for example, an extraction well located near the downgradient boundary of a contaminant plume. Many of the flowlines that will be generated by the pumping will extend downgradient before arcing back toward the well. Those flowlines may transport the plume through what was previously a background-quality portion of the aquifer (Figure 41). Such an occurrence may lead to the detection of contaminants in a nearby monitoring well. The primary concern is that this condition not be interpreted as plume advancement. The zone in which this can occur is relatively small, and the contamination will eventually be removed by the remedial well.

However, an extraction well placed upgradient of the downgradient plume boundary, such that all the flowlines associated with the well are within that boundary, may result in inadequate capture and control of the outermost portion of the plume. A stagnation point in the flowfield, where velocity is equal to zero, will form directly downgradient of the extraction well. The portion of the aquifer close to this point may not be effectively remediated, unless the flow rate of the well is changed (caus-

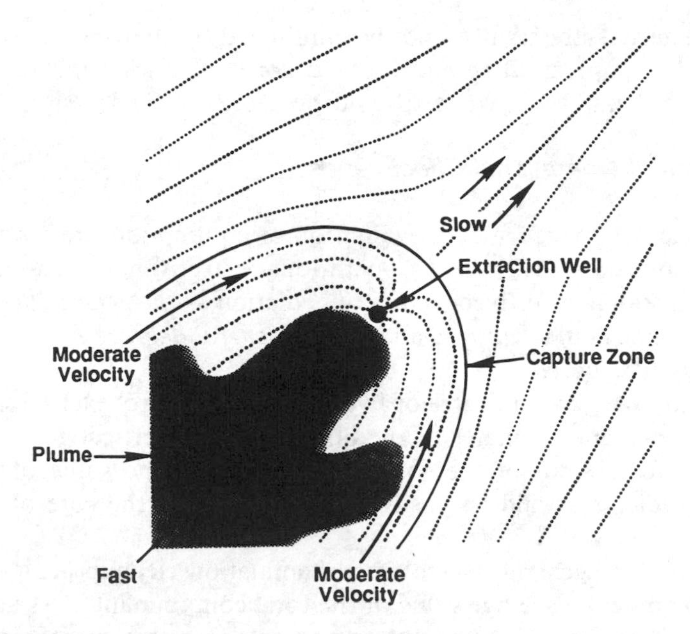

Figure 41. Flowline and velocity variations associated with locating an extraction well at the outer plume boundary.

ing the location of its stagnation point to move) or significant flowline changes are imposed by actions at other wells.

Occasionally, there may also be a need to relocate or add extraction or injection wells to remediate portions of the plume more efficiently. This is normally the case when regulatory compliance requirements or system performance objectives cannot be met by modifying pumping rates and schedules at existing wells.

An intuitively correct way to organize the design of monitoring networks must include mapping of the flowlines generated by the remediation well field in sufficient detail to characterize the capture zones of the extraction wells (and the flushing zones of the injection wells). For each remediation well, the location of the stagnation point and the maximum width of its capture zone must be identifiable. Of course, a different flowline plot is required for each change in such key variables as well flow rates and locations, natural hydraulic gradient, and flow direction. Provided that the estimates of key parameters such as hydraulic conductivity are reasonably accurate, these flowline maps should form a sound starting point for selecting monitoring-point locations.

There is, of course, a need to ensure that other nearby irrigation wells or industrial wells are accounted for when mapping the flowlines. Generally, the external stresses need not be eliminated by shutting the wells down, but their operation should be held constant during the period when data are being acquired for the purpose of well-field design.

4. Measures of Operational Effectiveness

At the heart of a performance evaluation of a pump-and-treat remediation is the need to evaluate its operational effectiveness. One may view the operational effectiveness of a remediation as *the general degree of hydrodynamic control exerted* and/or *the general degree of contamination cleanup achieved*.

To quantify the general degree of hydrodynamic control exerted, one may compute several estimates of the horizontal and vertical gradients along the outer bounds of the contaminant plume. The value of the hydraulic gradients should be positive inward, toward the core of the plume.

To quantify the general degree of contamination cleanup achieved, one may compute the average value of the total contaminant mass at all monitoring locations within the plume and compare this average with previous values. There are several ways to do this, and some are better than others.

The total contaminant mass at any monitoring point within the plume may be estimated most rigorously by analysis of chemical extracts of subsurface core samples collected at the monitoring point. The basis of this estimation procedure is that chemical extraction of the core samples recovers all contaminant residuals, regardless of the form or phase of the residuals. Obviously, the sampling technique, spatial coverage of the sampling efforts, and efficiency of the chemical extractions are quality-control considerations that must be dealt with properly for this technique to generate reliable estimates. If the remediation progresses satisfactorily, the estimated values should trend downward over time. Trend reversals may occur, however, especially where site characterization efforts have been inadequate to determine the true magnitude of the contamination problem (e.g., where DNAPLs have gone undetected or where specific portions of the site were inaccessible due to the presence of buildings).

The total contaminant mass at monitoring locations has often been estimated in the past from contaminant concentrations in groundwater samples. These values are then used, together with linear isotherm relationships and partitioning ratios, to infer the concentrations of contaminants on subsurface sediments from the measured concentration values of the same contaminants in local groundwater samples. Unfortunately, this approach may seriously underestimate the masses of contaminant residuals on subsurface sediments. For example, the assumptions on which the use of linear isotherm relationships depend are easily violated in the high-concentration, multicomponent environment at the core of a contaminant plume.

Estimation of the total contaminant mass, on the basis of contaminant concentration values in groundwater, may also be in error because remediation well fields cause increased flow velocities locally and induce invasion of the plume by uncontaminated water beneath and adjacent to the plume. Both of these factors serve to depress contaminant concentrations in the groundwater, at least until pumping ceases.

The general degree of hydrodynamic control or contaminant cleanup, as represented by averaged values of hydraulic gradient or total contaminant mass, is an informative and useful measure of operational effectiveness, but it does not address the spatial variations in control or cleanup. Therefore, a second way in which one may view the operational effectiveness of a pump-and-treat remediation is to examine the *spatial uniformity of the control exerted and/or the cleanup level achieved.* In this context, hydrodynamic control should be evident along the periphery of the plume at each location where gradients are estimated from water-level elevation measurements. Likewise, the total contaminant mass at each location of the within-plume monitoring points should not vary significantly from the average. Otherwise, the remediation may be deemed ineffective at such locations, even though by balance considerations, it will be especially effective at others. Variability is the rule rather than the exception in the data obtained from subsurface measurement networks. Both operators and regulators have to be prepared to make decisions despite the presence of considerable uncertainty in data interpretation.

Finally, operational effectiveness may be viewed in terms of the *persistence of the desired effects of the remediation*. Hydrodynamic control of plume movement should be maintained continuously, regardless of seasonal variations in the recharge and flow rate of the regional system. Concentration profiles should remain stable. However, they can be affected by long-term drainage of contaminated residuals from the unsaturated zone to the groundwater or by rising and falling water tables throughout the year. Concentration profiles can also be affected by increased infiltration due to aging of an on-site cap or due to settling and slumping of materials beneath the cap. Any of the foregoing may cause the desired effects of a pump-and-treat remediation to diminish over time.

5. Measures of Operational Efficiency

Operational effectiveness, as described in the previous subsection, may be a focus of both regulatory and operational concerns. Operational efficiency, on the other hand, is an economic concept that is of primary interest to the site operator. We will look at three measures of operational efficiency: minimizing total costs, maximizing contaminant removal per unit volume of groundwater pumped, and minimizing remediation time.

The *minimization of total costs required to reach and maintain remediation targets* is the most economically oriented view of operational efficiency. Total costs can be tracked on a quarterly or annual basis to provide a relatively simple picture of how the system operating cost is evolving. The total costs per year may decline continuously as the remediation progresses, but they may also be erratic because of the specific pattern of removal of the contaminants (e.g., hot spots in the plume may not respond initially). Unfortunately, the tracking of total costs also can be a misleading measure of operational efficiency unless costs are normalized for the effects of inflation, interest rates, depreciation of recoverable assets, and overall discounting.

Another means of evaluating the operational efficiency of a pump-and-treat remediation is to strive for *maximization of contaminant removal per unit volume of groundwater pumped and treated.* This approach focuses on the technical efficiency, from which the bottom-line economics may be inferred. The conventional assumption is that costs are inversely proportional to the contaminant levels of the treated water because it normally costs much less on a per unit-mass basis to remove contaminants from a concentrated waste stream than from a dilute waste stream. Hence, to evaluate the operational efficiency by this approach, the contaminant concentrations in the pumped waters must be tracked. Active management of the remediation may result in a trend of increasing concentrations in the pumped water, which reaches and levels off at a ceiling that represents the maximum concentrations that can be withdrawn and which eventually falls as the contaminant residuals are depleted from the subsurface. Hot spots and slowly mobilized contaminants such as DNAPLs may cause variations about this pattern, but these can be minimized by active management of the remediation.

Another possible approach for achieving operational efficiency would be to minimize the volume of contaminated water that will be produced over the lifetime of the remediation. Generally, treatment costs are proportional to some function of the discharge of the waste stream for a given contaminant concentration level. There are, therefore, economic attractions to minimizing the total volume pumped and treated. Minimization of pumping is also of interest to regulators where groundwater is in short supply. The *total volume of pumped and treated groundwater* is, therefore, an appropriate measure of operational efficiency.

Lastly, the time needed for completion of the remediation may be minimized. This is of interest to both regulators and the financial backers of remediations because they share sensitivity to public perceptions of progress at the site. The *time for completion of major phases of the remediation* is, therefore, an appropriate measure of operational efficiency.

The benefits of minimizing the time required to complete a remediation include (1) a reduction in remediation-oversight costs borne by the regulatory agencies, (2) a reduction in project management and labor costs and an increase in the value of salvageable assets, and (3) an improvement in the site owner's credibility with the public.

6. Interpretation of Chemical Time Series From Monitoring Points

The measures of operational effectiveness and operational efficiency of capture and containment remedial actions introduced in the previous subsections require evaluation and interpretation of the time series of contaminant concentration values obtained from monitoring wells. Because of the complexities of aquifer stratigraphy, natural flow patterns, plume geometry, and remedial drawdown cones, these time series may not be simple.

Figure 42 illustrates some of the contaminant arrival patterns that may be observed in chemical time series data collected during periods of steady pumping. No attempt will be made here to detail the litany of potential interpretations for each pattern illustrated, but some of the more likely possibilities are described.

Patterns (a), (b), (c), and (d) portray cases where there is an initial decrease in concentrations with pumpage. Pattern (a) might result from placing the well so that it penetrates a contaminant plume of very limited extent; contaminant levels decrease due to an increase in the availability of uncontaminated aquifer water as the distance from which the water is drawn into the well becomes greater and greater. It might also be that the well penetrates a substantial plume but is located close to one edge of the plume. Or it might be that a slow but steadily increasing contribution of additional uncontaminated water, such as from upconing or from leaky confined beds, is gradually reducing the ratio of contaminated-to-uncontaminated aquifer water drawn in to the well. Each hypothesis

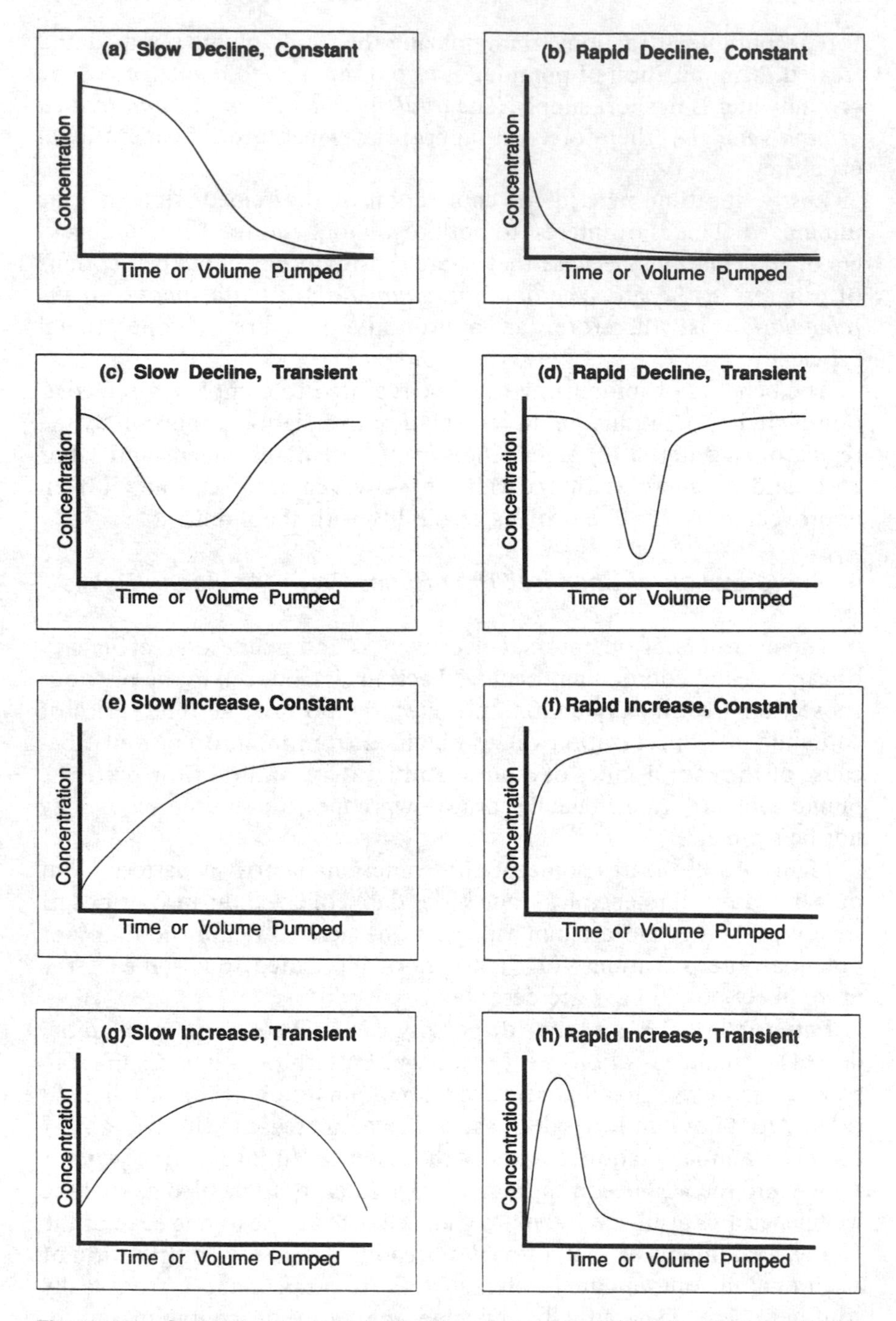

Figure 42. Contaminant-concentration time series in monitoring wells (after Keely, 1982).

could be strengthened or challenged by supplemental data (e.g., if pumping tests show the existence of leaky artesian conditions, the last hypothesis may be most supportable; if nearby wells are unable to penetrate the same plume, the first hypothesis may be correct; and so on).

A rapid decline of contaminant concentrations after a very brief period of pumping, as shown in pattern (b), should alert investigators to the possibility of inadvertent contamination of the well. This might occur from a poor surface seal, use of inappropriate lubricating agents, or leaching of contaminants from well-construction materials. Again, supplemental data (here, regarding well construction) would be of value in reducing the number of hypotheses that must be seriously considered.

Patterns (c) and (d) might suggest the possibility of intercepting flow from a temporary recharge source at some finite distance from a well that penetrates a contaminant plume. Or alternately, they might reflect intermittent loading of contaminants, as might occur in fractured crystalline rock aquifers, karstic limestone aquifers, or unconsolidated aquifers laced with large macropores. A very brief transient decline is likely to be dismissed as a relatively unimportant anomaly. A more gradual, transient decline is less likely to be ignored and a reasonable effort should be make to determine the cause of the anomaly.

Increasing trends of contaminant concentrations with continued pumpage, as illustrated by patterns (e) through (h), are also possible. A well that is laterally or vertically offset from a plume might exhibit pattern (e). The well would produce increasing concentrations of contaminants as it caused the plume to migrate toward it. Samples from nearby wells would be useful in verifying this hypothesis, both by levels of contaminants encountered and, in the case of nearby pumping wells, by the ability to influence contaminant concentrations when creating hydraulic divides through well interference.

Pattern (f), though similar in most respects to pattern (e), would be more apt to result from physical, chemical, or biological degradation of contaminants in the well itself. The decline in the concentration of iron in stagnant casing water has been documented. Analogous effects could influence organic contaminant concentrations. The contaminant concentrations would rise rapidly as pumping continues because the stagnant, affected waters in casing storage would be quickly voided and replaced by waters from the contaminated surrounding formation.

The implications of patterns (g) and (h) are that an isolated plume is being sampled. The isolation may be the result of separate sources or events, physical occlusion by clay strata, or intermittent release of the contaminants into the subsurface. Alternatively, the patterns could be caused by variations in the rate of clean water reaching the contaminant plume. Pattern (h) could also be the result of a combination of casing storage effects and poor well construction (leaky surface seal) or poor isolation of the aquifer of interest (communication of contaminated waters along the outside of the casing due to poor grouting). Supplemental data are essential to the interpretation of these patterns.

The effect of hydraulic isolation of a plume was noted in an investigation of two public water-supply wells that had become contaminated with tetrachloroethene, trichloroethene, and other volatile organics. During chemical time-series sampling of Well B, contaminants were routinely detected until Well A began pumping, beyond which point the contaminant levels in Well B rapidly declined to near nondetectable levels. Figure 43 illustrates this mechanism of contaminant isolation. Note that contaminant-arrival effects are not limited to the pumping wells, since local displacement of waters caused by pumping will be observed in nearby monitoring wells.

Regardless of whether chemical time-series sampling results in a pattern of chemical arrival that shows a definite trend or is completely random, the data collected are useful to the investigator in identifying the uncertainties in contaminant concentrations in the plume he or she is attempting to define. Such data can be used in deciding when a model has been adequately calibrated and can ultimately help reduce the amount of parameter uncertainty.

7. Strategies for Termination of Remedial Systems

Measures of the operational effectiveness and efficiency of a pump-and-treat remediation can be used to monitor the progress of the remediation, but some end points must be selected to bring it to termination. Such end points may involve absolute or relative measures of success and they may involve statistical considerations.

Absolute measures of success leading to the termination of the operational phase of a pump-and-treat remediation may be as simple as specifying the removal of a specific mass of contaminants. Alternatively, absolute measures may involve achieving specific contaminant concentration levels such as background quality or nondetectability. For practical reasons, the former is much less likely to be accepted by regulators than the latter. This is because it is difficult to guarantee the validity of initial estimates of contaminant residuals, so that preoperational decisions to remove only a specific mass or volume run the risk of terminating the remediation without adequate reduction of the potential risks. The latter form of absolute measure is much less likely to be acceptable

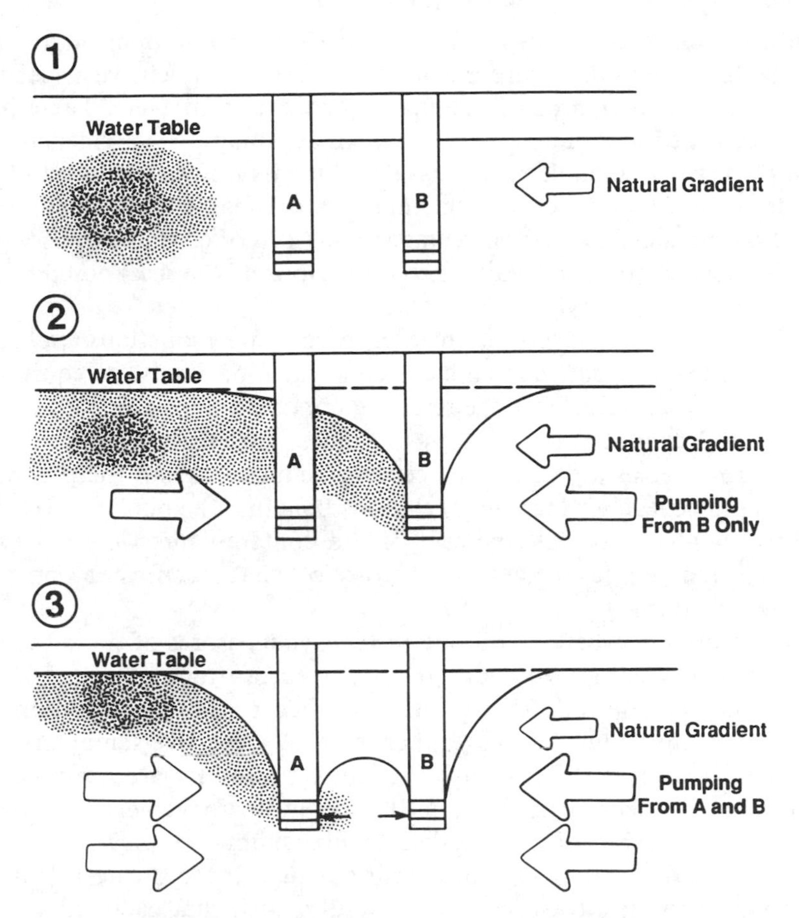

Figure 43. Pumping-well contamination by a downgradient source and plume isolation by hydraulic interference (after Keely, 1982).

to those responsible for the operation of the remediation because it may be difficult to achieve the specified concentration limit and virtually impossible to guarantee it *a priori*. A similar concern can be voiced on the part of the regulators: it is just as difficult to foresee what technological innovations may come along during the years of operation of the remediation, and they would not want to see termination pose a premature cutoff to what might otherwise be a more thorough remediation. Fortunately, absolute measures need not be chosen as stand-alone endpoints; contingencies for demonstrated limitations or technological innovations can be integrated into the definition of the absolute measures.

Relative measures of the success of the operational phase of a pump-

and-treat remediation are similar to some permit requirements for discharging waste to the environment: the process must achieve a specified percentage reduction of contaminant concentration levels. These have not been used as stand-alone criteria for termination of remediations, but they are often implemented as requirements for the treatment plant. There is merit, however, in applying relative measures to the contaminant plume, since specific percentage reductions of contaminant concentration levels can be translated into reductions of the risks posed by the plume.

For all practical purposes, however, percentage reductions applied to known preoperational contaminant concentration levels are equivalent to absolute measures since the resulting concentrations are readily computed and cannot vary. Perhaps the major incentive to phrasing the cleanup target as a specified percentage reduction from the preoperational plume levels is the public relations benefits: the public can readily grasp the idea that 99.99 percent of the contaminants were removed, whereas few people are likely to be aware of what specific concentration levels may mean.

Regardless of whether absolute or relative measures of the success of a pump-and-treat remediation are employed, a single round of water-level measurements or samples will not suffice to justify termination of a pump-and-treat remediation. Rather, several successive samplings on a monthly or quarterly basis are needed. The results are examined to determine whether or not the remediation has truly achieved the target objectives. Examination of the data sets inevitably reveals that the data values vary from one sampling/measurement event to the next. While it is possible to view these variations rigidly, such that each value from each sampling/measurement event must be at or below the target concentration level, this is an unduly restrictive approach which is not usually employed by regulatory personnel. More often a statistical approach is followed.

The evaluation of pretermination data sets with *statistical measures* entails plotting the contaminant concentration time series and fitting a nonlinear regression line to the plotted values (Figure 44). Here the form of the regression line is a matter of judgement. Two kinds of outcomes of the regression-line fitting may be acceptable to regulators as an initial indication that the remediation is ready for termination. The first outcome is that the regression line has a negative slope (successive values are generally trending downward) and it extends below the target concentration level.

The second outcome is more stringent. Some regulators require that the regression line extend below the target concentration levels until a

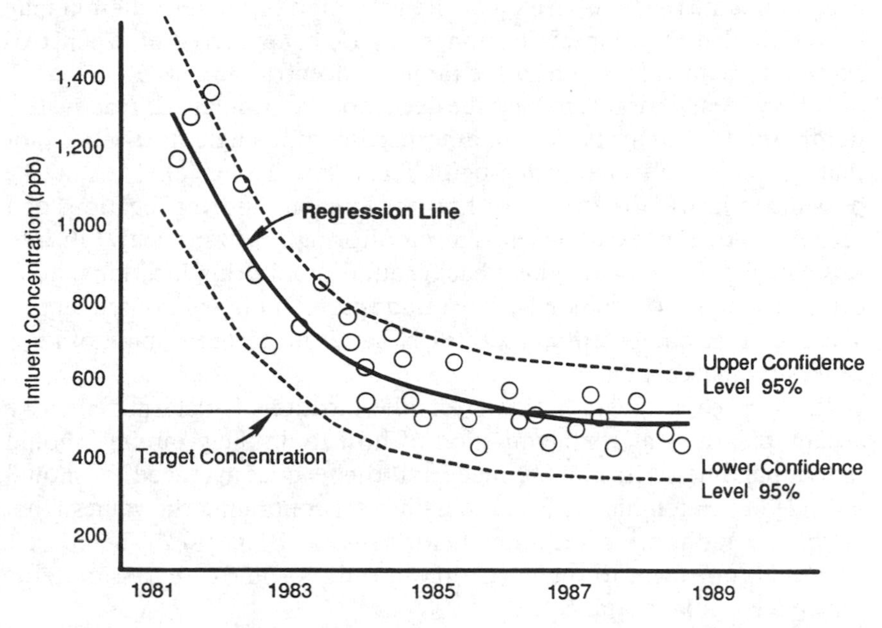

Figure 44. Example of concentration versus time data plot with fitted regression line.

zero-slope regression plot is achieved (signifying stabilization of the remediation at its best possible level). In such cases, the final result is that the remediation produces a better cleanup than targeted.

The slope of a fitted regression line may be an insufficient statistical measure by itself, since data values inevitably vary about this line. It may be necessary to perform statistical tests to ascertain that the slope of the fitted regression line is not greater than zero at some level of confidence. Quality-control considerations also require that a certain percentage of the data values fall within a specified confidence interval about the regression line. Figure 44 shows the 95-percent confidence interval that lies between the upper confidence level and the lower confidence level.

Following termination of the remediation, the sampling/ measurement events must be continued to ensure persistence of the effects of the remediation. A declining frequency of sampling/ measurement events is appropriate. Monthly sampling may be needed for the first 3 or 4 months immediately following termination, followed by quarterly sampling and then annual sampling. The sampling program then terminates at the end of the postoperational regulatory period. The data values obtained during these sampling/measurement events are used to update the time-series plot. Reactivation of the remediation may be required if the slope is no longer at or below zero, or when two successive data values exceed the target concentration.

It may seem desirable to base the decisions to terminate or reactivate a pump-and-treat remediation on examination of a single time-series plot that represents all monitoring-point locations. Technically, it may be possible to justify the grouping of results from monitoring locations of a certain kind. For example, on-site monitoring locations may fall into this category. At downgradient background-monitoring locations, however, grouping of results may have undesirable political consequences. The idea that one potential receptor is better protected than another is generally unacceptable.

In those cases where grouping of like locations is determined to be acceptable, there arises the question of how to do the grouping: should individual time-series plots be made and their slopes averaged, or should a single regression line be fitted to a data set containing the values from all of the monitoring locations belonging to the group? The former option is equivalent to the latter option only for linear regression plots that have similar variances.

If the former option is chosen, the confidence interval is computed for a data set that is composed of the slopes of the regression plots of data sets from the different monitoring locations. This computed value will be largely independent and insensitive to actual variations of the data in each of the data sets. If the latter option is chosen, the confidence interval is computed for a single regression line that is fitted to a super data set that contains the raw data sets of all monitoring locations. This confidence interval will be dependent on the actual variations of the raw data values in every one of the contributing data sets. It will also be larger than the confidence interval computed by the former option because it is based on all of the values in the raw data sets, not just their regression slopes. As a result, regulators may prefer the former option (smaller confidence interval), and remediation operators may prefer the latter option (larger confidence interval). Neither is more or less correct in terms of statistical considerations; the choice of which to use must be made on philosophical grounds.

The foregoing discussion presumes that a decision has been made regarding whether to prepare time-series plots of contaminant concentration levels versus time for each contaminant, groups of contaminants, or the total contaminant load. In addressing this point, the methods for computing confidence intervals may be considered in a manner that is parallel to the foregoing discussion. One may either compute the confidence interval of measurements of each contaminant at a given monitoring-point location and then average these results, or one may combine several contaminants' data values into a single data set and then compute its confidence interval. Note that these computations can be based on some *normalized* data (for example, the results of dividing the raw data values of each contaminant by its average value). Otherwise, the results will be biased toward the contaminant data sets with the highest concentration levels. Again, regulators may prefer the former option (a smaller 95-percent confidence interval), and operators may prefer the latter option (a larger 95-percent confidence interval). Philosophically, the grouping of the data sets from several contaminants at a given monitoring-point location is desirable only if justified by similarity of the contaminant transport behaviors and the health risks posed by the contaminants to be grouped.

B. Protocols for Performance Evaluations

The following paragraphs summarize the issues addressed in this chapter from the perspective of how they impact site protocols.

It is best to begin the selection of performance-monitoring criteria by reviewing the data-collection history at the site. This will determine if it is necessary to adopt specific spatial or temporal patterns of sampling to ensure continuity with previous efforts. This step may be as simple as reviewing past water-quality data and organizing a sampling episode to proceed from uncontaminated wells to contaminated wells, thereby minimizing the potential for cross-contaminating samples. One can expect that a year or more may elapse between the completion of the RI/FS and the effective start-up of the remediation and that preoperational programs of sample collection and analysis, quality control, and data interpretation will be required during this period. These will be needed to provide current information for use in the final stages of the remedial design. It will be necessary to select monitoring criteria appropriate to these data needs.

One can also expect that bench-treatability studies and pilot-plant demonstrations may be called for by design engineers prior to starting the construction of the remedial system. It is wise to conduct column studies at flow rates equivalent to those expected for different locations within the proposed remediation well field. In addition, overlay plots of all sampling points and sample types should be prepared to identify locations of inadequate or excessive sampling. It is best to prepare brief reports of results of sample collections and testing as soon as each result is ready, rather than wait for all items to be synthesized into a major report. In this way, the information will be most useful in the final phases of the remedial design and the early phases of remediation start-up.

It may be necessary to modify the frequency and locations for sampling during remedial operations. Specifically, one should focus on characterizing the flowlines that will result from operating the remediation well field and on tracking the transport of contaminants along the flowlines. It is necessary to anticipate possible flowline shifts due to partial shutdowns or operational failures. Adequate sampling coverage should also be provided beyond the horizontal and vertical bounds of the plume, such that flowlines that originate in the plume, but which then enter uncontaminated regions, can be tracked to ensure eventual capture by an extraction well. This will also allow a more accurate calculation of the ratio of uncontaminated groundwater volume to the total volume pumped.

Provisions should be made for measurements germane to active management of the operation. These measurements include flow rates and contaminant concentration levels at individual extraction wells, the treatment-plant inlet, key treatment units (e.g., air stripper, carbon filtration unit), and the treatment-plant discharge point. The frequency of chemical sampling should be reduced if stable trends of contaminant concentration levels are apparent. It is wise to provide contingencies for increased sampling in response to those periods when the data are highly erratic. The program of filling in gaps in the site characterization data base should be continued with opportunistic and periodic testing.

The program of evaluating depletion of the residual contamination by collection of core samples for chemical extraction and analysis should also be continued. This can be done by tracking the estimated mass of contaminants removed from the aquifer and comparing this estimate with the mass of contaminants removed by the treatment system. One should also plot and examine water-level elevation contours to determine if the plume is adequately contained and to identify stagnation zones for subsequent action.

As the concentration-time plots begin to flatten out, one should try varying the flow rates of a few extraction wells at a time to determine the potential for improvement by means of well-field scheduling changes. When all potential improvements have been attempted and exhausted and contaminant concentrations have reached or nearly reached target concentrations, preparations should be made for terminating the remediation. This is accomplished by increasing the sampling frequency to build a statistically significant data base for regulatory consideration.

Throughout the operational phase of the remediation, monitoring data should be utilized for management purpose as soon as they become available, and comprehensive annual reports should be prepared for regulatory oversight purposes.

The data collected during each sampling/measurement event must be examined for internal consistency prior to accepting the accuracy or relevance of anomalous values. If a single monitoring point indicates an unacceptable level of a particular contaminant, the concentration levels of other contaminants at the same location should be checked to see if they have approximately the same relationship (ratio) to the suspect contaminant as in the past; if not, it may be possible to reject the suspect value as internally inconsistent. If the concentration appears to be internally consistent, or if internal consistency cannot be judged due to the small number of contaminants associated with a plume (for example, predominantly TCE and nothing else), then the suspect value should be compared with the concentration levels found in nearby wells and viewed as a possible parent compound or daughter product. If poor agreement is indicated, it may be possible to reject the suspect value as locally inconsistent. If the suspect value is not rejected due to internal inconsistency or due to local inconsistency, the value should be retained. It will then be necessary to resample/remeasure at that location and at two or more adjacent locations (for control). The resulting data should then be evaluated in the same fashion as the initial data in order to confirm or disprove the conclusions drawn from the earlier analysis.

Regardless of the status (rejected or retained) of the first such indication of unacceptable contaminant levels, one must examine any successive indications to evaluate the potential loss of adequate control of the plume. If a problem does, in fact, exist, the well-field scheduling should be studied to determine how pumping schedules and rates may be altered to improve system performance. One can expect that the flowline pattern will shift with the implementation of a new schedule, so it is necessary to evaluate the significance of the shift in terms of the need for repositioning the sampling/measurement points. Subsequently, one must verify the operational viability of the modified well field by increasing the sampling/measurement frequency for a limited period. One can also utilize information obtained during the increased sampling/measurement period to assess the impacts of the correctional action in terms of changes in the rate of depletion of the contaminant residuals.

Once the time-series plots indicate that the target contaminant concentration levels have been reached, termination of the remediation can be proposed. If termination is granted by the regulators, the sampling program should be continued at the same frequency for up to 2 years to test the need to reactivate the remediation. It will probably be necessary to prepare brief quarterly reports for management and regulatory purposes during this period. Subsequently, it may be possible to reduce sampling frequencies to an annual basis. This final sampling program is then continued for some time to monitor the persistence of the effects of the remediation. The total duration of postremediation (also referred to as postclosure) monitoring will vary from site to site.

References for Chapter VII

Andersen, P. F., C. R. Faust, and J. W. Mercer, "Analysis of Conceptual Design for Remedial Measures at Lipari Landfill," *Groundwater*, Vol. 22, No. 2, 1984.

Bear, J., and A. Verruijt, *Modeling Groundwater Flow and Pollution*, D. Reidel Publishing Company, Boston, MA, 1987.

Cheng, Songlin, "Computer Notes—Trilinear Diagram Revisited: Application, Limitation, and an Electronic Spreadsheet Program," *Ground Water*, Vol. 26, No. 4, pp. 505–510, 1988.

Davis, J. C., *Statistics and Data Analysis in Geology,* 2nd Ed., John Wiley & Sons Inc., New York, NY, pp. 227–250, 1986.

de Marsily, G., *Quantitative Hydrogeology: Groundwater Hydrology for Engineers*, Academic Press Inc., Orlando, FL, 1986.

El-Kadi, A., "Applying the USGS Mass Transport Model (MOC) to Remedial Actions by Recovery Wells," *Ground Water*, Vol. 26, No. 3, 1988.

Freeze, R. A., and J. A. Cherry, *Groundwater*, Prentice Hall, Inc., Edgewood Cliffs, NJ, 1979.

Gilbert, R.O., *Statistical Methods for Environmental Pollution Monitoring,* Van Nostrand Reinhold Company, New York, 1987.

Huntsberger, D. V., and P. Billingsley, *Elements of Statistical Inference*, 4th Ed., Allyn and Bacon, Inc., Boston, MA, 1977.

Huyakorn, P. S., *Testing and Validation of Models for Simulating Solute Transport in Ground Water: Development and Testing of Benchmark Techniques,* IGWMC Report No. GWMI 84–13, International Ground Water Modeling Center, Holcolm Research Institute, Indianapolis, IN, 1984.

Javandel, I., C. Doughty, and C. F. Tsang, *Groundwater Transport: Handbook of Mathematical Models*, AGU Water Resources Monograph No. 10, American Geophysical Union, Washington, D.C., 1984.

Konikow, L. F., "Predictive Accuracy of a Ground-Water Model—

Lessons from a Postaudit," *Ground Water*, Vol. 24, No. 2, pp. 173–184, 1986.

Krabbenhoft, D. P., and M. P. Anderson, "Use of a Numerical Ground-Water Flow Model for Hypothesis Testing," *Ground Water*, Vol. 24, No. 1, 1986.

Keely, J. F., "Chemical Time-Series Sampling," *Ground Water Monitoring Review*, Vol. 2, No. 4, pp. 29–38, 1982.

Mercer, J. W., and C. R. Faust, *Ground-Water Modeling*, National Water Well Association, Dublin, OH, 1981.

Rogers, R. J., "Geochemical Comparison of Ground Water in Areas of New England, New York, and Pennsylvania," *Ground Water*, Vol. 27, No. 5, pp. 690–712, 1989.

SAS Institute Inc., *SAS® Introductory Guide for Small Computer, Release 6.03 Edition*, SAS Institute Inc., Cary, NC, 1988a.

______, *SAS/GRAPH® User's Guide, Release 6.03 Edition*, SAS Institute Inc., Cary, NC, 1988b.

______, *SAS/STAT® User's Guide, Release 6.03 Edition*, SAS Institute Inc., Cary, NC, 1988c.

SAS Institute Inc., *SAS/ETS® User's Guide, Version 6, First Edition*, SAS Institute Inc., Cary, NC, 1988d.

______, *SAS/OR® User's Guide, Version 6, First Edition*, SAS Institute Inc., Cary, NC, 1989a.

______, SAS/QC® *User's Guide, Version 6, First Edition*, SAS Institute Inc., Cary, NC, 1989b.

Steinhorst, R. K., and R. E. Williams, "Discrimination of Groundwater Sources Using Cluster Analysis, MANOVA, Canonical Analysis, and Discriminant Analysis," *Water Resources Research*, Vol. 21, No. 8, pp. 1149–1156, 1985.

Taylor, J. K., *Quality Assurance of Chemical Measurements*, Lewis Publisher Inc., Chelsea, MI, 1987.

Tufte, E. R., *The Visual Display of Quantitative Information*, Graphics Press; Cheshire, CT, 1983.

U.S. Environmental Protection Agency, *Statistical Analysis of Ground-Water Monitoring Data at RCRA Facilities (Interim Final Guidance)*, Office of Solid Waste, Waste Management Division, Washington, D.C., 1989.

Usunoff, E. J., and A. Guzman-Guzman, "Multivariate Analysis in Hydrochemistry: An Example of the Use of Factor and Correspondence Analysis," *Ground Water*, Vol. 27, No. 1, pp. 27–34, 1989.

van der Heijde, P. K. M., Y. Bachmat, J. D. Bredehoeft, B. Andrews, D. Holtz, and S. Sebastian, *Groundwater Management: The Use of*

Numerical Models, 2nd Edition, AGU Water Resources Monograph No. 5, American Geophysical Union, Washington, D.C., 1985.

van der Heijde, P. K. M., and P. Srinivasan, *Aspects of the Use of Graphic Techniques in Ground-Water Modeling*, IGWMC Report No. GWMI 83-11, International Ground Water Modeling Center, Holcolm Research Institute, Indianapolis, IN, 1983.

Walton, W. C., *Practical Aspects of Ground Water Modeling*, 2nd Ed., National Water Well Association; Dublin, OH, 1985.

Wang, H., and M. P. Anderson, *Introduction to Groundwater Modeling: Finite Difference and Finite Element Methods*, W. H. Freeman and Company, San Francisco, CA, 1982.

Watson, Ian., "Contamination Analysis—Flow Nets and the Mass Transport Equation," *Ground Water*, Vol. 22, No. 1, pp. 31-37, 1984.

CHAPTER VIII

Examples of Remedial Well Field Design and Optimization

A. Introduction

1. Purpose

The previous chapters presented thorough discussions of the techniques that can be used to design, monitor, and manage a high-efficiency pumping system for aquifer remediation. The tool identified as having the most potential to aid a hydrologist in designing effective remediation systems is the aquifer management modelling capability of AQMAN, coupled with an optimization package such as MINOS. When applied to a site for which a calibrated groundwater flow model has been developed, AQMAN allows a hydrologist to quickly examine a variety of remediation systems and select a design which achieves the required hydrologic objectives with minimal pumping rates. Using AQMAN and MINOS together with an integer programming package such as MINT (Chapter V.F.1), one can restrict the number of wells that are included in a particular remedial system. Concerns regarding restricted well numbers or well locations can also be addressed through trial-and-error AQMAN and MINOS runs.

In this chapter, the optimization approach described in Chapter V will be applied to a problem that is loosely based on actual field data. Limited modifications of the actual data have been made for instructive purposes. The purpose of this chapter is to provide a simple example that illustrates in detail the steps involved in the application of AQMAN and MINOS to the design of a remedial well field. The remedial strategies examined in this chapter are patterned after those presented in Chapter V and include the following specific alternatives:

- Simple hydraulic gradient control,
- General hydraulic gradient control (plume boundary and interior pumping),
- Hydraulic gradient control with injection,
- Flow reversal with injection, and
- Screening plus removal with velocity constraints.

Emphasis is placed on describing the steps involved in setting up, executing, and comparing optimization results produced by AQMAN and MINOS runs. For simplicity, the scope of the design simulations is limited to linear optimization; quadratic optimization is not used.

2. Site and Model Description

The site chosen as the basis for this demonstration chapter is an industrial-airfield complex that overlies a 5-to 30-foot-thick unconfined aquifer. The aquifer has been contaminated by landfill leachate, waste liquids, and leakage from industrial process pipes. This latter component is suspected to be the source of a TCE plume that has formed in the uppermost unconfined aquifer. The plume is shown, together with the modelled hydraulic head map, in Figure 45 and is approximately 2,000 feet long and 700 feet wide.

The unconfined aquifer comprises alluvial clay and sandy gravel. It is bounded on the northwest by a lake and on the north-northeast by an area of high bedrock where the aquifer pinches out. On the east, the aquifer is bounded by an intersecting river. Hydrogeologic information on the southern and western extent of the aquifer is limited.

As required by AQMAN, the USGS two-dimensional flow code (Trescott et al., 1976) was used to model groundwater flow at the site. The model was based on a 77-column by 64-row grid with prescribed head and no-flow boundaries (Figure 46). A no-flow boundary was assigned to the north and northeast portion of the grid where the aquifer pinches out. Prescribed head boundaries were assigned on the northwest where the aquifer intersects the lake and on the east where the aquifer intersects the river (see Figure 45). Although the presence of a groundwater divide suggests a no-flow boundary on the west (Figure 45), a prescribed head boundary based on observed water levels was assigned west of the groundwater divide. This was done to maximize the distance between the west boundary and the area of interest surrounding the plume and to enable the divide to shift positions under conditions of pumping. For lack of a better alternative, a prescribed head boundary based on observed water levels was also chosen for the southern boundary.

In the vicinity of the plume, a 50-foot grid spacing was used to allow accurate simulation of local hydraulic gradients. Grid spacing was increased to 400 feet over the majority of the model domain. In the extreme eastern portion of the grid, spacing was increased to 2,000 feet. Incremental increases in column and row width do not exceed 50 percent of the width of the adjacent column or row.

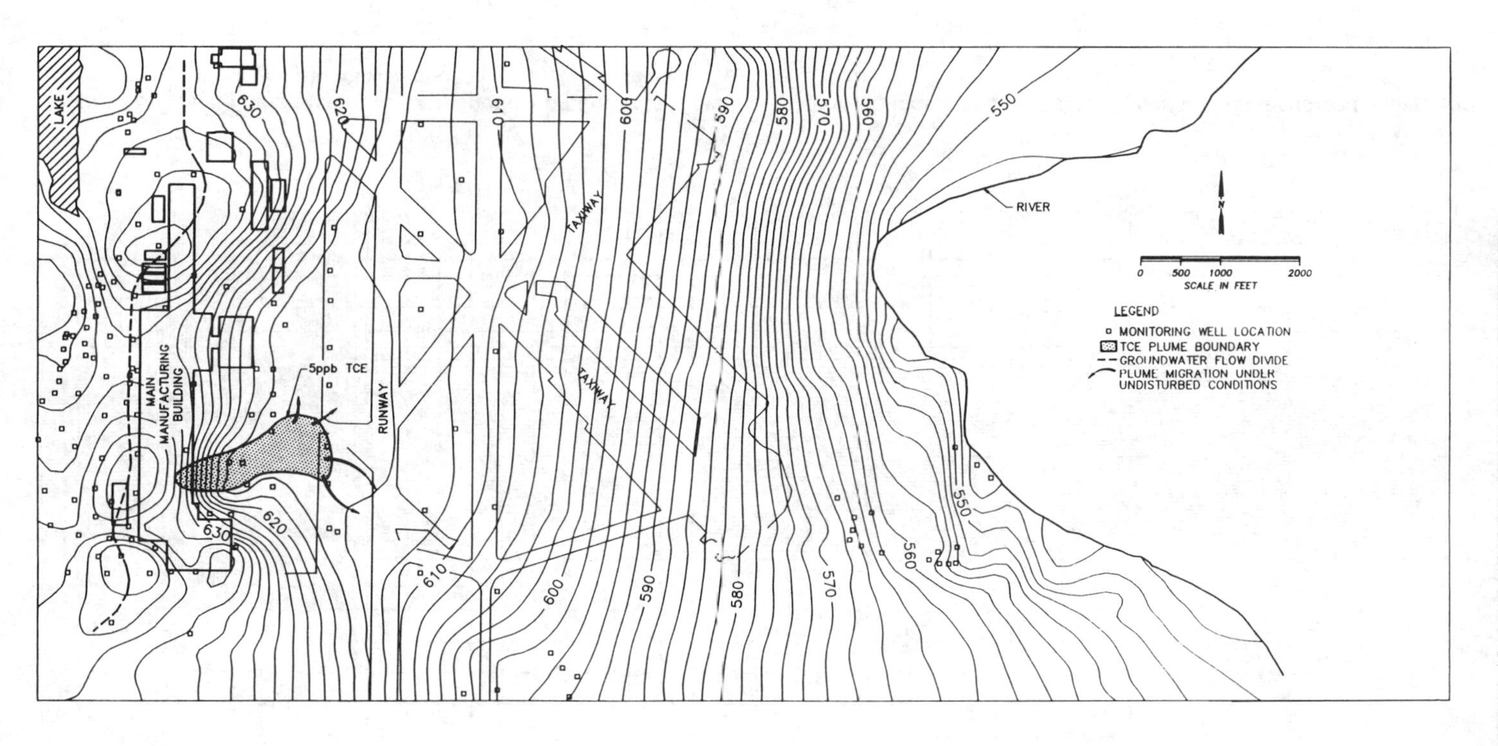

Figure 45. Map of field site showing simulated water-table elevation contours, TCE plume, buildings, runway, and taxiway.

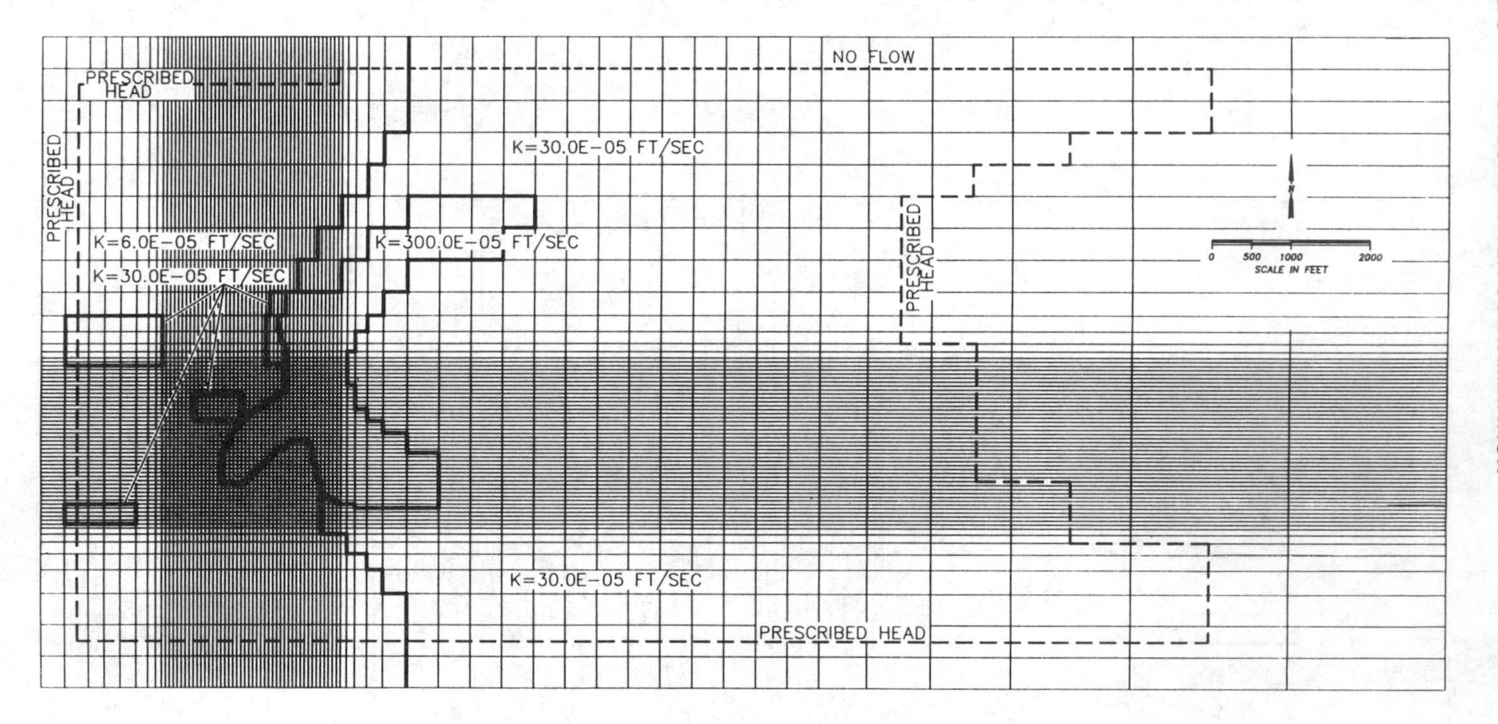

Figure 46. Finite difference grid for numerical groundwater flow model showing locations and types of model boundaries and initial hydraulic conductivity values.

Input parameters for the numerical model included hydraulic conductivity, aquifer bottom elevations, and recharge rates. A specific yield was not needed because a steady-state simulation was used. Based on slug tests, aquifer tests, and lithologic logs from monitoring-well and soil borings, three different values of K were assigned to specific sections of the model grid (Figure 46). These values served as the basis for further model calibration. Aquifer bottom elevations were assigned by superimposing the model grid on a bedrock contour map generated from lithologic logs (Figure 47). Lastly, a uniform recharge rate of 5×10^{-9} inches per second (in/s) was initially assigned to most of the area on the basis of recorded precipitation rates and empirical estimates of runoff and evapotranspiration. In the vicinity of the main industrial complex in the western portion of the model domain, recharge rates were increased to a maximum of 4.5×10^{-7} in/s to reflect leakage from the plant determined by an approximate plant water-balance calculation. This water balance showed that leakage from the facility was in the range of tens to hundreds of thousands of gallons per day. These recharge rates in the vicinity of the main building were also determined during calibration.

Figure 45 shows the contour map for the simulated water-table elevations produced by the calibrated model of groundwater flow under natural conditions (no pumping). The approximate flow direction and migration path for the plume is shown near its downgradient boundary. At this point, the first 2 of the 10 remediation design steps presented in Chapter V.A.2 were completed. It should be noted that the calibration phase showed the simulated hydraulic heads to be quite sensitive to uncertainties in hydraulic conductivity and recharge values. However, for the illustrative purposes of this chapter, the calibration shown in Figure 45 was considered acceptable. In practice, one would need to determine if additional data would be needed to further improve model accuracy.

3. Development of Management Objectives

The third item identified in the 10 steps from Chapter V.A.2 is the development of management objectives and constraints. In practice, this step is driven by a variety of factors associated with the contamination and remediation problem. For the purposes of this chapter, each combination of an objective function and a set of constraints will be regarded as a distinct pumping strategy. The following menu of objectives and constraints will form the basis for formulating alternative pumping strategies.

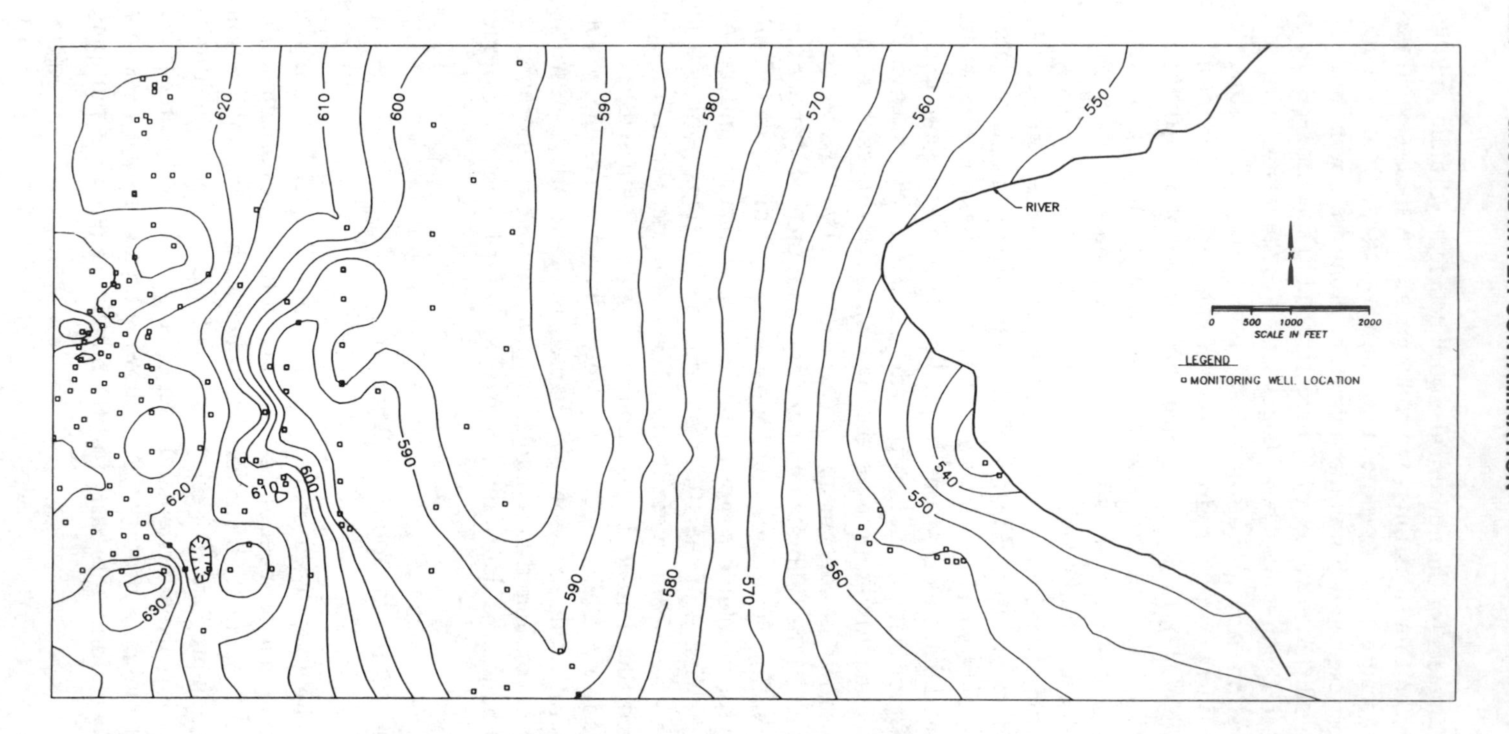

Figure 47. Contour map of bedrock elevations used to assign nodal values for aquifer bottom elevation.

Objective Function

- Minimize the sum of pumping (withdrawal) rates;

Constraints

- Contain the contaminant plume (no spreading);
- Restrict the total volume pumped to allow use of a low-cost, low-capacity treatment system;
- Restrict individual pumping rates to allow the use of small, inexpensive pumps;
- Restrict the number of active wells to a "reasonable" limit;
- Restrict drawdown in pumping wells to be less than a prescribed percentage of the saturated thickness (e.g., 30 percent);
- Restrict the total injection rate to be less than or equal to the total pumping rate;
- Reverse flow in low-level contamination areas towards contamination "hot spot";
- Screen regional flow to reduce volume of clean water in the treatment stream; and
- Constrain inward velocity of plume boundary to exceed some minimum velocity.

In formulating the following simulation/optimization problems, various combinations of the previous constraints will be selected to produce several distinct pumping strategies. An optimal solution will then be sought for each strategy. In this context, the term "optimal" is used with respect to a particular set of design considerations and constraints. Each successful pumping strategy will then be characterized by an optimal solution.

4. Remediation Modelling Considerations Using the Trescott Code

Several important modelling considerations are worth noting when using the Trescott code in conjunction with AQMAN. The most important consideration concerns the AQMAN Unit 15 input file that is derived from the Trescott code input file. Because AQMAN is written to solve only the 2–D confined flow equation (Equation 66), unconfined problems must be converted to confined flow approximations before application of AQMAN. Table 24 shows the structure of the Trescott code input file for the unconfined flow system under consideration. Table 25 shows the AQMAN Unit 15 input file that is based on the Trescott input file with modifications to represent a confined flow approximation of the unconfined flow system. In Table 25, a confined

Table 24. Structure of the Trescott Code Input File for Simulation of the Natural Unconfined Flow System. Note that the Input File is Constructed to Represent an Unconfined Aquifer System.

```
************UNCONFINED FLOW MODEL INPUT FOR TRESCOTT CODE ************
          Blank row or continuation of title
WATE                 RECH SIP  CHEC <-------+         HEAD Print head matrix
    64DIML      77DMW         0NW      1000ITMAX
   Blank row                                |
     1NPER      10KTH    .0001ERR     .01EROR                      LENGTH10
   0.44HMAX      1FACTX        1FACTY        This line specifies unconfined
   Three blank lines                         aquifer simulation with
        :                                    recharge, SIP solution, and
   Starting head matrix, STRT(I,J)           mass balance calculation.
        :
   Storage coefficient matrix, S(I,J), (-1's and 0's to define prescribed
   head boundaries)
        :
   Hydraulic conductivity matrix, PERM(I,J)
        :
   Aquifer bottom elevation matrix, BOTTOM (I,J)
        :
   Recharge matrix, QRE(I,J)
        :
   Grid spacing values for X direction, DELX(J)
        :
   Grid spacing values for Y direction, DELY(I)
        :
   Group IV data (not required in absence of pumping wells)
        :

                                             Text in italic is not part of the
                                             actual file. UPPERCASE ITALIC
                                             TEXT denotes variable names
                                             for actual values.
```

simulation is indicated by the lack of the value WATE (for the input variable WATER) on line three of the Unit 15 file. The use of a transmissivity matrix instead of hydraulic conductivity and bottom-elevation matrices also indicate a confined simulation is planned.

A confined flow approximation of the unconfined flow system is easily obtained by first developing a calibrated model of the water-table aquifer using only the Trescott model. Nodal transmissivities are subsequently defined as the product of input K values and simulated saturated thickness values. The latter are calculated as the difference between simulated head values and input values for aquifer bottom elevations. If a problem involves a combined artesian and water-table aquifer, equivalent transmissivities are calculated only for the water-table portions of the model grid.

When converting an input file for a stand-alone unconfined flow simulation to a confined flow approximation for use with AQMAN, it

Table 25. Structure of the AQMAN Unit 15 Input File for the First Iteration of the Simple Hydraulic Gradient Control Optimization Problem. Note that the Input File is Constructed to Represent a Confined Approximation of an Unconfined Aquifer.

```
************CONFINED FLOW MODEL INPUT, FIRST AQMAN ITERATION ************
        Blank row or continuation of title
                 RECH SIP  CHEC <-----------------      HEAD Print head matrix
   64DIML      77DMW        7NW       1000ITMAX  |
Blank row                                        |
   1NPER     10KTH     .0001ERR      .01EROR     |               LENGTH10
 1.63HMAX     1FACTX      1FACTY                 This line specifies unconfined
Three blank lines                                aquifer simulation with
   :                                             recharge, SIP solution, and
Starting head matrix, STRT(I,J)                  mass balance calculation.
   :
Storage coefficient matrix, S(I,J), (-1's and 0's to define prescribed
head boundaries)
   :
Transmissivity matrix, T(I,J), obtained by subtracting aquifer bottom elevations from water
table elevations obtained from the unconfined aquifer solution and multiplying the result by
the nodal conductivity value
   :
Recharge matrix, QRE(I,J)
   :
Grid spacing values for X direction, DELX(J)
   :
Grid spacing values for Y direction, DELY(I)
   :
Group IV data (not required for AQMAN runs)
   :

                                                 Text in italic is not part of the
                                                 actual file. UPPERCASE ITALIC
                                                 TEXT denotes variable names
                                                 for actual values.
```

may also be necessary to change the value of β', the parameter used during matrix solution in the Trescott model. This parameter is read as HMAX in the Trescott input (Table 24 and Table 25). For problems such as this example in which the strongly implicit procedure (SIP) is used for the matrix solution, HMAX is defined as the solution parameter β'. When running the Trescott code for unconfined simulations, use of $\beta' = 0.44$ produced the most rapid convergence. Using the same value for β' during confined simulations as part of AQMAN runs resulted in several trials that failed to converge in less than 1,000 iterations. After experimenting with different values, $\beta' = 1.63$ was found to produce consistent convergence in approximately 130 iterations.

Another input parameter that must be changed when converting from the Trescott input file (Table 24) to the AQMAN Unit 15 file (Table 25) is NW, the number of pumping wells. For stand-alone runs of the Trescott code, NW was set equal to zero because there were no pumping wells in the unconfined aquifer. When converting the Trescott input file to the Unit 15 file structure, NW must be set equal to the number of potential pumping and injection wells that are included in the optimization problem. The total number of pumping and injection wells included in an AQMAN run is specified as NWLS in the AQMAN Unit 14 input file.

B. Simple Hydraulic Gradient Control

1. Problem Formulation

The first remediation design problem to be examined is an example of simple hydraulic gradient control. As outlined in Chapter V, the objective is to prevent further migration or expansion of the contaminant plume. For this example, this will be accomplished by extracting contaminated water from within the plume. The number of possible locations for pumping wells is limited to seven. Table 26 presents the complete formulation for the simulation-optimization problem.

Only two constraint types were specified for this problem. Hydraulic gradient constraints are used to ensure that groundwater flow is directed toward the interior of the plume. Drawdown constraints are used at the pumping wells to prevent excessive drawdowns that could lead to inefficient well operation or pump damage.

The location of possible pumping wells, hydraulic gradient constraints, and the plume boundary are shown in Figure 48. To ensure containment of the TCE plume, hydraulic gradient constraints were specified at every pair of finite difference nodes along the downgradient boundary of the plume. Under ambient conditions, inward hydraulic

Table 26. Formulation of Simply Hydraulic Gradient Control Optimization Problem

Minimize Z = Total Pumping Rate (seven pumping wells possible)	$\sum_{i=1}^{7} Q_i$
Subject to	
1. Hydraulic gradients directed inward toward the plume around its entire boundary.	$H_{out_j} - H_{in_j} \geq 0$ $j = 1–46$
2. In-well drawdowns restricted to 30 percent of the saturated thickness, b.	$H_i \geq$ Bottom Elevation + 0.7(b) $i = 1–7$

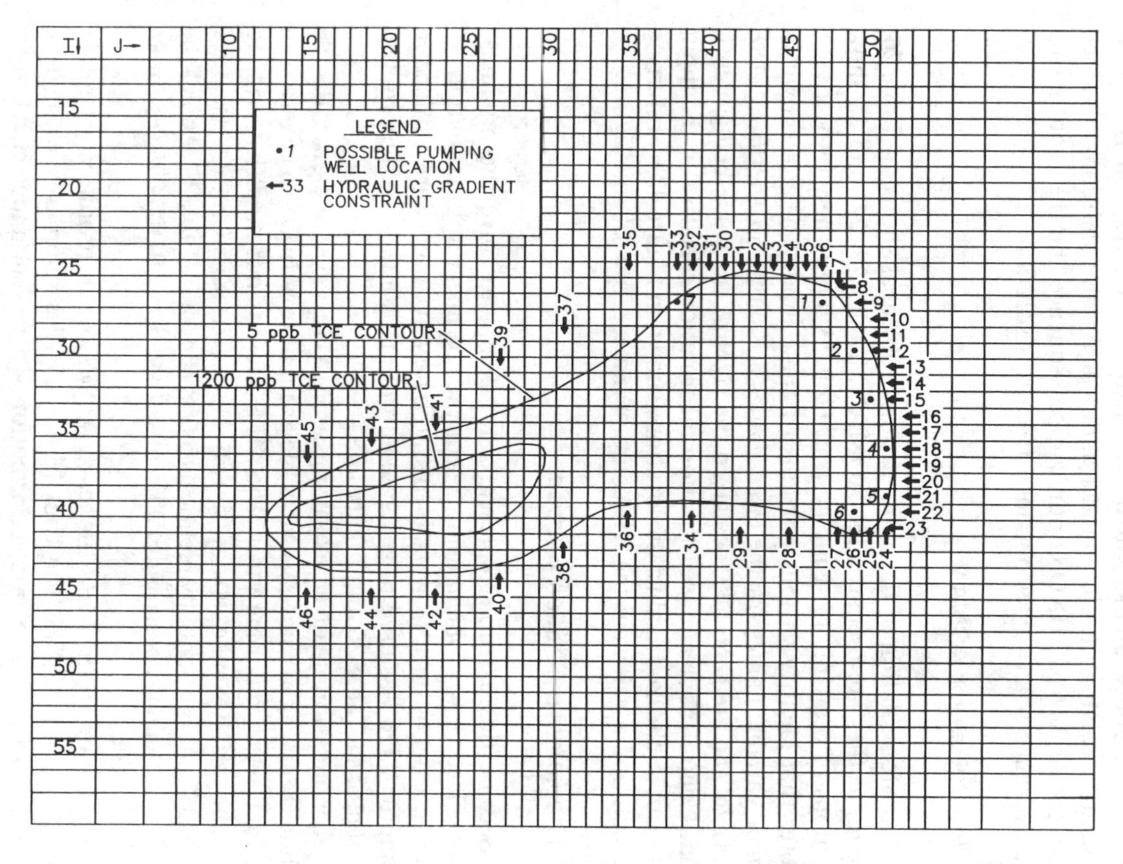

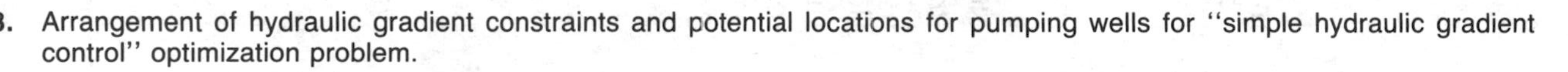
Figure 48. Arrangement of hydraulic gradient constraints and potential locations for pumping wells for "simple hydraulic gradient control" optimization problem.

gradients existed around the rest of the plume boundary (Figure 45). Therefore, gradient constraints were only placed at every fourth pair of grid nodes around the rest of the plume. After a solution to the optimization problem is obtained, attainment of inward hydraulic gradients around the entire plume will be verified. The verification is accomplished by solving the flow model under pumping conditions obtained from the AQMAN/MINOS optimization sequence and then contouring the hydraulic head values obtained from this solution.

2. Input Files for AQMAN

After formulation of the optimization problem as shown in Table 26, the input files are prepared for AQMAN. Tables 25, 27, and 28 show portions of the Unit 15, Unit 14, and Unit 13 input files, respectively. Notes that identify various portions of these input files are shown in *italic*. Variable names for the input values are shown in *UPPER CASE ITALIC* and are generally located to the right of the value being identified. Control-node and well-location information specified in the input files can be checked against the locations shown in Figure 48 by referring to the I and J values in the left column and top row of this figure.

3. The Iterative Solution Sequence

Because this example represents an attempt to solve the nonlinear unconfined flow problem using a sequence of linear approximations (a sequence of confined flow problems), the iterative procedure outlined in Chapter V.F.6 was used to generate an optimal solution to the problem presented in Table 26 and Figure 48.

Each iteration involved executing AQMAN to generate an MPS file (Unit 18) and then using this MPS file as input to MINOS to generate an optimal set of pumping rates. The MPS file was not modified prior to running MINOS and is not shown here. The SPECS file used for MINOS runs is essentially identical to that shown in Table 17, except the BOUNDS line has been omitted because there is no BOUNDS section in the MPS file.

In addition to the MPS file, each AQMAN run produces a Unit 16 file and a Unit 17 file. The Unit 16 and 17 files summarize error messages and input control node information. To prevent undetected errors, these two files should be checked after the first AQMAN run of each new problem formulation.

For each iteration, the completion of an AQMAN run was followed by the execution of MINOS. Output from MINOS included a new set of

Table 27. AQMAN Unit 14 Input File for the Simple Hydraulic Gradient Control Optimization Problem

```
     CASE
LINE   GRADCONTROL NNAME
          7 NWLS      99 NCNTR      1 NNPER     1.CDELT        46NGRAD
          1. TIMPER
         24. TIMINC(N)
          0 NKEYQ
         27 ILOCW(I)  47 JLOCW(I) 0.75 XRAD(I)   0 UNITQ(I) -.001
         30           49          0.75           0          -.001
         33           50          0.75           0          -.001  Information pertaining
         36           51          0.75           0          -.001  to wells
         39           51          0.75           0          -.001
         40           49          0.75           0          -.001
         27           38          0.75           0          -.001
         27 ILOCC(J)  47 JLOCC(J)   1 KEYWL(J)   0 KEYGRD(J)  0 KDEFHD(J)
         30           49            1            0            0
         33           50            1            0            0  Information pertaining
         36           51            1            0            0  to wells at which
         39           51            1            0            0  hydraulic head is
         40           49            1            0            0  constrained
         27           38            1            0            0
Control {24 ILOCC(J)  42 JLOCC(J)   0 KEYWL(J)   1            0 KDEFHD(J)
Pair 1  {25           42            0            1            0
Control {24           43            0            1            0  Information pertaining
Pair 2  {25           43            0            1            0  to control nodes that
         24           44            0            1            0  make up hydraulic
          :                                                       gradient control pairs
      Same control node information for 87 additional
      control nodes entered here
          :  CONHD(JN)
      588.0    G CONTYP(JN)
      588.0    G
      588.0    G       Limiting value and constraint
      588.0    G       type for constraints on hydraulic
      588.0    G       head at pumping wells
      588.0    G
      588.0    G
          0 CONHD(JN)
          0
          0            Dummy values for control nodes
          0            without individual hydraulic
          0            head constraints. CONTYP(JN)
          :            not required.
          :
      Dummy values for 87 additional control nodes
      entered here.
          :
```

Text in *italic* is not part of the actual file. *UPPERCASE ITALIC TEXT* denotes variable names for actual input parameters and data.

Table 28. AQMAN Unit 13 Input File for the Simple Hydraulic Gradient Control Optimization Problem

```
Control
Pair 1 \24 ILOCG1(K)  42 JLOCG1(K)  25 ILOCG2(K)  42 JLOCG2(K)  0 KDEFGR(K)
Control/24            43            25            43            0
Pair 2  24            44            25            44            0
        24            45            25            45            0
        24            46            25            46            0
        24            47            25            47            0
        25            48            26            48            0
        26            49            26            48            0
        27            50            27            49            0
        28            51            28            50            0
        :
   Same control pair information for 36
   additional control pairs entered here
     :                          GCON(KN)
          50.0 GFACT(KN)    0.    G GRATYP(KN)
          50.0              0.    G
          50.0              0.    G     Separation distance (50.0 feet in
          50.0              0.    G     grid area near plume) between nodes
          50.0              0.    G     of each control pair, and limiting
          50.0              0.    G     values (0.) and constraint type
          50.0              0.    G     for hydraulic gradient constraints.
          50.0              0.    G
          50.0              0.    G
          50.0              0.    G
     :
   Same gradient constraint information for 36 additional contraints
   entered here. Dummy value can be entered and GCON and GRATYP omitted
   if control pair is used for velocity definition and KDEFGR=1
     :

   Note:  Because GCON is multiplied by GFACT to convert the gradient
   constraint into a difference in head constraint and GCON is zero,
   GFACT could have been set equal to 1.0 [see Equation (90)].
```

Text in *italic* is not part of the actual file. *UPPERCASE ITALIC TEXT* denotes variable names for actual values.

pumping rates that was used to prepare a new set of Group IV data for the Trescott model. An unconfined simulation was run and new hydraulic head values for all grid nodes were obtained. These head values were the new water-table elevations, which together with the aquifer-bottom elevations define the unconfined saturated thickness, b. A new transmissivity matrix was then prepared by multiplying b x K, and the optimization step was repeated. The results of this procedure, presented in Table 29, show that successive iterations converged to less than a 0.08 percent change in five iterations.

Table 29. Pumping Rates for Sequential Iterations of the Optimization Problem Using a Confined Flow Approximation During the AQMAN/Minos Optimization Step

Optimization Run	Pumping Rate in Cubic Feet per second (cfs)								
	Q_1	Q_2	Q_3	Q_4	Q_5	Q_6	Q_7	Total Pumping Rate (cfs)	Percent Change
1	0.0200	0.0123	0.0090	0.0038	0.0022	0.0060	0.0141	0.0674	
									8.4
2	0.0199	0.0121	0.0089	0.0038	0.0021	0.0059	0.0095	0.0622	
									−1.1
3	0.0200	0.0122	0.0089	0.0038	0.0021	0.0059	0.0100	0.0629	
									0.2
4	0.0200	0.0121	0.0089	0.0038	0.0021	0.0059	0.0100	0.0628	
									<0.08
5	0.0200	0.0121	0.0089	0.0038	0.0021	0.0059	0.0100	0.0628	

Table 30. Simulated Hydraulic Head at the Radii of Pumping Wells and Average Hydraulic Head for Corresponding Finite Difference Cells

Well No.	Head at Well	Average Cell Head
1	610.22	610.35
2	610.21	610.28
3	610.18	610.24
4	610.19	610.21
5	610.18	610.19
6	610.13	610.17
7	610.63	610.70

4. Optimal Solution

The pumping rates produced from the fifth iteration were used to prepare a final unconfined flow input file for the Trescott model. The Group IV data at the end of the file included a radius of 0.75-foot for each well (variable named RADIUS). Non-zero values for well radii cause the code to output drawdowns and hydraulic heads at the wells. These data are important because the hydraulic head at a pumping well will always be lower than the average head for a cell. Contour maps prepared using hydraulic heads at the well locations will be more accurate than those based solely on average cell heads. Table 30 provides a comparison of in-well heads and average cell heads for this example. For this example, none of the constraints on hydraulic heads at the pumping wells were binding.

The water-table contour map for steady-state conditions under the design pumping rates is shown in Figure 49. The figure shows that the hydraulic gradients and flowlines are directed inward across the entire plume boundary. The groundwater divide that separates the flow captured by the remedial wells is located just downgradient of the plume. This divide extends far to the north of the plume boundary, indicating that a substantial volume of uncontaminated groundwater will be captured by the remedial well field.

Figure 49 also shows that pumping at Well 7 may not be necessary to prevent further uncontrolled migration of the plume. The fact that Constraint 33 is binding suggests that pumping at Well 7 is necessary only to reverse a small northward component of flow in this area. Additional simulations did in fact show that the plume could be captured without pumping at Well 7. Pumping from only the downgradient boundary wells simply permits a small amount of plume expansion in the area of Well 7. This example illustrates that very conservative constraints such as "no plume expansion" may not be practical. For the example shown

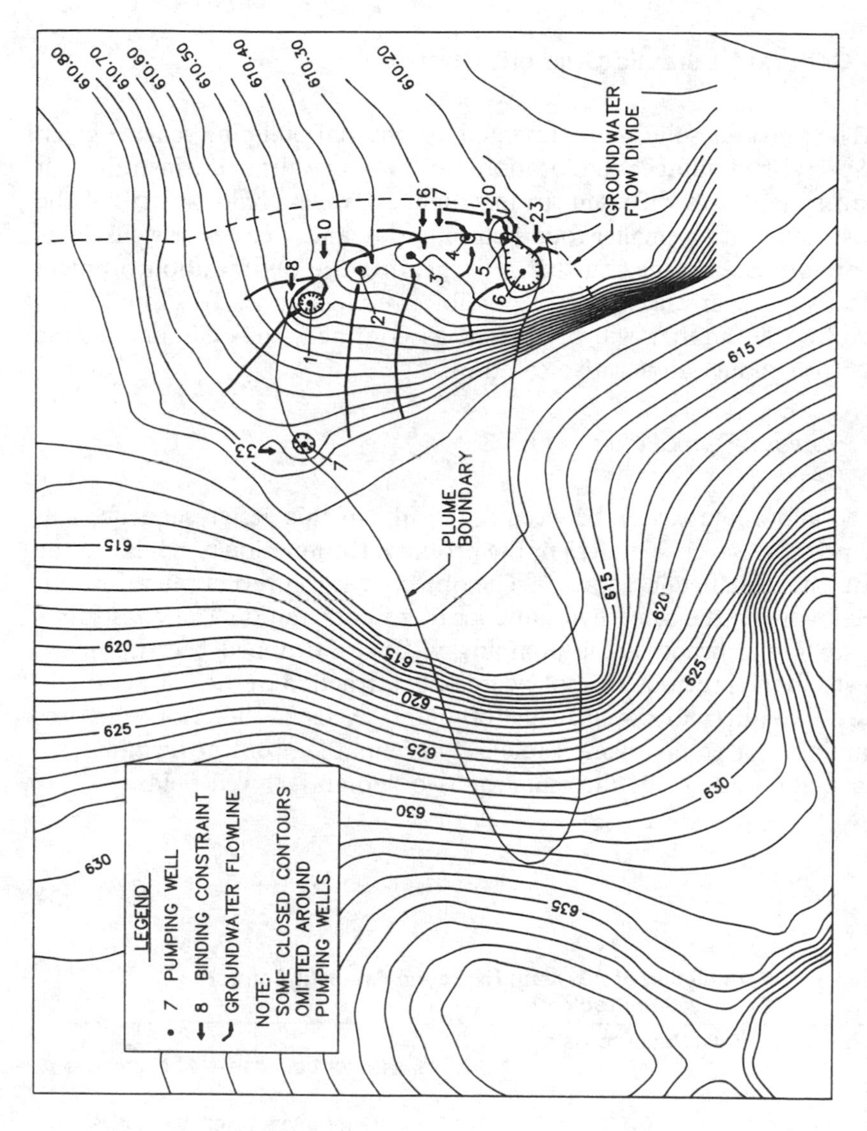

Figure 49. Steady-state hydraulic head map produced by the optimal pumping system for the "simple hydraulic gradient control" problem.

in Figure 51, a constraint that precluded migration east of the downgradient boundary would be equally protective of public health and would eliminate the need to install and operate Well 7.

C. General Hydraulic Gradient Control

The previous solution represented an optimal pumping scheme given the prescribed number and location of pumping wells and constraints. It is possible that a different arrangement of wells could satisfy all the constraints with a smaller total pumping rate and even fewer wells. This possibility is examined further by repeating the optimization problem with two new arrangements of possible well locations. To simplify this exercise, the aquifer will be treated as confined, thus eliminating the need to conduct an iterative solution.

1. Problem Formulation

To allow comparison between the results of this design sequence and the previous seven-well design, the problem formulation is similar to the problem presented in Table 26. The objective is to prevent plume expansion by using the same hydraulic gradient constraints. The constraints on hydraulic heads at the pumping wells are dropped because drawdowns in the previous example were only a fraction of the aquifer thickness. Two alternative well fields (boundary wells and interior wells) are examined for comparison. Table 31 presents the problem formulation, and Figures 50 and 51 illustrate the two "potential" well fields.

Table 31. Formulation of "General Hydraulic Gradient Control" Optimization Problem

Minimize Z = Total Pumping Rate	$\sum_{i=1}^{n} Q_i$	n = 26 for boundary-well alternative
		n = 21 for interior-well alternative
Subject to		
1. Hydraulic gradients directed inward toward the plume around its entire boundary.	$H_{out_j} - H_{in_j} \geq 0$	j = 1–46

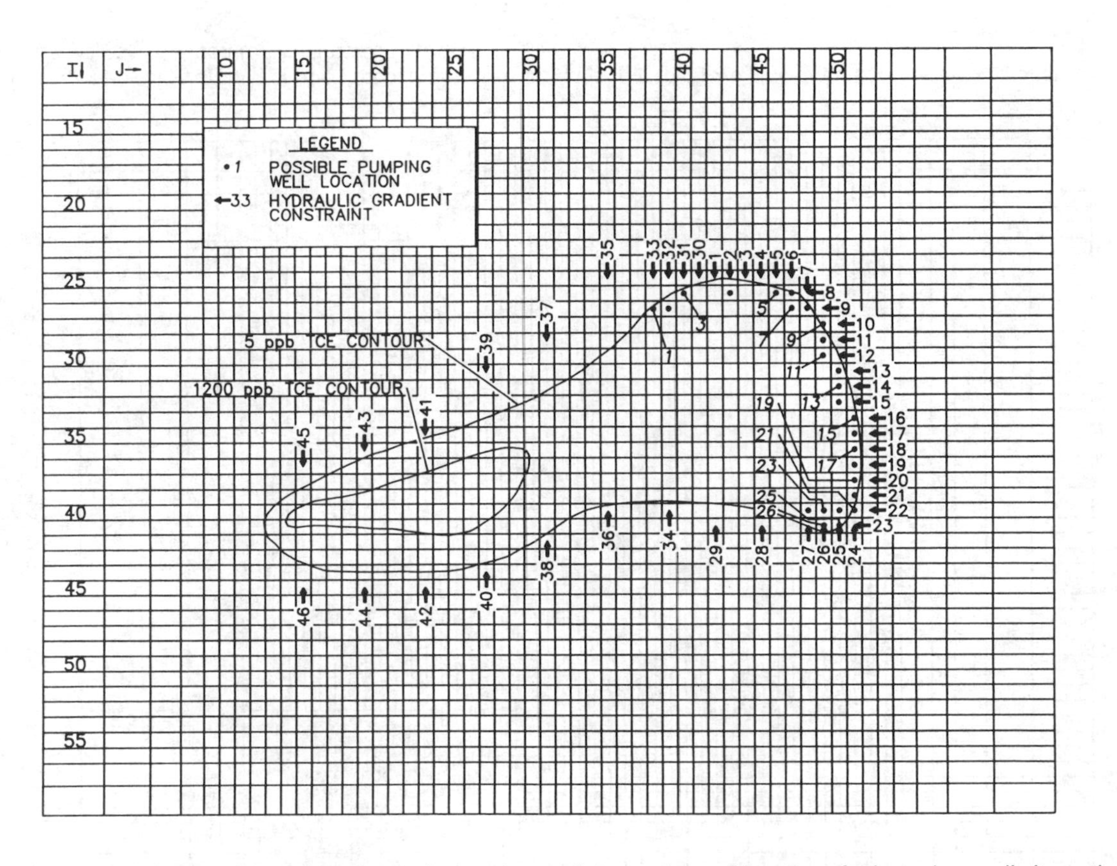

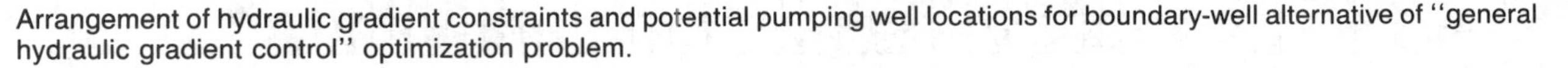

Figure 50. Arrangement of hydraulic gradient constraints and potential pumping well locations for boundary-well alternative of "general hydraulic gradient control" optimization problem.

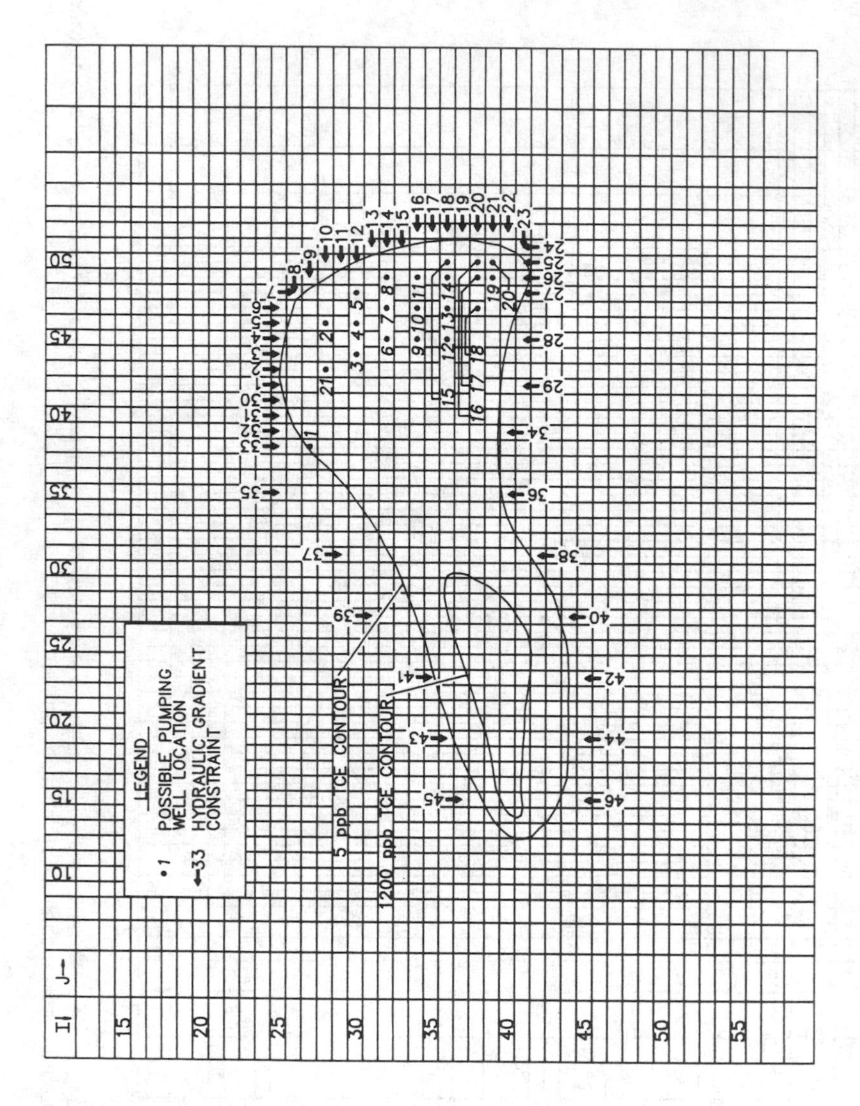

Figure 51. Arrangement of hydraulic gradient constraints and potential pumping well locations for interior-well alternative of "general hydraulic gradient control" optimization problem.

2. Problem Solution

The boundary-well alternative includes 26 potential well locations (Figure 50), and the interior-well alternative includes 21 potential well locations (Figure 51). The MPS files produced by AQMAN for the two alternatives will, therefore, contain 26 and 21 decision variables, respectively. Using the unaltered MPS file for the boundary-well alternative as input to MINOS generated an optimal solution in which 17 of the 26 wells are pumped at a total pumping rate of 0.0648 cubic foot per second (cfs). Compared to the seven-well system of Figure 48 (Q_{total} = 0.0679 cfs), this represents a 0.0031 cfs reduction in the total pumping rate and a 10-well increase in the number of active wells. The optimal solution for the interior-well alternative involved pumping from eight wells at a total rate of 0.0684 cfs, so it offered no improvement over the earlier case. These two solutions included no restrictions on the number of active pumping wells and are referred to as the "unrestricted solutions."

To obtain solutions in which the number of possible pumping wells is restricted, one could use mixed-integer programming, as described in Chapter V.F.1. Because a mixed-integer programming package was not installed on the computer system used for these simulations, the trial-and-error method was used. Using a BOUNDS section that was added to the MPS file, selected decision variables were removed from the objective function by using equality constraints to set the decision variables equal to zero. The use of a BOUNDS section is described in Chapter V.D.6.f. An example BOUNDS section from a modified MPS file for the boundary-well alternative is shown in Table 32. For the trial shown in Table 32, pumping was permitted at Wells 1, 7, 11, 14, 16, 19, and 22.

In an effort to find the well configuration that provided the smallest total pumping rate and a reasonable limit on the number of wells, repeated trials were conducted for both the boundary-well and the interior-well alternatives (Table 33). Each trial considered a new combination of wells and used a newly modified MPS file for input to MINOS. The number of wells permitted in the trials was arbitrarily restricted to seven or less. For comparison, a solution was also obtained for each alternative with all decision variables included in the objective function (Trials 1 and 12). It is worth noting that all of the trials in Table 33 were generated from only two AQMAN runs—one for the boundary-well alternative and one for the interior-well alternative.

As expected, the lowest total pumping rate for each alternative is obtained when decision variables for all possible wells are included in the objective functions (Trials 1 and 12). Restricting the number of

Table 32. Modified MPS File Showing Bounds Section Used to Set Select Decision Variables Equal to Zero

```
NAME           GRADCONTROL
ROWS
 L  DIF01001
 L  DIF01002
        :
     ROW names for 44 additional
     constraints appear here.
        :
 N  OBJ                                  ┌─Objective function cost coefficient
COLUMNS                                  ▼ for decision variable Q1.
    Q01001     OBJ          0.10000E-02
    Q01001     DIF01001   -0.38866E-03   DIF01002   -0.28738E-03
        :
     Response coefficients for columns Q01001 through
     Q01026 and cost coefficients for decison variables
     Q01002 through Q01026 appear here.
        :
    Q01026     DIF01045   -0.49198E-03   DIF01046   -0.17174E-03
RHS
    RHS        DIF01001   -0.54163E-02   DIF01002   -0.26018E-02
        :
     RHS values for constraints DIF01003 through DIF01046
     appear here.
        :
BOUNDS
 FX BOUND      Q01002       0.0
 FX BOUND      Q01003       0.0       BOUNDS section added
 FX BOUND      Q01004       0.0       to MPS file. Note that
 FX BOUND      Q01005       0.0       decision variables for
 FX BOUND      Q01006       0.0       wells 1, 7, 11, 14, 16,
 FX BOUND      Q01008       0.0       19, and 22 are not included
 FX BOUND      Q01009       0.0       in the BOUNDS section
 FX BOUND      Q01010       0.0
 FX BOUND      Q01012       0.0
 FX BOUND      Q01013       0.0
 FX BOUND      Q01015       0.0
 FX BOUND      Q01017       0.0
 FX BOUND      Q01018       0.0
 FX BOUND      Q01020       0.0
 FX BOUND      Q01021       0.0
 FX BOUND      Q01023       0.0
 FX BOUND      Q01024       0.0
 FX BOUND      Q01025       0.0
 FX BOUND      Q01026       0.0
ENDATA
```

Text in italic is not part of the actual file. Bold type indicates additions to the MPS file.

Table 33. Results of Multiple Well-Field Design Trials Based on 2 AQMAN Runs and 21 Minos Runs. Results Include the Number of Wells or Decision Variables in each Objective Function and the Number of Active Wells Actually Used in the Design Solution.

	Trial	Number of Wells in Objective Function	Number of Wells Used in Solution	Total Q (cfs)
Boundary-Well Alternative				
	1	26	17	0.06476
	2	7	7	0.07215
	3	7	7	0.07080
	4	7	7	0.06964
Well locations	5	7	7	0.06869
are different	6	7	7	0.06933
for each trial	7	7	7	0.04749
	8	7	7	0.06771
	9	7	7	0.06749
	10	5	5	0.07269
	11	6	6	0.07139
Interior-Well Alternative				
	12	21	8	0.06838
	13	5	5	0.06940
	14	5	5	0.06910
	15	5	5	0.06968
Well locations	16	5	5	0.06884
are different	17	7	5	0.07121
for each trial	18	7	6	0.07102
	19	7	5	0.06944
	20	7	7	0.06893
	21	7	7	0.06847

pumping wells to seven or less for the boundary-well alternative produced a minimum pumping rate of 0.06749 cfs (Trials 7 and 9). Because this value was less than the overall minimum of 0.06838 cfs for the interior-well alternative, Trials 7 and 9 were chosen as the optimum well field designs, given the potential well locations in Figure 50 and Figure 51.

3. *Optimal Solution*

Given the possible well locations shown in Figures 48, 50, and 51 and restricting the number of active wells to seven or less, the minimum total pumping rate that will satisfy the hydraulic gradient constraints is 0.06749 cfs for Trial 7 or 9 from Table 33. This value is based on the confined flow approximation. Applying the iterative solution sequence used for the previous problem formulation would lead to an even smaller value for the total pumping rate.

Using the individual pumping rates for the seven active wells of Trial

9 as input to the Trescott model, the steady-state hydraulic head map in Figure 52 was generated. The individual pumping rates, hydraulic head values, and drawdowns for each well are shown in Table 34. As shown by the hydraulic head contours and flow lines in Figure 52, the optimal set of pumping rates should be successful in preventing further migration of the TCE plume.

Table 34. Pumping Rates, Hydraulic Heads, and Drawdowns for Individual Wells in Optimal Solution to "General Hydraulic Gradient Control" Problem

Well	Q (cfs)	Hydraulic Head (ft)	Drawdown (ft)
1	0.0142	608.43	6.37
2	0.0205	608.12	6.38
3	0.0127	608.11	6.29
4	0.0068	608.12	6.18
5	0.0044	608.12	6.08
6	0.0035	608.10	6.00
7	0.0055	608.08	6.02

D. Hydraulic Gradient Control With Injection

1. Problem Formulation

Suppose that as part of a larger remedial strategy for the entire site, it was necessary to design an interim remedial action that would prevent further migration of the TCE plume and would require only limited start-up funding. This objective would require minimizing pumping rates and restricting the number of remedial wells. In addition, an upper limit might be imposed on the total pumping rate to allow an inexpensive low-capacity treatment system to be used for treating the contaminated water withdrawn from the aquifer. Small portable treatment systems of this type are available for $5,000 to $10,000 with capacities that range from 10 to 50 gallons per minute (gpm).

Table 35 presents the formulation of the optimization problem for this example. The total pumping rate is restricted to 25 gpm (0.055 cfs), and the total number of wells is restricted to eight. Individual pumping rates are restricted to 15 gpm (0.033 cfs) to allow the use of small, inexpensive pumps.

The solution for the previous problem formulation required a total pumping rate of 30.3 gpm (0.06749 cfs) to achieve gradient control. To meet the 25 gpm constraint on total pumping for this problem, injection

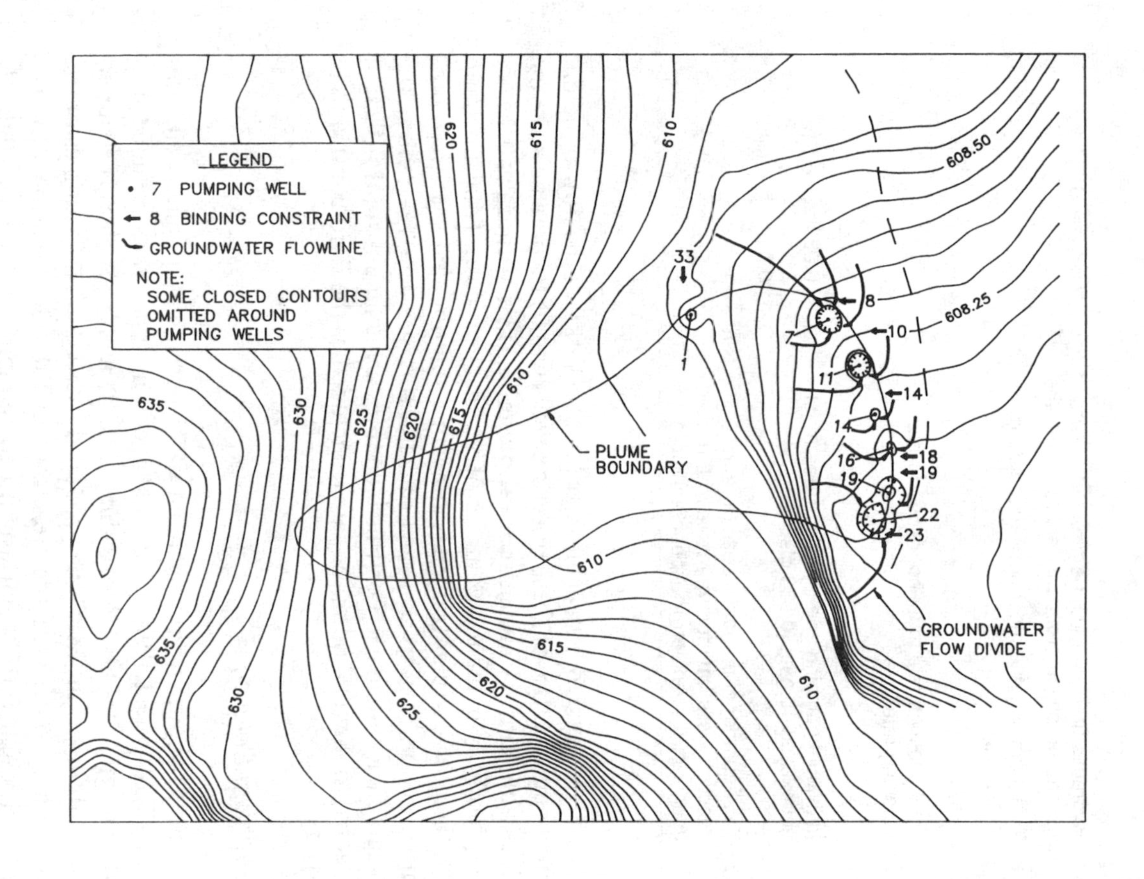

Figure 52. Steady-state hydraulic head map produced by optimal pumping system for "general hydraulic gradient control" problem.

Table 35. Formulation of Optimization Problem for Hydraulic Gradient Control with Injection

Minimize Z = Total Pumping Rate	$\sum_{i=1}^{n} Q_i$
Subject to	
1. Hydraulic gradient must be directed inward toward the plume around its entire boundary.	$H_{out} - H_{in} \geq 0$
2. Total pumping rate must not exceed 25 gpm (0.055 cfs).	$\sum_{i=1}^{n} Q_i \leq 0.055$
3. Total injection rate must not exceed total pumping rate.	$\sum_{i=1}^{n} Q_i - \sum_{j=1}^{m} Q_j \geq 0$
4. Individual pumping and injection rates must not exceed 15 gpm (0.033 cfs).	$Q_i \leq 0.033 \quad i = l, n$ $Q_j \leq 0.033 \quad j = l, m$
5. Total number of pumping wells (n) and injection wells (m) must not exceed 8 (mixed integer or trial-and-error experimentation).	$n + m \leq 8$

wells must be used to help achieve gradient reversal at the downgradient plume boundary. Use of these **injection** wells will be favored because the optimal solution will minimize total **pumping** (i.e., withdrawal) by injecting possibly unlimited volumes of water at no cost. Total injection must, therefore, be constrained to prevent a solution that specified enormous injection rates as a means of achieving gradient reversal at the downgradient boundary. For this example, a maximum total injection rate equal to the total pumping rate is included as the third constraint in Table 35.

2. Problem Solution

Figure 53 shows the arrangement of 21 potential pumping well locations and 20 potential injection well locations that were included in a single run of AQMAN. Constraints 2, 3, 4, and 5 (Table 35) required manual editing of the MPS file produced by AQMAN. Examples of the modifications to the MPS file are shown in Table 36. Decision variables Q01001 through Q01021 correspond to pumping Wells 1 through 21, and decision variables Q01022 through Q01041 correspond to injection Wells 22 through 41.

The first modification of the MPS file is associated with the objective

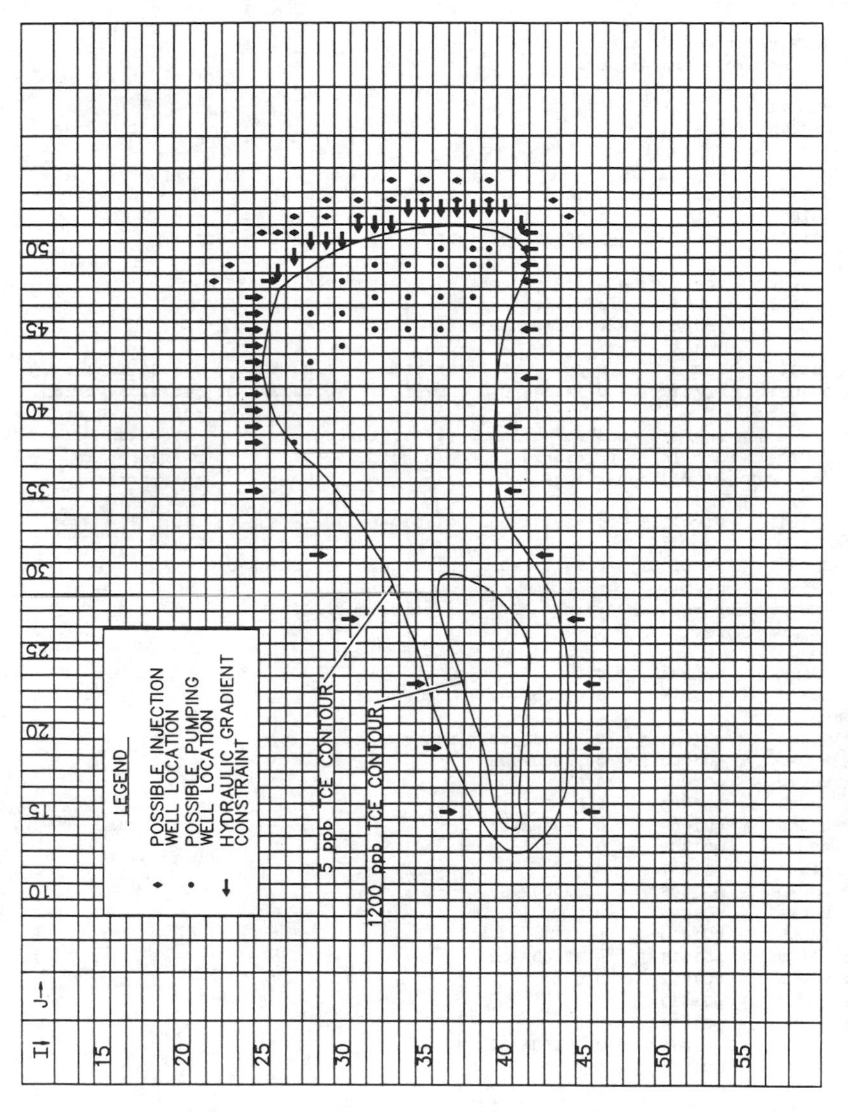

Figure 53. Arrangement of hydraulic gradient constraints and potential locations for pumping and injection wells for the "hydraulic gradient control with injection" problem.

Table 36. Sections of Modified MPS File for "Hydraulic Gradient Control with Injection" Problem

```
NAME          GRADCONTROL
ROWS
 L  DIF01001
 L  DIF01002
      :
    ROW names for 44 additional
    constraints appear here
      :
 L  QLIMIT
 G  RECHBAL
 N  OBJ
COLUMNS
    Q01001     OBJ          0.10000E-02    RECHBAL     1.0
    Q01001     QLIMIT       1.0
    Q01001     DIF01001   -0.38866E-03     DIF01002   -0.28738E-03
      :
    Response coefficients for columns Q01001 through
    Q01024 and cost coefficients for decision variables
    Q01002 through Q01024 appear here.
      :
    Q01025     OBJ          0.0            RECHBAL    -1.0          Q01025 is the
    Q01025     DIF01001     0.12488E-04    DIF01002    0.22232E-04  decision variable
      :                                                             for the second
    Response coefficients for columns Q01025 through                injection well
    Q01041 and cost coefficients for columns Q01026                 permitted in
    through Q01041 appear here.                                     this trial.
      :
    Q01041     DIF01045     0.48523E-03    DIF01046    0.16455E-03
RHS
    RHS        DIF01001   -0.54163E-02     DIF01002   -0.26018E-02
      :
    RHS values for constraints DIF01003 through DIF01046
    appear here.
      :
    RHS        RECHBAL      0.0
    RHS        QLIMIT       0.55000E+02
BOUNDS
 UP BOUND      Q01001       0.33000E+02 ┐
 UP BOUND      Q01002       0.33000E+02 │
 FX BOUND      Q01003       0.0         │
 FX BOUND      Q01004       0.0         │
 UP BOUND      Q01005       0.33000E+02 │
 FX BOUND      Q01006       0.0         │
 FX BOUND      Q01007       0.0         ├─Pumping wells
 UP BOUND      Q01008       0.33000E+02 │
 FX BOUND      Q01009       0.0         │
 FX BOUND      Q01010       0.0         │
 UP BOUND      Q01011       0.33000E+02 │
 FX BOUND      Q01012       0.0         │
 FX BOUND      Q01013       0.0         │
 FX BOUND      Q01014       0.0         │
 UP BOUND      Q01015       0.33000E+02 │
 FX BOUND      Q01016       0.0         │
 FX BOUND      Q01017       0.0         ┘
```

Text in italic is not part of the actual file. Bold type indicates lines added to the MPS file to incorporate constraints 2, 3, 4, and 5 of Table 34. The MPS file shown here is for the trial that permitted pumping at wells 1, 2, 5, 8, 11, 15, and 20 and injection at wells 23, 25, 29, 33, 37, and 40.

Table 36. Sections of Modified MPS File for "Hydraulic Gradient Control with Injection" Problem (Continued)

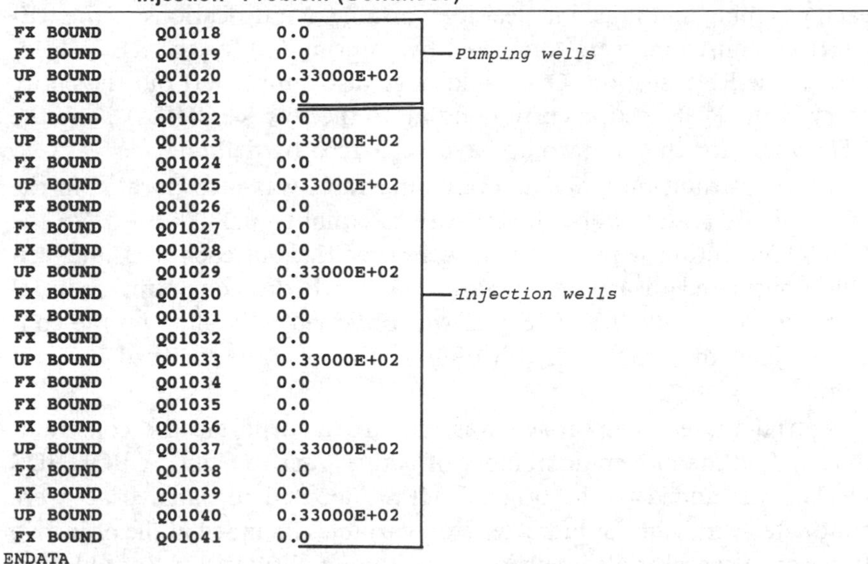

```
 FX BOUND     Q01018     0.0
 FX BOUND     Q01019     0.0            Pumping wells
 UP BOUND     Q01020     0.33000E+02
 FX BOUND     Q01021     0.0
 FX BOUND     Q01022     0.0
 UP BOUND     Q01023     0.33000E+02
 FX BOUND     Q01024     0.0
 UP BOUND     Q01025     0.33000E+02
 FX BOUND     Q01026     0.0
 FX BOUND     Q01027     0.0
 FX BOUND     Q01028     0.0
 UP BOUND     Q01029     0.33000E+02
 FX BOUND     Q01030     0.0            Injection wells
 FX BOUND     Q01031     0.0
 FX BOUND     Q01032     0.0
 UP BOUND     Q01033     0.33000E+02
 FX BOUND     Q01034     0.0
 FX BOUND     Q01035     0.0
 FX BOUND     Q01036     0.0
 UP BOUND     Q01037     0.33000E+02
 FX BOUND     Q01038     0.0
 FX BOUND     Q01039     0.0
 UP BOUND     Q01040     0.33000E+02
 FX BOUND     Q01041     0.0
ENDATA
```

Text in italic is not part of the actual file. Bold type indicates lines added to the MPS file to incorporate constraints 2, 3, 4, and 5 of Table 34. The MPS file shown here is for the trial that permitted pumping at wells 1, 2, 5, 8, 11, 15, and 20 and injection at wells 23, 25, 29, 33, 37, and 40.

function. As written by AQMAN, the MPS file includes "cost coefficients" equal to 0.10000E–02 for pumping wells and –0.10000E–02 for injection wells. Decision variables for the injection wells (Q01022 through Q01041) were eliminated from the objective function by changing their cost coefficients to 0.0.

Constraint 2 (Table 35) limits total pumping and was incorporated by adding the constraint row QLIMIT (named arbitrarily). A coefficient of 1.0 was then added in the COLUMNS section for each managed pumping well (does not include injection wells). The constraint limit of 0.055 cfs was entered in the RHS section by adding the value 0.55000E + 02 for the QLIMIT constraint. Note that the constraint on total pumping, 0.055 cfs, was scaled up by dividing by the absolute value of the unit stress (0.001 cfs) when entered in the MPS file.

Another constraint row, RECHBAL (also named arbitrarily), was added to the ROWS section for the constraint on the total injection rate (see Constraint 3 in Table 35). The COLUMNS section was then modified by adding a coefficient of 1.0 for the RECHBAL constraint row to each pumping and injection decision variable. Modifications of the RECHBAL constraint were completed by entering 0.0 for the RECHBAL row in the RHS section. One could have also simply left out the entire entry in the RHS section corresponding to the row RECHBAL, and the RHS value for this row would have been zero by default.

The constraint on individual pumping rates was incorporated in the MPS file by adding upper bound values equal to 0.33000E+02 to the BOUNDS section. Upper bounds were specified for each decision variable considered in a particular trial. As with the constraint on total pumping, the value 0.33000E+02 was obtained by scaling up the constraint limit of 0.033 cfs by dividing by the absolute value of the unit stress.

A trial-and-error approach was used to incorporate the constraint that limited the maximum number of active wells to eight. A BOUNDS section was added to the original MPS file, and repeated trials were conducted with eight or more decision variables included in the objective function. Several trials involving more than eight decision variables resulted in optimal solutions with only eight active wells.

For optimal solutions with a maximum of eight active wells, total pumping rates ranged from 0.05188 to 0.05437 cfs. The mass-balance constraint on total pumping and injection rates was binding in every solution. It is likely that using an integer programming package would lead to an optimal solution with a smaller total pumping rate than the 0.05188 cfs obtained for the minimum in this problem.

3. Optimal Solution

The solution with total pumping = 0.05188 cfs was obtained using an MPS file that included seven possible pumping wells (1, 2, 5, 8, 11, 15, and 20) in the objective function and six possible injection wells (23, 25, 29, 33, 37, and 40) in the injection balance constraint. Three active pumping wells and five active injection wells were specified in the solution. The contour map of the steady-state water table for this remedial design is shown in Figure 54. Also shown are the groundwater flowlines, the location of the resulting groundwater flow divide, and the location of the binding constraints. The figure indicates that hydraulic gradient control of the plume is achieved.

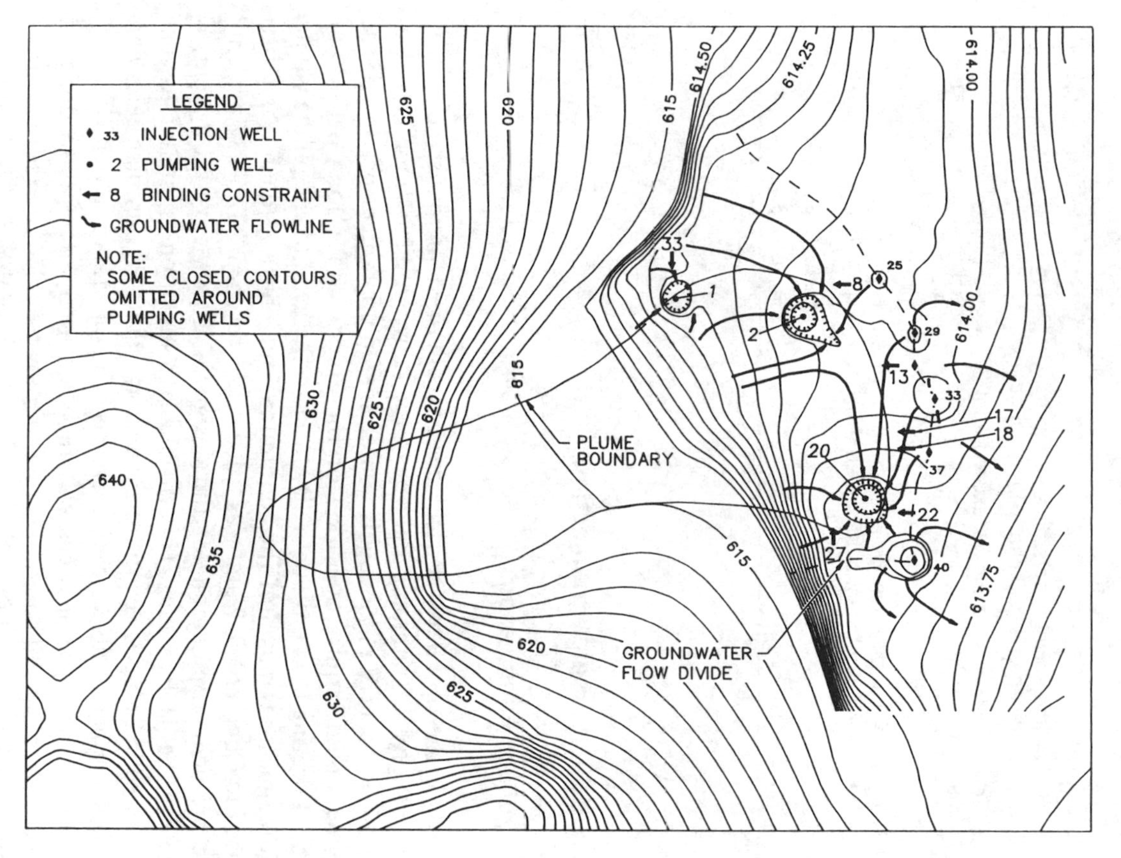

Figure 54. Steady-state hydraulic head map produced by optimal pumping system for "hydraulic gradient control with injection" problem.

Table 37. Formulation of Optimization Problem for "Flow Reversal with Injection"

Minimize Z = Total Pumping Rate	$\sum_{i=1}^{4} Q_i$
Subject to	
1. Hydraulic gradient directed inward toward the plume around the entire plume boundary.	$H_{out} - H_{in} \geq 0$
2. Total injection rate must not exceed total pumping rate.	$\sum_{i=1}^{4} Q_i - \sum_{j=1}^{21} Q_j \geq 0$
3. Drawdown in pumping Wells 1 through 4 must not exceed 30 percent of saturated thickness.	$H_1 \geq 611.7$ $H_2 \geq 611.8$ $H_3 \geq 612.1$ $H_4 \geq 613.1$

E. Flow Reversal With Injection

1. Problem Formulation

All of the optimal well-field designs obtained previously have the disadvantage of pumping only near the downgradient boundary. Removal of the TCE in the highly contaminated upgradient portion of the plume would require operation of these systems for long periods of time. Locating the pumping wells closer to the head of the plume would allow quicker removal of the contamination. This approach was introduced in Chapter V.F.2 and was titled "flow reversal." Table 37 presents the problem formulation for the design of a flow reversal system to capture the TCE plume. This problem formulation is similar to that presented in Table 35, except that the constraints on pumping rates and on the number of active wells have been dropped.

The use of injection wells was required after attempts to design a flow-reversal system based only on pumping wells led to excessive drawdowns (below the aquifer bottom) in the pumping wells. The mass-balance constraint on total pumping and injection rates was necessary to prevent enormous injection rates. The addition of drawdown constraints at the pumping wells ensures that wells and pumps operate efficiently.

2. Problem Solution

Figure 55 shows the locations of potential pumping and injection wells considered in the design of the flow-reversal system. As in the previous design problems, a single AQMAN run was used to generate an MPS file that was manually edited to permit examination of systems involving various combinations of wells. For completeness, the first optimization trial was conducted with no constraint on the total injection. An optimal solution was obtained with a total pumping rate of 0.03690 cfs from Well 1 only. The total injection rate was 2.87 cfs (1,288 gpm) for 11 active wells. In terms of operating and well-installation costs, this large injection rate involving so many active wells was undesirable. Including Constraint 2 on the total injection rate led to a feasible solution with a total pumping rate of 0.05097 cfs from Well 1. Total injection was equal to total pumping, and 10 injection wells were active. Four of the 10 injection wells had rates less than 0.0020 cfs. This result suggests that the number of active injection wells could be reduced without a significant increase in pumping and injection rates.

Additional trials were then conducted with a BOUNDS section added to the MPS file to restrict the number of potential injection wells to five. Feasible solutions were obtained with total pumping rates ranging from 0.05161 to 0.06329 cfs. Well 1 was the only active pumping well in all the solutions.

The injection constraint was also binding in all the solutions. Compared to the 0.05097 cfs design (10 active injection wells), the 0.05161 cfs design represents a 1.2-percent increase in total pumping rate and a 50-percent reduction in the number of active wells. In terms of logistic and economic considerations, the latter design based on a pumping rate of 0.05161 cfs represents the most desirable flow reversal system.

3. Optimal Solution

The contour map of steady-state hydraulic head elevations generated by the 0.05161 cfs design is shown in Figure 56. Well 1 is the only active pumping well, and Wells 10, 14, 18, 22, and 23 are active injection wells. Drawdown at Well 1 is 3.18 feet. An additional 0.5 foot of drawdown could be developed before the drawdown constraint at Well 1 would become binding. This suggests that flow reversal might be possible using pumping wells located near the center of the 1,200 parts per billion (ppb) contour. Such a system would have the advantage of both quicker removal of the highly contaminated mass of groundwater and quicker

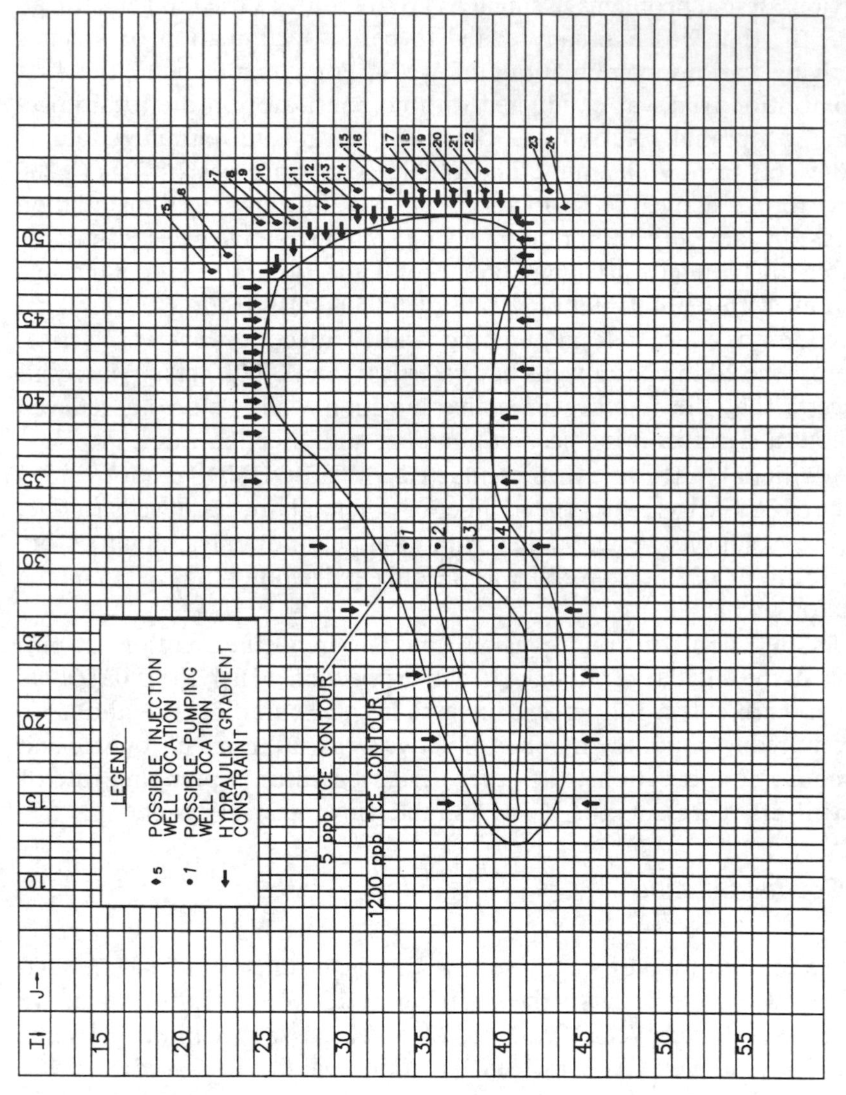

Figure 55. Arrangement of hydraulic gradient constraints and potential pumping and injection wells for "flow reversal with injection" problem.

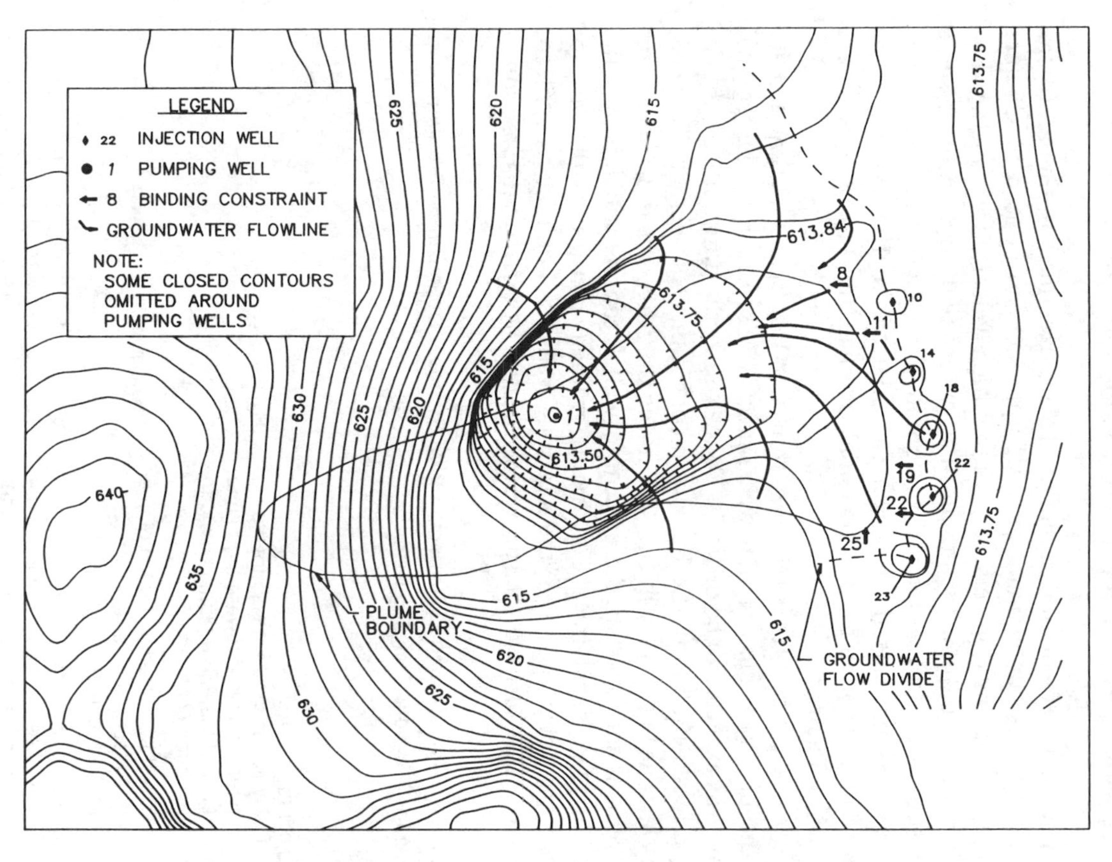

Figure 56. Steady-state hydraulic head map produced by optimal pumping system for "flow reversal with injection" problem.

removal of any contaminants adsorbed on the porous media in the high-concentration region of the plume.

F. Screening Plus Removal With Velocity Constraints

The final remediation technique to be examined is similar to the screening plus contaminant-removal approach introduced in Chapter V.F.3. Constraints on retarded contaminant velocities are included in this example to allow the design of a system that could achieve some measure of control on the rate of contaminant removal.

Table 38 presents the problem formulation for this example. The locations of constraints and potential removal and screening wells are shown in Figure 57.

Because the aquifer is relatively thin with low hydraulic conductivity in the vicinity of the screening wells, successful screening of the regional flow was expected to be difficult. To maximize the possibility of obtaining a feasible solution, the drawdown constraint (Number 4 in Table 38) was intended to be very liberal. Requiring only 50 percent of the aquifer to remain saturated at the screening well locations will increase the likelihood of satisfying the hydraulic gradient constraints that are located just downgradient of the screening wells. It should be recognized, however, that such extreme drawdowns could be difficult to maintain in

Table 38. Formulation of Screening Plus Removal Optimization Problem

Minimize Z = Total Pumping Rate	$\sum_{i=1}^{28} Q_i$
Subject to	
1. Hydraulic gradient constraints directed inward toward the plume around the entire plums boundary (47 such constraints).	$H_{out} - H_{in} \geq 0$
2. The flow upgradient of the plume boundary must be reversed (25 reversal constraints).	$H_{downgradient} \geq H_{upgradient}$
3. Retarded contaminant migration velocities towards the contaminant removal wells must exceed either 50 feet per year (ft/yr) or 100 ft/yr (see Figure 57).	$V_{retarded_j} \geq 50$ ft/yr j = 1–3 $V_{retarded_j} \geq 100$ ft/yr j = 3–6
4. Drawdown at the screening wells must not exceed 50 percent of the saturated thickness, b.	$H_i \geq H_{min_i}$ i = 19–31

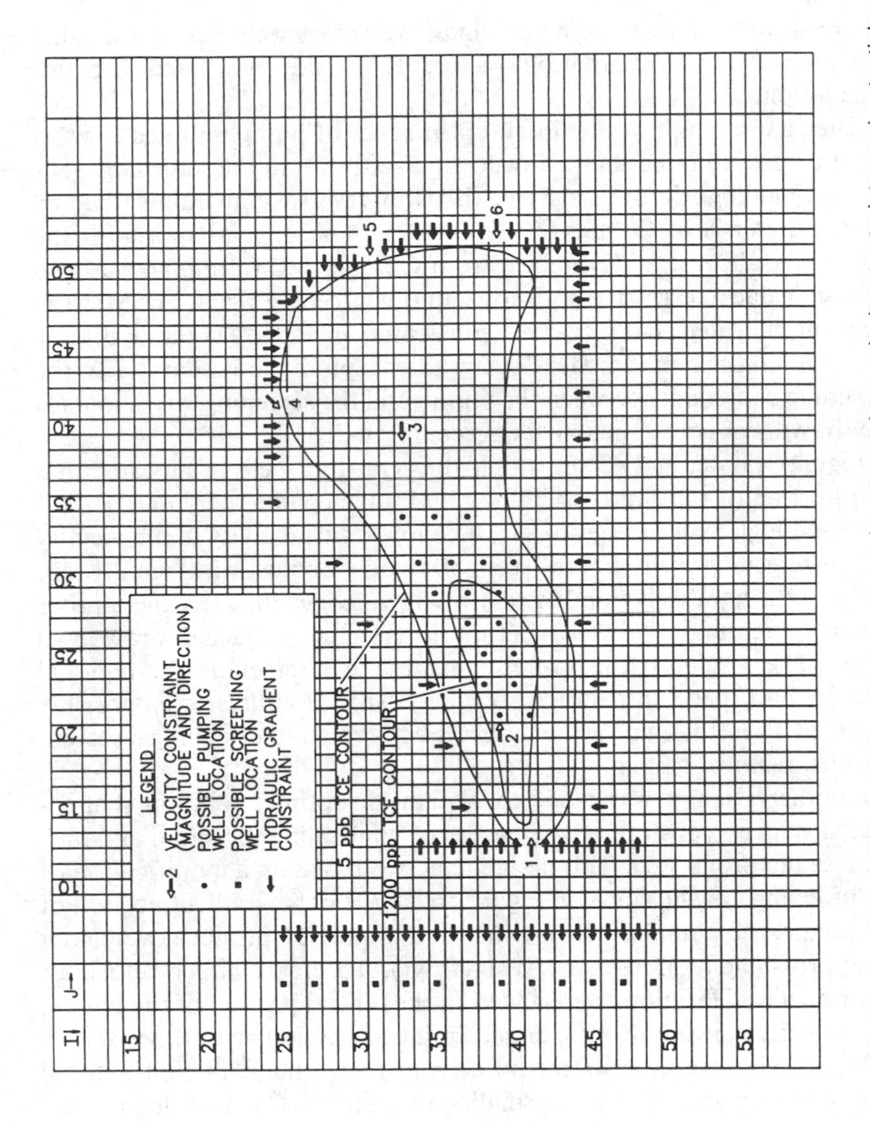

Figure 57. Locations of hydraulic gradient control constraints, screening constraints, velocity constraints, and potential removal and screening wells for "screening plus removal" problem.

a practical application. Drawdowns equal to 50 percent of the saturated thickness would also dictate using the iterative solution approach to more accurately account for the nonlinear nature of the resultant flow system.

Constraints on retarded contaminant velocities were developed using Equation (93). A retardation factor, $R_f = 12$, was used for this calculation.

The first attempt to obtain an optimal set of pumping rates for this example proved infeasible. Drawdown constraints in the screening wells became binding without satisfying the flow-reversal constraints directed from Column 8 to Column 7 (Figure 57). A second attempt with the flow-reversal constraints directed from Column 7 to Column 6 was also infeasible due to excessive drawdown in the screening wells. The MINOS solution indicated that infeasible constraints included (1) the hydraulic gradient constraints at the downgradient plume boundary, (2) the screening constraints between the plume and the screening wells, and (3) the drawdown constraints at the screening wells.

Figures 45, 46, and 47 show that the screening wells are located in a thin portion of the aquifer with low hydraulic conductivity (6.0×10^{-5} ft/s) and high hydraulic gradients. The low transmissivity in the vicinity of the screening wells creates steep drawdown cones with small radii. Because the aquifer is thin, these drawdown cones intersect the aquifer bottom (creating dry wells) before intersecting adjacent drawdown cones. The results of this exercise suggest that screening the regional flow at this site would require very closely spaced wells or even possibly a trench. Therefore, no further effort was devoted to designing a screening plus removal system.

Another attempt was made to obtain an optimal solution that included velocity constraints as a means of controlling the speed at which the cleanup progressed. This design trial was based on a modification of the pumping system shown in Figure 56. In the downgradient portion of the plume, the pumping scheme shown in Figure 56 produced velocities of approximately 55 feet per year (ft/yr). To create larger velocities, additional well locations were added in the central portion of the plume, and velocity constraints were added in locations similar to those shown in Figure 57. Several unsuccessful attempts were made to find optimal solutions with each trial using smaller magnitudes for the velocity constraints. In the downgradient portion of the plume, velocity constraints ranged from 75 to 100 feet per day (ft/day). In the upgradient portion of the plume, velocity constraints ranged from 50 to 75 ft/day. In each attempt, drawdown constraints of the form ($s \leq 0.30 \times b$) became binding before several of the velocity constraints were met. Only the velocity constraints at the downgradient plume boundary (near the injection wells) were satisfied.

These results indicate that the low conductivity sediments at the site are simply not conducive to rapid removal of the contaminated groundwater. A likely alternative for an acceptable remedial action is some degree of "perpetual care" using a low-volume pumping system for a long period of time. During operation, this system would prevent the spread of contamination and slowly remove it.

G. Conclusions

All the pumping systems designed in this chapter are solutions to optimization problems that included the same objective function: minimize the total pumping rate. The differences between the final designs resulted from different constraint sets and different potential locations for pumping and injection wells. In spite of these differences, useful comparisons can be made between the different pumping systems. Table 39 summarizes the number of active wells and the total pumping and injection rates for six different systems.

Basing comparisons on the pumping and injection rates from the confined flow approximation, the boundary-well alternative is the most desirable system if the main concern is minimizing the combined pumping and injection rates. This alternative requires installing seven wells and is nearly identical to the simple hydraulic-gradient control and interior-well alternatives. The disadvantage of these three systems is the long travel distances and travel times required for contaminants to move from the upgradient hot spot to the pumping wells. The larger total pumping rates for the first three strategies in Table 39 would also require a treatment system with approximately 30-percent greater capacity than is required for the two strategies that use pumping and injection.

The fifth strategy in Table 39 could be considered the most attractive for several reasons, including the following:

- it requires the smallest number of wells,
- it requires the smallest treatment capacity,
- it has the shortest average distance between the plume boundary and the pumping well, and
- it has the shortest distance and travel time between the plume hot spot and the pumping well.

Table 39. Summary of the Required Number of Wells and Total Pumping and Injection Rates for the Alternative Remedial Strategies

Pumping Strategy	No. of Active Pumping Wells	No. of Active Injection Wells	Total Rate (gpm)	
			Pumping	Injection
Simple Hydraulic Gradient Control	7	0	30.5[a]/28.2[b]	NA[c]
General Hydraulic Gradient Control				
—Boundary-Well Alternative	7	0	30.3[a]	NA
—Interior-Well Alternative	7	0	30.7[a]	NA
Hydraulic Gradient Control With Injection	3	5	23.3[a]	23.3[a]
Flow Reversal With Injection	1	5	23.2[a]	23.2[a]
Screening Plus Removal With Injection	—Infeasible—			

[a]Pumping and injection rates based on single solution ofthe optimization problem using a confined flow approximation of the unconfined flow system.
[b]Pumping rate based on iterative solution of the optimization problem to represent unconfined flow conditions.
[c]Not applicable.

The only apparent disadvantage of this system is that the required combined pumping and injection rate is nearly twice the pumping rate of the first three alternatives.

The fourth strategy in the table can safely be dismissed as a viable alternative because it requires the largest number of wells, the largest combined pumping and injection rate, and large travel times for upgradient contaminants.

Once a decision has been made regarding which pumping strategy to use, it would be prudent to conduct a pilot-scale design test before full construction of the remedial system. The pilot-scale test could include installing one or two of the planned remedial wells along with several observation or monitoring wells. The test would not necessarily involve any additional expense because the remedial action wells are required as part of the system and the monitoring wells are commonly required for performance assessment and operational management. Once this initial group of wells is installed, a medium-term pumping and treatment test can be conducted. Drawdown data collected from the pumping and monitoring wells can be compared with numerical model predictions. An updated model could then be developed and used to validate or modify the system design. Once the hydrologist or engineer feels that the numerical model is predicting the aquifer response with reasonable accuracy, construction of the remainder of the remedial system can proceed.

Reference for Chapter VIII

Trescott, P. C., G. F. Pinder, and S. P. Larson, *Finite-Difference Model for Aquifer Simulation in Two Dimensions With Results of Numerical Experiments*, U.S. Geological Survey Techniques of Water Resources Investigations, Book 7, Chapter C1, 1976.

APPENDIX A

Decision Analysis With a Risk-Cost-Benefit Objective Function

A. Introduction

In Chapter IV.C.3, a design option was described that involves coupling simulation and decision analysis. In that chapter, decision analysis was differentiated from optimization analysis and compared with it. Decision analysis is less general than optimization in that it identifies the best alternative from a discrete set of alternatives rather than the optimal alternative over all possible alternatives. However, the objective function used in decision analysis is in a sense more general than that used in optimization in that it includes the risks associated with a design as well as the costs and benefits. Decision analysis is well suited to a risk-based philosophy of engineering design.

The approach has been espoused by Massmann and Freeze (1987a, 1987b), but thus far their applications have been directed toward the design of new waste-management facilities rather than remedial action. Research on the latter topic is currently under way.

Computer programs are also under development for eventual documented release but are not yet available. For this reason, we cannot recommend the decision analysis approach as an "on-the-shelf" alternative to optimization. We feel, however, that a somewhat more detailed outline of the philosophy and methodology is in order and that outline is provided in this appendix. The emphasis will be on concepts and methodology. Detailed theoretical treatments will be referenced but not presented.

1. Uncertainty, Failure, and Risk

The process of engineering design is often described as a sequence of decisions between alternatives under conditions of *uncertainty*. This is a particularly apt definition for the types of engineering projects that arise in a hydrogeological context. In engineering projects that require a knowledge of the hydrogeological environment, uncertainty as to the system's properties and expected conditions is far greater than in most traditional engineering practice. Not only is there uncertainty as to the parameter values needed for design calculations, there is even uncertainty about the very geometry of the system being analyzed. The uncertainties of lithology, stratigraphy, and structure introduce a level of complexity to geotechnical and hydrogeological analyses that is unknown in most other engineering disciplines.

Consider the design of a remedial pump-and-treat well network for the purpose of migration control in a contaminated aquifer. Uncertainties of three types may plague the design process. First, as noted above, the geological setting may be uncertain; that is, there may be uncertainty about the geometry of the geological boundaries and the continuity or connectivity of the aquifers and aquitards present. Second, within each aquifer or aquitard, there is uncertainty as to the exact spatial distribution of the values of hydrogeological parameters such as porosity and hydraulic conductivity. Third, there is uncertainty as to the exact location of the boundaries of the contaminant plume that is to be captured by the well network.

When uncertainty arises, it is no longer possible to ensure that a particular engineering design will meet its objective with certainty. In the risk-balancing approach to design, it is common to use the probability of failure, Pf, as the measure of design uncertainty. This probability of failure, when multiplied by the dollar consequences of failure, constitutes the *risk*, which has units of dollars. The risk can be included with the direct benefits and costs in a risk-cost-benefit objective function within an economic decision-making framework.

Utilizing the concept of a probability of failure requires a definition of *failure*. For a migration- control network, failure involves the spread of contamination into previously uncontaminated areas of the aquifer. This could occur by contaminants slipping downstream through the well network or by lateral migration of contaminants across the presumed capture-zone boundaries. In a regulated environment, failure would usually be identified by a downstream contamination incident that violates a performance standard at a regulatory compliance point. Among the possible compliance points are regulatory monitoring wells, downstream production wells, or surface-water sampling points in streams, ponds, or lakes. Performance standards usually take the form of maximum concentration limits for particular chemical species.

2. Complexity of the Design Environment

Engineering design in a hydrogeological context is carried out in a complex technical, economic, legal, and political environment.

The *geological environment* is often heterogeneous and complex. The *hydrogeological conditions* are usually uncertain. The engineering project, itself, involves many interdependent components. Decisions between alternative courses of action are based on *economic decision-making*. They are often subject to constraints that arise from the *legal regulatory framework*. The entire process is carried out in an *adversarial environment* in the political arena.

In this day and age, engineering designs for hydrogeological projects that consider only the technical factors are not suitable. The design framework outlined in this appendix provides the necessary integration of the technical and social aspects of the decision-making milieux. It includes a decision model, a simulation model, and an uncertainty model.

The *decision model* permits the comparison of alternative engineering designs. It is an economic analysis based on a risk-cost-benefit objective function. It includes the impact of legal and political issues through the risk term.

The *simulation model* is used to represent the expected performance of the hydrogeological component of the system. It can be an analytical solution or a numerical model of the hydrogeological system at the site. Most often it will probably be a finite-difference or finite-element model of flow and transport. The hydrogeological simulation model is utilized in a stochastic mode; its purpose is to predict the probability of failure, which is a component of the risk term in the decision model.

The need for stochastic simulation arises in order to take into account the uncertainty in the hydrogeological system. The *uncertainty model* provides the necessary quantitative description of uncertainty. For uncertainty in geological boundaries or plume boundaries, the uncertainty model can be based on search theory; for uncertainty in parameter values, it is based on geostatistical concepts.

The remainder of this appendix describes these three components of the decision-analysis framework in more detail.

B. Decision Model

1. Objective Function

The technical objective of a remedial well-network design is to create a capture zone that prevents migration past a specified location and that encompasses the entire plume and source area. Alternatives are differentiated on the basis of a set of decision variables that include the number, location, depth, pumping rates, and pumping schedules of extraction and injection wells.

From the perspective of the owner-operator of the remedial facility, we can define an objective function as the net present value of the expected stream of benefits, costs, and risks taken over an engineering time horizon and discounted at the market interest rate (Crouch and Wilson, 1982). If an objective function Z_j is defined for each j = 1 . . . .N alternatives, then the goal is to maximize Z_j:

$$Z_j = \sum_{t=0}^{T} \left[\frac{1}{(1+i)^t} \right] [B_j(t) - C_j(t) - R_j(t)] \tag{A-1}$$

where
Z_j = objective function for alternative j [$],
T = planning horizon [years],
i = annual discount rate [decimal fraction],
$B_j(t)$ = benefits of alternative j in year t [$],
$C_j(t)$ = costs of alternative j in year t [$], and
$R_j(t)$ = risks of alternative j in year t [$].

The risks, R(t), in (A–1) are defined as the expected costs associated with the probability of failure:

$$R(t) = Pf(t)Cf(t)\gamma(Cf) \tag{A-2}$$

where
Pf(t) = probability of failure in year t [decimal fraction],
Cf(t) = costs associated with failure in year t [$], and
γ (Cf) = normalized utility function [decimal fraction, $\geq$ 1].

In Equation (A–1) the C(t) term represents the cost to the decision maker and the B(t) term the benefits. If the decision maker is the owner-operator of a remedial facility at the design stage and a set of alternative conceptual designs are under consideration, then the costs would be the capital costs and operational costs of constructing and operating the remedial facility under each design, and the benefits would be any revenues that might accrue. The probabilistic costs, Cf(t), that appear in the risk term are those costs that would be incurred only in the event of failure. They might include regulatory penalties, loss of good will in the community, possible closure of the waste management facility that is the source of the plume (if such is the case), and the additional costs of further remedial action. In many remedial situations there are no direct benefits to accrue to the decision maker and the decision model reduces to a risk-cost minimization.

Figure A–1 (adapted from Crouch and Wilson, 1982) summarizes the application of a decision model to a set of alternatives.

2. Utility

The utility function, $\gamma(Cf)$, in Equation (A–2) allows one to take into account the possible risk-averse tendencies of some decision makers.

The concept of utility encompasses the second and third terms on the right-hand side of Equation (A–2) (Crouch and Wilson, 1982). Utility theory holds that decision makers often do not make decisions on the basis of the expected cost associated with failure. Rather, they exhibit their risk-averseness by multiplying these expected costs by a factor that is thought to depend on the level of expected failure costs relative to the net worth of the decision maker. A risk-averse decision maker will set γ in (A–2) greater than one; a risk-neutral decision maker will set $\gamma = 1$. Smaller owner-operators who do not have a large net worth are the most likely to use a risk-averse utility function. Larger companies are more likely to take a risk-neutral approach.

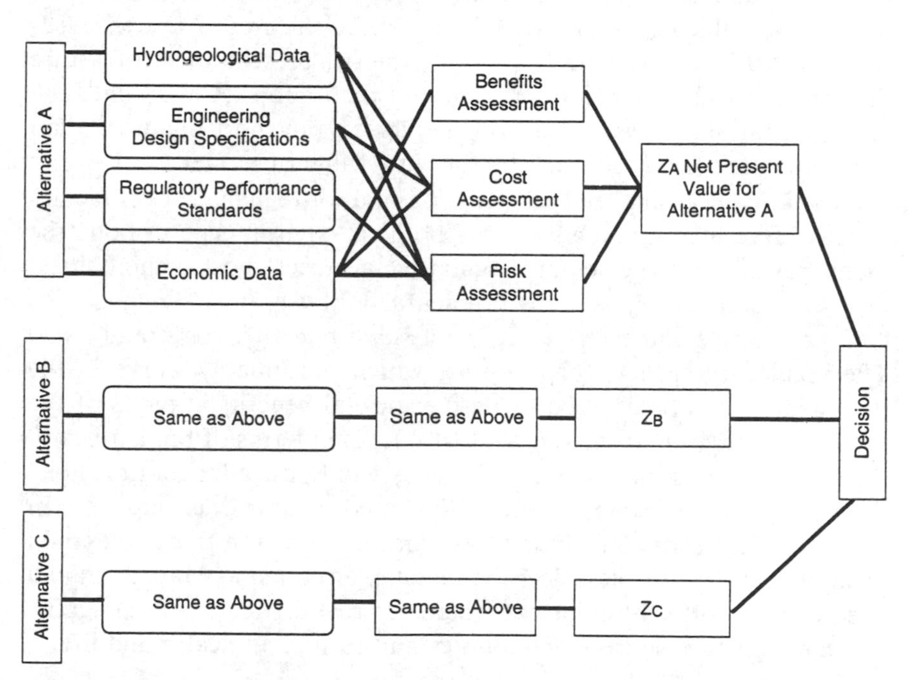

Figure A–1. Application of a decision model to a set of alternatives (adapted from Crouch and Wilson, 1982).

Risk aversion should not be confused with risk perception. People often do not have a realistic estimate of their risk due to actual or perceived threats. This causes them to overestimate the expected value of Cf. The more experienced a company is in the waste management industry, the more likely they are to have accurate risk perceptions.

Having raised the issues of risk-aversion and risk-perception, we now dismiss them. We will assume from here on that the decision makers come from large experienced companies with a realistic view of the Cf terms, and that they are willing to set the γ-term in Equation (A–2) equal to unity for all Cf.

3. Time Horizon and Discount Rate

In this appendix, the decision-analysis framework is presented from the perspective of the owner-operator of a waste management facility. From this perspective, the pertinent time-horizon for decision-making is a relatively short one, on the order of 20 to 50 years, and the pertinent discount rate is the market interest rate on borrowed money.

The discount rate and the time horizon are not really independent in a net-present-value economic analysis. It can be shown (cf. Dieter, 1983) that for a discount rate of 10 percent, the net present value of future dollars approaches zero for periods more than about 50 years into the future. A rational decision maker will not use a time horizon longer than the economic period indicated by the time value of money.

The siting, licensing, and construction of a waste management facility usually takes place in an adversarial political environment. Among the many players in the game are regulatory agencies, host communities, insurance companies, and environmental lobby groups. Each has its own perspective and each can define its own benefits, costs, and risks. The regulatory agency, for example, which presumably works in the public interest, would probably define societal benefits in terms of the value of clean groundwater and societal risks in terms of human health and environmental protection. An agency would use a longer time horizon and a lower discount rate than that used by an owner-operator. In principle, the decision-analysis framework described in this book could be applied by any of these decision makers, although as Massmann and Freeze (1987a, 1987b) point out, there are real difficulties in assigning dollar values to resource commodities and to human health and life.

4. Integrated Design Process

The design process at a remedial facility involves a sequence of at least four steps: (1) siting, (2) site investigation, (3) design of the remedial facility, and (4) design of the monitoring network. Each of these steps involves a decision process of deciding between alternatives. How many holes will be drilled during site investigation? How many wells are needed for capture? How many monitoring points are needed?

In the traditional approach, there has been a tendency to treat each of these decision processes independently. They are carried out sequentially, and decisions in the later steps are postponed until the results arising from the decisions made in earlier steps are available.

It is possible to conceive of a more integrated decision process in which the set of alternatives are lined out in such a way that each alternative covers the entire design process. This allows the owner-operator to assess *economic trade-offs* between the various steps. For example, would it be better to use minimal site investigation and a conservative design, or would it be better to carry out a detailed site investigation in the hopes of buying reduced construction costs. The owner-operator would like to know how to partition his resources among the competing requirements of site investigation, remediation, and monitoring.

It must be emphasized that this integrated approach does not require that all decisions be reached at the beginning before site investigation results can be assessed. On the contrary, as will become clear below, it is logical to carry out a sequentially iterative decision process whereby decisions are constantly reassessed and updated as additional site information becomes available. But at each of these decision points, it is the fully integrated alternatives that are assessed.

5. Bayesian Design Philosophy

Decision analysis is carried out in a Bayesian framework. The implications of this statement are best seen from a statistical perspective, where there have always been two philosophical camps: those who espouse classical statistics, and those who espouse Bayesian statistics. Classical statistics requires the development of a probability density function based on measured data. Estimates of the mean and variance of the data set, or the use of such statistics to test hypotheses, must await the existence of sufficient data to allow the form of the probability density function to be established and to ensure that the desired level of confidence in the estimates has been reached. When additional data become available, the data are used to enlarge the data set, and the summary statistics are recalculated with a higher level of confidence.

With a Bayesian statistical approach, a *prior estimate* of the form of the probability density function and its summary statistics is made. This prior estimate may be based on limited early data, or it may be based on subjective information that is available in the form of experience and personal judgment, even before any measurements are taken. When additional data become available, they are used to update the prior estimates of the statistics to *posterior estimates* using Bayes theorem. The posterior estimates are influenced both by the new data and by the prior estimates. Under sparse data conditions, the Bayesian statistical estimates could be quite different from the classical statistical estimates for the same data set. As the data set becomes larger, the two sets of estimates converge.

In a hydrogeological context, data sets for porosity or hydraulic conductivity are often very sparse, and a classical statistical analysis often seems to suggest that little has been learned from early measurement programs. It seems right and proper to allow the experience gained at other sites or the implications that can be gleaned from "soft" data at the site, such as lithologic descriptions or geophysical surveys, to play a role in reducing uncertainty. In a Bayesian context, this type of information provides the basis for the subjective prior estimates that may still exert considerable influence on posterior estimates following data collection.

In an engineering context, the Bayesian approach fits perfectly into the sequential design framework described above, whereby the design engineer iterates between analysis and measurements as the project progresses. If the analysis takes the form of a hydrogeological simulation model, the Bayesian approach supports the establishment of a preliminary model early in the project, followed by continual updating as field data become available.

C. Simulation Model

1. Steady-State Flow; Advective Transport

The purpose of the simulation model is to provide an estimate of the probability of failure, Pf, for use in the decision model.

The simulation model may take the form of an analytical solution, or it may take the form of a finite-difference or finite-element model. In either case, it provides a predictive analysis of groundwater flow and/or contaminant transport in the hydrogeological environment. The simulation model will be invoked for all alternatives under consideration. These may include a no-action alternative as well as several alternative remedial pump-and-treat well networks.

Most capture-zone analyses for the design of migration-control well networks are carried out with two-dimensional steady-state flow models that assume the primacy of advective transport. They are based on the steady-state flow equation (Chapter II) with suitable boundary conditions prescribed. Input requirements include (1) the boundaries of the flow domain; (2) the boundary conditions on the boundaries; (3) the location of boundaries between geological layers, units, or zones; (4) the spatial distribution of hydraulic conductivity, K(x,y), and porosity, n(x,y), within each layer, unit, or zone; and (5) the locations of extraction and injection wells and their pumping rates.

Output is in terms of the head distribution, h(x,y). However, simulation models can easily be programmed to provide equipotential output, stream-function output, particle-tracking output, or travel-time output. One or more of these options are provided by the finite-difference model, MODFLOW; the particle-tracking model, GWPATH; and the analytical models, RESSQ and DREAM. All four models are described in Chapter IV.D.

If one of these models is to be used in the design of a remedial well network, the location of the plume boundaries must also be known. The strength of the source, or the pattern of concentrations within the plume, are not required as long as it is assumed that compliance is required downstream of the plume and that the performance standard for the contaminant under analysis is greater than the ambient concentration and less than the concentrations in the plume.

2. Monte Carlo Simulation

All of the input parameters listed above are subject to uncertainties. The methods of addressing these uncertainties are described in Chapter D of this appendix. Here, we consider only how such uncertainties can be taken into account in stochastic simulations.

In stochastic analysis, uncertainty in the input parameters is specified in the form of a probability density function (pdf) or by the first two moments of such a distribution. There are three basic approaches used to propagate these uncertainties through the hydrogeological simulation model to estimate uncertainties of output variables: (1) first-order analysis (Dettinger and Wilson, 1981), (2) perturbation analysis (Gelhar, 1984), and (3) Monte Carlo analysis. We limit our discussion here to the latter, which is the most general and most widely used approach.

With Monte Carlo simulation, a large number of equally likely realizations of each parameter field are generated and the hydrogeological simulation model is run for each realization. The method requires that the full pdf be specified for each input parameter. The pdf of the output variables is obtained from a statistical analysis of the output from the Monte Carlo runs. Peck et al. (1988) discuss some of the issues involved in Monte Carlo analysis of groundwater systems. The method is widely used because of its generality and simplicity. It can be used with analytical or numerical solutions, in infinite or bounded domains, and for any level of input uncertainty. Its Achilles heel lies in the computer-intensive nature of the endeavor. For complex transport problems in large heterogeneous domains, Monte Carlo analysis may be computationally infeasible. However, for steady, saturated, advective transport in two-dimensional systems on the scale of an engineering site, Monte Carlo simulation is a feasible proposition.

In most advective transport analyses, the input parameter with the largest uncertainty is the hydraulic conductivity, K. It must be treated as an autocorrelated spatial stochastic process. The available tools for the generation of heterogeneous hydraulic-conductivity realizations in a two-dimensional domain are discussed later in this appendix. Monte Carlo simulation can also be used to assess the impact on the probability of failure of uncertainties in geologic boundaries or plume boundaries.

3. Unconditional and Conditional Simulations

Stochastic simulations identify uncertainty in predicted output variables due to uncertainty in input parameters. This uncertainty can be reduced by measurements of the input parameters.

Unconditional simulations are simulations that are not conditioned on any measured input. In the Bayesian framework, they are usually simulations designed to produce prior estimates of output uncertainty based on subjective estimates of input values made before measured data become available.

Conditional simulations are simulations conditioned on the values of measured input parameters and on the specific locations of those measurements. At the points of measurement of K, for example, uncertainty in K is reduced to zero (or to the level of measurement error). Between the points of measurement, uncertainty in K remains. Stochastic conditional simulation provides the transfer function that predicts the uncertainty in hydraulic heads and, therefore, in gradients and plume migration directions.

The available tools for updating uncertainties under the influence of additional input measurements are discussed later in this appendix. One of these tools is kriging, and applications of conditional simulation are often associated with discussions of kriging (cf. Clifton and Neuman, 1982).

D. Uncertainty Models

Predictions of hydraulic-head patterns or plume-front travel times at field sites are subject to large uncertainties because of the uncertainty in input parameters. There may be uncertainties in the geological configuration, plume boundaries, and hydrogeological parameter values. Of these, the first two can be addressed with search theory and the third with the geostatistical concepts of stochastic process theory. Search theory is simple in concept and we will discuss it first. Stochastic process theory is complex, and a full description of it is outside the scope of this appendix. What follows should be sufficient to glean an understanding of the concepts that underlie these two methodologies. The interested reader is directed to Freeze et al. (1990) and Massmann et al. (1991) for a more complete presentation.

1. Boundary Uncertainty; Search Theory

In this appendix, a simple but realistic geological environment is used to illustrate geological uncertainty. Consider the question of aquitard continuity, as illustrated in Figure A–2. If the horizontal aquitard shown on that figure is continuous, then any contaminant plume that might migrate past a remedial facility will probably be constrained to the upper aquifer horizon. If it is not continuous, and particularly if there is pumping in the lower aquifer, the plume may be drawn through the holes in the aquitard. The probability of this occurrence could have a large impact on the probability of failure and the risk term associated with specific alternative designs for a remedial facility.

Uncertainty as to the presence of holes in the aquitard would be reduced by the drilling and logging of boreholes. Search theory is a method by which the geological data obtained from such boreholes can be used to assess the reduction in uncertainty. Search theory treats the drillhole intersection with the aquitard as a yes-no situation. Either the aquitard is present at the location of the drillhole or it is absent, in which case a hole of some dimension exists. The first applications of search theory in a geological context were in the mineral exploration field for the design of drilling patterns in search of ore bodies. Drew (1979), for

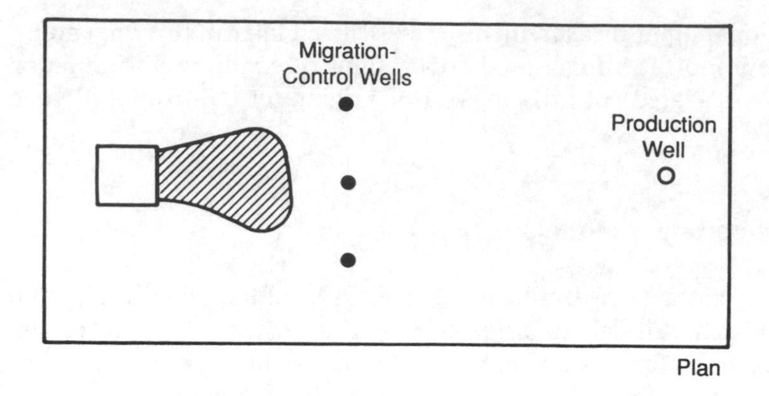

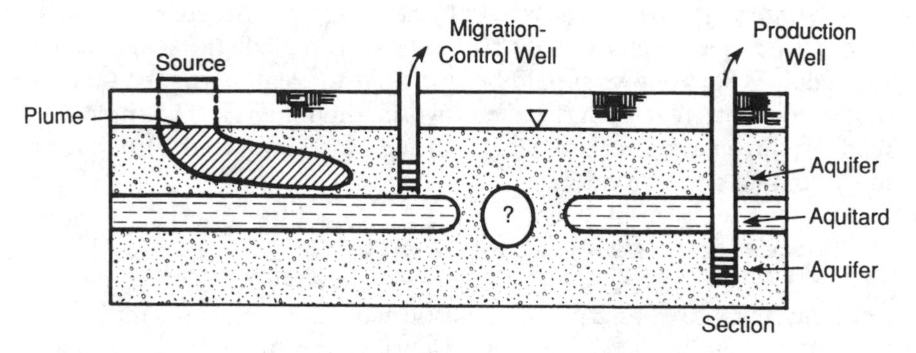

Figure A–2. Example of aquitard continuity.

example, provides an analysis of the optimum drilling pattern and hole spacing for elliptical targets of various sizes. Savinskii (1965) provides probability tables for locating elliptical targets with a rectilinear grid of drillholes. These analyses and tables are relevant for hydrogeological exploration.

Consider a circular target of diameter, a, and a square drillhole grid of spacing, d (Figure A–3a). The tables referenced above provide the probability that the target exists even though it is missed by the grid. As the grid spacing is reduced (Figure A–3b), this probability decreases for all target diameters. If the target is an aquitard hole, pattern drilling can be used to reduce the probability of its existence (i.e., to reduce the uncertainty associated with its possible presence).

Suppose, for example, that it is decided that an aquitard hole 60 meters in diameter would produce a failure at a particular site. A square-

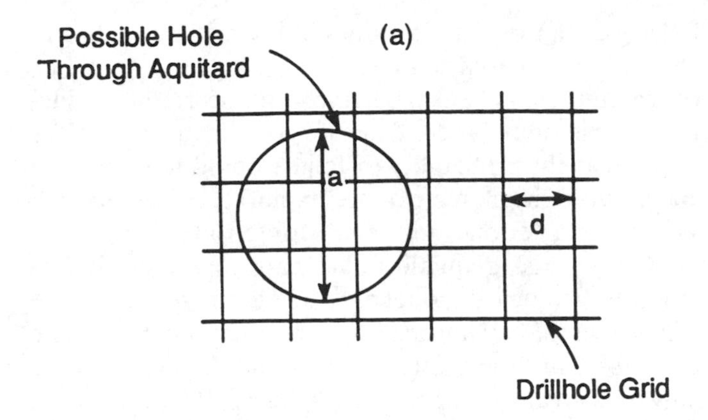

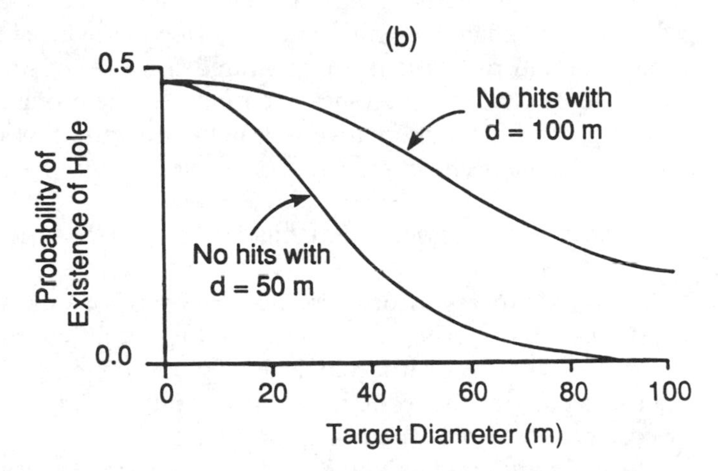

Figure A–3. Use of search theory to reduce uncertainty as to possible presence of holes through an aquitard.

grid drilling program at a spacing of 100 meters has failed to encounter a hole of any size. However, according to Figure A–3b, the probability of the existence of a 60-meter diameter hole is still greater than 30 percent. With a probability of failure, Pf, greater than 30 percent, and a large potential cost, Cf, associated with such a failure, the risk (which is the product of Pf and Cf) could be very large. If a denser drilling program is used, with a grid spacing of 50 meters, the probability of encountering a hole is reduced to less than 10 percent, and the risk is reduced fivefold. The question of which drillhole spacing is better depends on whether the risk reduction that is achieved by the denser grid exceeds its additional cost, and this will depend on the relative values of the drilling costs and the costs associated with a potential failure.

The concepts of search theory are also pertinent to uncertainty reduction with respect to plume boundaries. Capture-zone analysis is quite sensitive to plume extent so the probability of failure ought to be equally sensitive to the uncertainties in plume extent. Exploratory drilling will reduce these uncertainties. Prior and posterior Monte Carlo simulation can identify the degree of risk reduction that can be achieved. For example, assume that the geological model and the parameter values are all known with certainty so that all uncertainty is bound up in the plume size. A prior distribution as to the center of mass and diameter can be assumed and equally likely realizations can be generated. These can then be used for a set of prior unconditional Monte Carlo simulations carried out for each alternative remedial well network. After a grid drilling program, the updated distribution, reflecting the reduced uncertainty, is calculated, and realizations are generated for a set of posterior, conditional Monte Carlo simulations. In both cases, the probability of failure is determined for all alternative well networks, and this is fed into the risk term in the decision model.

2. Parameter Uncertainty; Stochastic Process Theory

Stochastic process theory provides the tools for quantifying the uncertainty in the spatially distributed values of input hydrogeological parameters within a hydrogeological unit. It also includes the methodology for calculating the reduction in uncertainty that can be achieved by a specific proposed measurement program.

The hydraulic conductivity K(x,y) and the porosity, n(x,y), have been identified as the hydrogeological parameters of interest for steady saturated advective transport in an aquifer. For the purposes of discussion in this chapter, we have selected the hydraulic conductivity as the uncertain parameter. The same conceptual framework would hold for porosity or for any of the other hydrogeological parameters that might arise from more complex transport formulations.

Most data sets that have been gathered for hydraulic conductivity exhibit a lognormal distribution. The presentation is therefore easier to follow if we define the log hydraulic conductivity, Y, such that:

$$Y = \ln K. \tag{A-3}$$

The parameter Y is normally distributed. The population of Y has a mean μ_y, and a variance, σ_y^2. An estimate of the population mean, $\hat{\mu}_y$, and

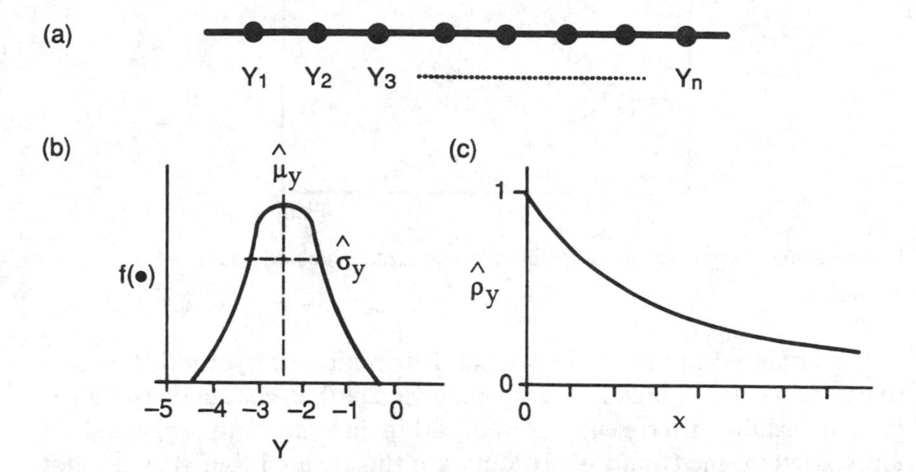

Figure A–4. (a) One-dimensional sequence of log hydraulic conductivity measurements; (b) Probability density function of measured values; and (c) Autocorrelation function of measured values.

variance, $\hat{\sigma}^2$, can be obtained from the sample mean, $\bar{Y}$, and the sample variance, S_y^2, where $\bar{Y}$ and S_y^2 are calculated from a set of measurements of Y. The parameter, S_y^2 (or σ_y^2), is one measure of the heterogeneity of an aquifer.

Consider a one-dimensional sequence of log hydraulic conductivity measurements: $\{Y_1, Y_2, \ldots Y_n\}$ as shown in Figure A–4a. Y is normally distributed with mean μ_y and standard deviation, σ_y (Figure A–4b). It is common to find that the observed values of Y_i are autocorrelated. Values separated by short distances will be highly correlated, and those separated by long distances only weakly correlated or not correlated at all. The function that displays the drop in correlation with distance is called the *autocorrelation function*, $\rho_y(x)$, (Figure A–4c). It is often found to be exponential in form:

$$\rho_y(x) = \exp[-|x|/\lambda_y] \tag{A–4}$$

where λ_y is an exponential decay parameter known as the *correlation length*. It is the distance over which $\rho_y(x)$ decays to a value of e^{-1}. A full geostatistical description of the heterogeneity of Y is given by the three-parameter *stochastic process* defined by the autocorrelated normal distribution, Y:N $[\mu_y, \sigma_y, \lambda_y]$. It is assumed for the purposes of our ongoing discussions that the stochastic process is *stationary*; that is, that the values of μ_y, σ_y, and λ are constant over space in the domain of interest.

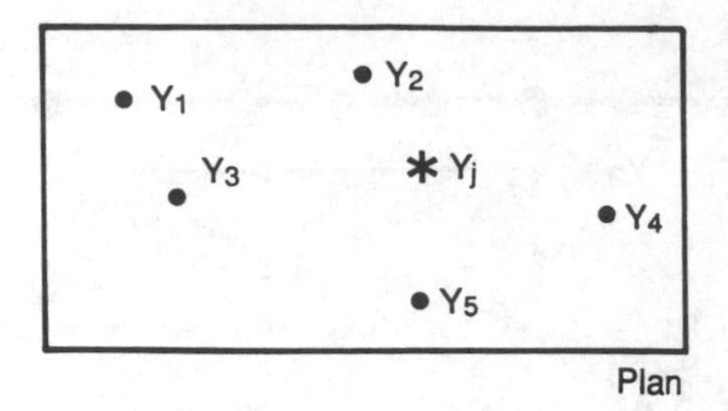

Figure A–5. Two-dimensional region of a horizontal confined aquifer.

Uncertainty as to the exact spatial distribution of hydraulic conductivity values arises because of our knowledge that conductivity values tend to exhibit heterogeneity, even within individual hydrogeological units. Heterogeneity and uncertainty are thus related, but they are not the same thing. *Heterogeneity* is in the geology, whereas *uncertainty* is in the mind of the analyst. The concept that allows us to integrate these issues, through the description of the K(x,y) field as a random realization of a stochastic process, is known as *ergodicity*.

Consider a two-dimensional region of a horizontal confined aquifer (Figure A–5) within which n measurements of Y are available: $\{Y_1, Y_2, \ldots Y_n\}$. Now suppose that we wish to estimate the value of Y_j at some position j that lies between (and well away from) any of the measured Y_i. If we assume that the most likely estimate of Y_j is $\bar{Y}$ (the mean value of the measured Y_i), and if we assume that our uncertainty about the value of Y_j is normally distributed with standard deviation, S_y (the standard deviation of the measured Y_i), then we are accepting the *ergodic hypothesis* (that is, we are assuming that the pdf of the analysts' uncertainty is the same as the pdf of the aquifer heterogeneity).

Given the above framework for the description of heterogeneity, and given an acceptance of the ergodic hypothesis, we now turn to the question of the uncertainty in K at unmeasured points.

Our uncertainty is greatest before we take any measurements at all; but in a classical statistical framework, we would have no way of estimating the values of μ_y, σ_y, or λ_y prior to taking measurements. In a Bayesian framework, however, we make subjective *prior estimates* of these parameters. We do so by using our knowledge of the geological environment at the site, our interpretation of any available "soft" data at the site, our past experience with respect to available data at similar sites, precedent, and common sense.

Consider now a one-dimensional field of Y values: $\{Y_1, Y_2, \ldots Y_n,\}$ as in Figure A–4a, where the locations of the Y_i now represent points at which we wish to estimate the expected value and uncertainty of Y_i prior to taking of measurements. Let $\{\mu_i\}$ be the vector of expected values, and let $\{\tau_{ii}\}$ be the vector of variances that describe our uncertainty. Prior to taking measurements and under the ergodic hypothesis, all the expected values, $\{\mu_1, \mu_2, \ldots \mu_n\}$, in the $\{\mu_i\}$ vector will be equal to $\hat{\mu}_y$, a subjective prior estimate of the mean of the stochastic process that describes the log hydraulic conductivity field; and all the uncertainty values $\{\tau_{11}, \tau_{22}, \ldots \tau_{nn}\}$ will be equal to $\hat{\sigma}_y^2$, the subjective prior estimate of the variance of the stochastic process that describes the log hydraulic conductivity field.

Given $\{\mu_i\}$ and $\{\tau_{ii}\}$ and an autocorrelation function like (A–4), it is possible to generate an infinite number of equally likely *realizations* of the log hydraulic conductivity field. There are several methods for doing so. The two most popular are Cholesky decomposition (cf. Clifton and Neuman, 1982) and the turning bands approach (cf. Mantoglou and Wilson, 1982). The interested reader is directed to these references for details.

In the above description, the uncertainty vector $\{\tau_{ii}\}$ has been given a double subscript because in reality it consists of the diagonal terms of the autocovariance matrix $\{\tau_{ij}\}$. This matrix figures prominently in the mathematical presentation of stochastic process theory.

Prior realizations are not conditioned on any measured Y values; they are unconditional realizations suited to unconditional simulation. Assume that some measurements of log hydraulic conductivity now become available at one or more points in the aquifer. For the purpose of explanation, we return to our one-dimensional sequence of points, $\{Y_1, Y_2, \ldots Y_n\}$, and assume that the measurement Y_4 has become available. This measurement allows us to update our prior vector of expected values, $\{\mu_i\}$, to a posterior vector, $\{\mu_i'\}$, and our prior vector of uncertainties, $\{\tau_{ii}\}$, to a posterior vector, $\{\tau_{ii}'\}$. At $i = 4$, the posterior expected value, μ'_4, will equal Y_4, and if there is no measurement error, the posterior uncertainty, t_{44}', will be equal to zero. Because of the autocorrelation properties of the log hydraulic conductivity field, the influence of this measurement will propagate through the vector of expected values and the vector of uncertainties, changing the expected values and reducing the uncertainty in a region around the point of measurement, the size of which depends on the correlation length. The quantitative posterior values can be calculated using the Bayesian updating procedures outlined by Hachich and Vanmarcke (1983) and Massmann and Freeze (1987a). The mathematics are not presented here.

At this stage it is possible to generate a set of equally likely conditional realizations using similar programs and procedures to those used to generate unconditional realizations. One can then apply Monte Carlo simulation on the conditional realizations to produce estimates of the output hydraulic head field and its uncertainty. For each remedial alternative, the probability of failure can be determined, the risk calculated, and the objective function assessed. Freeze et al. (1990) and Massmann et al. (1991) describe the entire framework for application of this methodology to waste management projects involving the hydrogeological environment.

The parameter uncertainty model using Bayesian updating as described thus far could easily be replaced by a parameter uncertainty model based on kriging. *Kriging* provides an alternate way to address the interpolation problem outlined in Figure A–5 where we wish to estimate the expected value, Y_j, and the uncertainty associated with the estimate, at points between a set of measurement points, $\{Y_1, Y_2 \ldots Y_n\}$. Kriging can be used with conditional simulation in the same way as Bayesian updating to predict uncertainties in output hydraulic head fields. Kriging is closely related to Bayesian updating, both in spirit and in the nature of the calculations. However, there are differences in the jargon and in the assumptions and philosophy that underlie the methods. These differences are summarized and explained in Freeze et al. (1988). de Marsily (1986) provides a sound treatment of kriging theory in a hydrogeological context.

3. Data Worth

One of the most valuable features of this decision framework is its ability to assess the worth of a proposed site-investigation program prior to actually taking the measurements. This is possible because the uncertainty reduction that can be attained by a measurement program depends only on the number and location of the measurements not on their measured values. The uncertainty reduction feeds into the risk term in the risk-cost-benefit objective function. A proposed site-investigation program should only be carried out if the value of the risk reduction it will achieve is greater than the cost of carrying it out. At any stage, it is possible to compare the cost-effectiveness of further exploratory drilling relative to the cost of a more conservative purge-well network as a means to reducing risk.

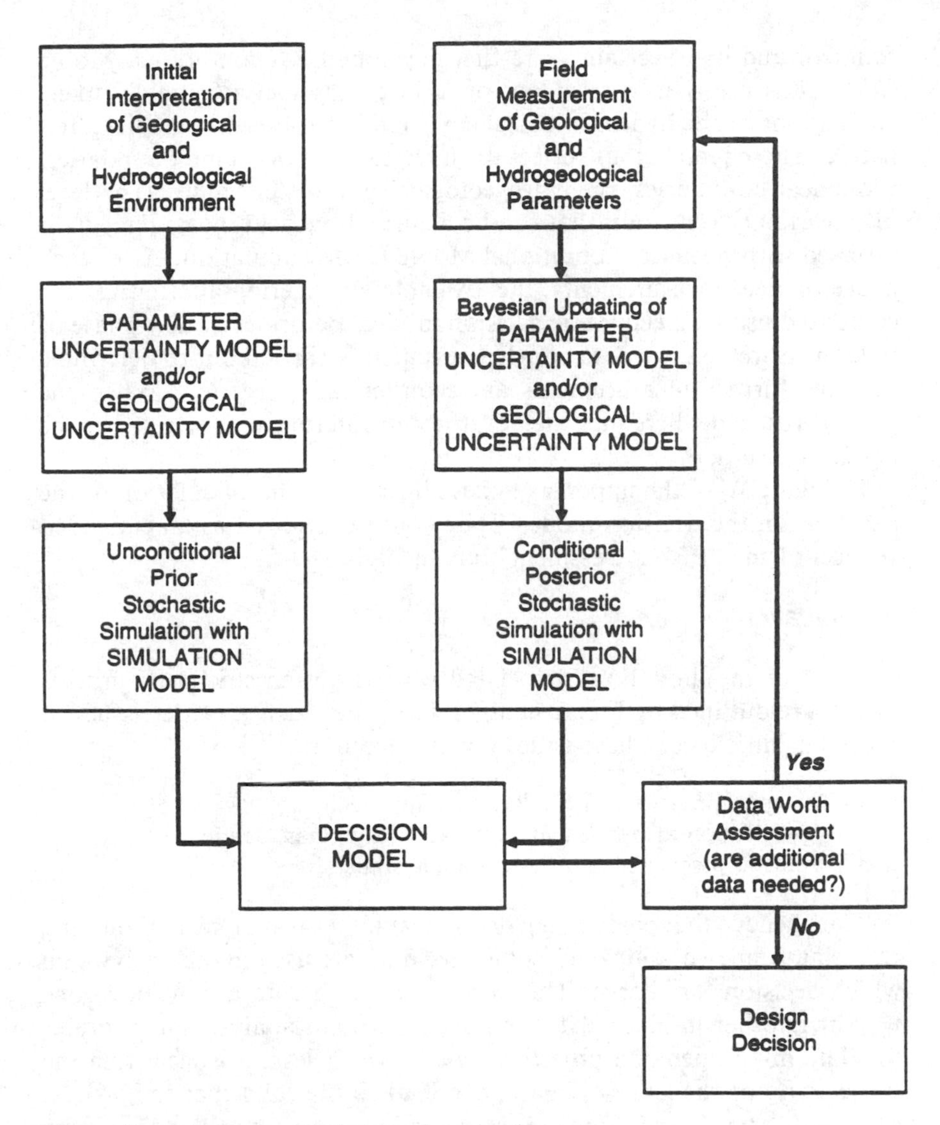

Figure A–6. Summary of decision framework.

E. Summary and Limitations of Decision Framework

1. Summary

Figure A–6 summarizes the decision framework presented in this appendix. It involves an iterative process between analysis and field measurement. A subjective prior interpretation of the hydrogeological environment and its uncertainty are first developed. Unconditional Monte Carlo runs are then carried out on a large number of equally likely realizations of the hydrogeological environment for each remedial alternative. These realizations reflect the uncertainties in plume boundaries, geological boundaries, or hydrogeological parameter values. The level of uncertainty reduction that can be achieved by field measurements is assessed with posterior conditional Monte Carlo simulation. After each phase of field measurements, the available engineering alternatives are compared using an economic decision model. Before each new phase of field measurements, a data worth assessment is carried out to determine whether further measurements are economically justified. When the point is reached where they are not, the best alternative is selected as the design decision.

In Figure A–6, the upper six boxes all relate to the calculation of the risk term in the decision model. They could be viewed as an expanded version of the "Risk Assessment" box in Figure A–1.

2. Limitations

Some of the limitations associated with Bayesian stochastic process theory are outlined by Freeze et al. (1988). They identify three issues of special significance. These issues revolve around

1. The appropriateness of the Bayesian approach,
2. Identification of trends and zonations in hydraulic conductivity, and
3. Problems associated with small-sample statistics.

They believe that engineering decision-making is best carried out in a Bayesian framework in which subjective data are used in the analyses on which decisions are based. They argue that such data are included one way or another in all analyses but that Bayesian analysis incorporates the data in an open and objective way. Nevertheless, it is clear that the subjectivity of the prior estimates can lead to bias and that the process could easily be abused. Some scientists and engineers may feel that using subjective prior estimates is unscientific or unprofessional, and there is some question as to public and political acceptance.

The identification of boundaries between hydrogeologic units and trends within hydrogeologic units may prove difficult, especially in fractured rocks involving complex, discontinuous, structural features that may exert considerable influence on flow and transport. Current geostatistical methods do not sufficiently emphasize the uncertainty spawned by potential geological misinterpretations.

The potential weakness in the application of uncertainty models at engineering sites lies with the problems that arise when data networks are sparse and sample sizes are small. Among the steps where limitations due to small-sample statistics may arise are the following: (1) testing the hypothesis of a lognormal distribution for hydraulic conductivity, (2) testing for stationarity of the hydraulic-conductivity field, (3) determining the most appropriate form of the autocorrelation function and estimating the values of its parameters, and (4) obtaining significant uncertainty reduction in the Bayesian updating step. Decision analysis highlights the weakness of sparse data networks by identifying the existence of a large risk. It also identifies the best decision given the level of data available, regardless of whether that level is sparse or full.

The discussion in this appendix assumes that parameter uncertainty is limited to a single parameter, the hydraulic conductivity. In reality, there is uncertainty in all input parameters, and some of them such as porosity and conductivity may be jointly distributed. To date, the hydrogeological decision-analysis literature has not addressed the complex issues associated with multiple-parameter uncertainty.

Those readers whose interest is piqued by this brief development of the emerging ideas of hydrogeological risk analysis are directed to the very few papers that have appeared thus far (Massmann and Freeze, 1987a, 1987b; Freeze et al., 1990; Massman et al., 1991; Ben-Zvi et al., 1988; Marin et al., 1989).

References for Appendix A

Ben-Zvi, M., B. Berkowitz, and S. Kesler, "Pre-posterior Analysis as a Tool for Data Evaluation: Application to Aquifer Contamination," *Water Resources Management,* Vol. 2, pp. 11–20, 1988.

Clifton, P.M., and S.P. Neuman, "Effects of Kriging and Inverse Modelling on Conditional Simulation of the Avra Valley Aquifer in Southern Arizona," *Water Resources Research*, pp. 1215–1234, 1982.

Crouch, E.A.C., and R. Wilson, *Risk/Benefit Analysis*, Ballinger, Boston, MA, 1982.

de Marsily, G., *Quantitative Hydrogeology,* Academic Press, 1986.

Dettinger, M.D., and J.L. Wilson, "First-Order Analysis of Uncertainty In Numerical Models of Groundwater Flow, 1. Mathematical Development," *Water Resources Research*, Vol. 17, pp. 149–161, 1981.

Dieter, G.E., *Engineering Design: A Materials and Processing Approach*, McGraw-Hill, New York, NY, 1983.

Drew, L.J., "Pattern Drilling Exploration: Optimum Pattern Types and Hole Spacings When Searching for Elliptical Shaped Targets," *Mathematical Geology*, Vol. 11, pp. 223–254, 1979.

Freeze, R.A., L. Smith, G. de Marsily, and J.W. Massmann, "Some Uncertainties About Uncertainty," Proc. AECL/DOE '87 *Conference on Geostatistical, Sensitivity, and Uncertainty Methods for Groundwater Flow and Radionuclide Transport Modelling*, San Francisco, CA, pp. 231-260, 1988.

Freeze, R.A., J. Massmann, J.L. Smith, A. Sperling, and B. James, "Hydrogeological Decision Analysis: 1. A Framework," *Ground Water,* Vol. 28, pp. 738-766, 1990.

Gelhar, L.W., "Stochastic Analysis of Flow in Heterogeneous Media," in *Fundamentals of Transport Phenomena in Porous Media*, edited by J. Bear and Y. Corapcioglu, NATO Advanced Study Institute, Series E., No. 82, pp. 673-717, 1984.

Hachich, W., and E. Vanmarcke, "Probabilistic Updating of Pore Pressure Fields," *Journal of Geotechnical Engineering Division, Procedures of American Society of Civil Engineers*, Vol. 109, pp. 373-385, 1983.

Mantoglou, A., and J. Wilson, "The Turning Bands Method for Simulation of Random Fields Using Line Generation by a Spectral Method," *Water Resources Research*, Vol. 18, pp. 1379-1394, 1982.

Marin, C.M., M.A. Medina, and J.B. Butcher, "Monte Carlo Analysis and Bayesian Decision Theory for Assessing the Effects of Waste Sites on Groundwater. I. Theory," *Journal of Contaminant Hydrology,* Vol. 5, pp. 1-13, 1989.

Massmann, J., and R.A. Freeze, "Groundwater Contamination From Waste Management Sites: The Interaction Between Risk-Based Engineering Design and Regulatory Policy, 1. Methodology," *Water Resources Research*, Vol. 23, pp. 351-367, 1987a.

———, "Groundwater Contamination From Waste Management Sites: The Interaction Between Risk-Based Engineering Design and Regulatory Policy, 2. Results," *Water Resources Research*, Vol. 23, pp. 368-380, 1987b.

Massman, J., R. A. Freeze, L. Smith, T. Sperling, and B. James, "Hydrogeological Decision Analysis 2. Applications to Ground-Water Contamination," *Ground Water*, Vol. 29, pp. 536-548, 1991.

Peck, A., S. Gorelick, G. de Marsily, S. Foster, and V. Kovalesvsky, "Consequences of Spatial Variability in Aquifer Properties and Data Limitations for Groundwater Modeling Practice," *International Association of Science Hydrology*, No. 175, 1988.

Savinskii, I.D., *Probability Tables for Locating Elliptical Underground Masses with a Rectilinear Grid*, Consultants Bureau, Plenum Press, 1965.

APPENDIX B

Nonlinear Optimization Methods for Groundwater Quality Management Under Uncertainty

A. Introduction

Chapter V presented optimization techniques that were both linear and deterministic. Although not yet ready for "on-the-shelf" applications at field sites, there have been many recent developments in nonlinear optimization for groundwater quality management and in the integration of uncertainty into such analyses. This appendix discusses the basic formulations and approaches that have been developed to integrate simulation with nonlinear optimization for aquifer management modelling under uncertainty. Much of this material has only recently appeared in the journal literature cited throughout this appendix. Formulations and some results are discussed for both deterministic and stochastic simulation-optimization problems.

B. Nonlinearities

Groundwater management problems are frequently nonlinear because of the coupling that exists between the groundwater flow equation and the contaminant transport equation (see Chapter II). They are coupled in three ways. First, they are coupled through the velocity vector components, V, which appear in the transport equation but which are obtained through Darcy's law. Second, they are coupled through the hydrodynamic dispersion tensor, D, which is a function of the groundwater velocity. Third, they are coupled through the fluid source/sink term that serves to add or extract both water and contaminant to or from the aquifer.

For problems involving simulation alone, flow and transport simulation is often treated as a linear system. Once hydraulic heads are computed using the flow equation, then the velocity components and dispersion coefficients are readily calculated. With these coefficients known, the transport equation is a linear system and solution is straightfoward.

Unfortunately, in combined simulation-management problems, the velocity components and dispersion coefficients cannot be defined because the source/sink terms are decision variables. The transport equation is then a nonlinear equation, containing products of unknown velocities and dispersion coefficients multiplied by unknown concentrations or their derivatives. The inherent coupling between the flow and transport equations is the source of the nonlinearities. The important point is that management formulations which seek to determine optimal well locations and pumping/recharge rates are inherently nonlinear even though contaminant transport simulation based on the flow and transport equations is "feed-forward" linear.

C. Model Uncertainty

A major limitation of the simulation and management formulations presented in Chapters IV and V of this book is that our models are simplifications and, by their nature, generate uncertain results. Hydraulic heads and concentrations are predicted deterministically by our models, but in fact they are random variables. The parameters in both the flow and transport equations are never known but are estimated and represent averaged or effective values over some volume. The greatest source of uncertainty in both simulated hydraulic head and concentration is believed to be hydraulic conductivity. Analyses of field data indicate that hydraulic conductivity can vary by an order of magnitude or more over the distance of a few meters (Bakr, 1976). Porosity variations tend to be smaller. Unfortunately, we cannot measure this spatial variability over the entire domain of our field problems. Treating transport simulation using stochastic partial differential equations, where spatial variability in hydraulic conductivity controls variability in heads and concentrations, has been the topic of extensive recent research (see overviews by Dagan, 1986; and Gelhar,1986). Dealing with flow and transport in a stochastic framework in simulation-management models is a current research topic on which little work has yet been done.

D. Remediation Design — Deterministic Formulation

If all the parameters in the groundwater flow and solute transport equations are perfectly defined, we can formulate a deterministic remediation design problem as a nonlinearly constrained optimization problem. Consider a typical contaminant remediation problem involving a

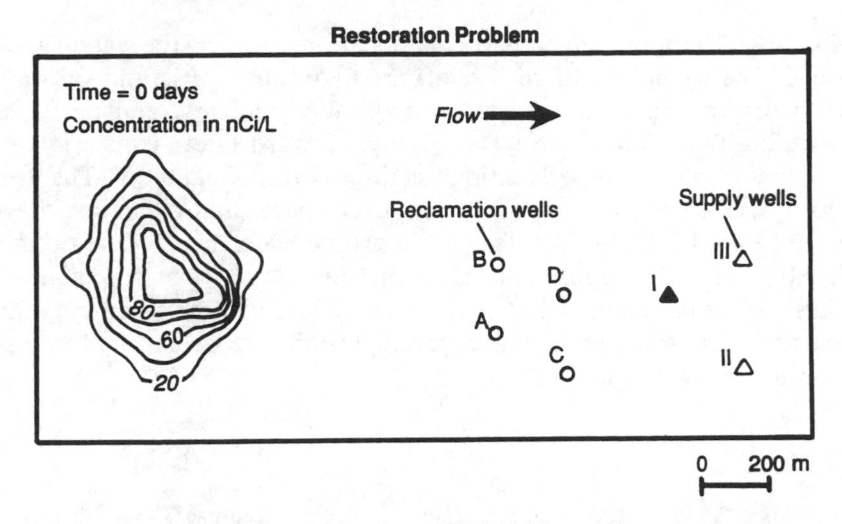

Figure B–1. Typical problem of aquifer restoration showing a tritium plume, potential reclamation wells, and threatened water-supply wells.

plume of tritiated water, which, if left unchecked, would migrate toward downgradient supply wells and begin to destroy their water quality after 3 months. The restoration problem is displayed in Figure B–1, which shows the contaminant plume, the reclamation wells, and the water-supply wells. Two mitigation strategies are possible. One is to intercept the plume with a series of pumping wells that remove the contaminated water. A second strategy is to inject clean water into the plume as it moves by and dilute it in situ. The second strategy is mathematically similar to the first, except that it involves management of recharge rates rather than pumping rates. For brevity, only the first strategy will be addressed. This strategy appears quite similar to that dealt with in Chapter V. There is an important difference, however, in that now some of the contaminated water is allowed to escape from the contaminated zone. Contaminant escape is permitted as long as it is a trace amount and water-quality standards are still met throughout time at the supply wells.

A variety of objective functions are possible, but the simplest economic surrogate for minimizing treatment costs is to simply minimize the total pumping rate (Lefkoff and Gorelick, 1986). Often the size (and cost) of the aboveground contaminated water treatment facility is dictated by the total rate at which water is pumped. The first set of essential constraints is to restrict contaminant concentrations to always be less than or equal to a specified water-quality standard of 20 nanocuries per liter (nCi/L) for tritium, at particular locations, namely the water supply wells. The second set of constraints are hydraulic restrictions on water level drawdowns, local velocities, hydraulic gradients, and particular pumping rates. These are rather straightforward linear constraints involving simple response functions as discussed in Chapter V. The decision variables are the pumping rates over space (and over time, if we were to consider transient changes in groundwater head). The optimal location of wells is approximated by providing a series of potential well sites for the design model to select from. Those values of pumping that are non-zero in the final solution are the preferred wells. The formulation in this case reduces to

OBJECTIVE:

Minimize the sum of the pumping rates	Minimize $Z = \sum_{i=1}^{n} q_i$

CONSTRAINTS:

Water-Quality Constraints (NONLINEAR)

The plume must be contained so that concentrations do not exceed specified values (e.g., water-quality standard) at critical check-point locations (e.g., water-supply wells).	$f(\mathbf{q}) = csim_j \leq c^*_j$

Hydraulic Constraints (LINEAR)

The simulated hydraulic head at each pumping well must be at least a specified minimum value.	$g(\mathbf{q}) = hsim_k \geq h^*_k$
The pumping rate for any well must not exceed a specified maximum value.	$q_k \leq q^*_k$

Nonnegativity Constraints

The values of the concentrations, heads, and pumping rates must not become negative.	$csim \geq 0$ for all values $hsim \geq 0$ for all values $q \geq 0$ for all values

where q_i = pumping rate decision variables;
$csim_j$ = simulated values of concentration and are a nonlinear function of pumping, $f(\mathbf{q})$;
c^*_j = water-quality standards;

$hsim_k$ = simulated values of hydraulic head and are a linear function of pumping, g(**q**), and

h^*_k = lower limits of local hydraulic head values.

The function g uses the flow equation to simulate hydraulic heads, drawdowns, velocities, and hydraulic gradients that are restricted by the vector **g***. These are linear restrictions and because they are easily handled using the response matrix approach described in Chapter V, they will not be discussed further in this appendix. The simulated concentrations are given by the nonlinear function, f. This function uses a transport model to simulate the concentrations over time and space. The finite element simulation model SUTRA (Voss, 1984) was modified and used to solve the above problem. A nonlinear optimization procedure (Gorelick et al., 1984) uses SUTRA as an independent module to help search for an optimal pumping strategy. The search is controlled by a nonlinear optimization procedure that serves as the basis for MINOS/ Augmented (Murtagh and Saunders, 1983) as well as NPSOL (Gill et al., 1986).

Transport simulations provide two types of information in the nonlinear search. First is evaluation of the function f(**q**) that describes the concentration distribution through time and space due to pumping. Second, the transport simulation provides information used to construct the Jacobian. The Jacobian is a matrix of sensitivities that give the change in concentration with respect to a change in pumping. These sensitivities are used to guide the search during the course of optimization. Hundreds to thousands of transport simulations are required during the course of targeting an optimal solution. NPSOL, which uses a sequential quadratic programming algorithm, was found to be more efficient for problems of this nature because it requires far fewer function and Jacobian evaluations to reach an optimal solution (Wagner and Gorelick, 1987). Because the function evaluations, f(**q**), are expensive, methods requiring fewer evaluations are preferred.

The optimal solution (not necessarily a global optimum) to this problem is to pump only at Well D (see Figure B–1) with no pumping at the remaining wells. The optimal pumping rate is 290 liters per second (L/s). Figure B–2 shows concentration histories at supply Well I for the cases of no reclamation and reclamation. Figure B–3 shows the simulated plume as it migrates through the system under managed and unmanaged conditions. Contaminant levels never exceed the water-quality standard, c*, under the optimal remediation case. In our experience, simulation-management model solutions of this type tend to select only one or a few

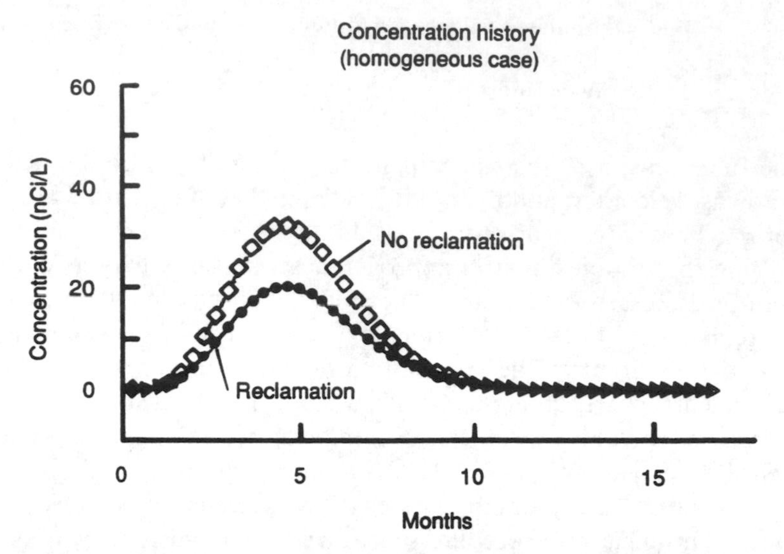

Figure B–2. Concentration histories showing tritium breakthrough for the case of no reclamation and optimal reclamation; note the water-quality standard is 20 nCi/L.

wells. For this reason, integer restrictions on the number of wells are not generally required.

The size of the problem represented by the above formulation is noteworthy. Our goal is to determine a pumping scheme such that contaminants never exceed the water-quality standard, c^*. Therefore, contaminant transport simulation is required for the entire time frame of the problem. For each time step, the concentration constraints over space (at the supply wells) must be met. We must calculate concentrations using sufficiently small time steps so that we reduce discretization error and guarantee that the plume peak is recognized and constrained by our pumping (or injection) strategy. To simulate the system with this fine "concentration history constraint mesh" may require solutions of the transport equation for tens to hundreds of time steps. For this rather small demonstration problem of 312 finite element nodes with 80 ten-day timesteps, 24,960 sparse nonlinear constraints describe transport alone over the entire temporal and spatial domain. Although simulation must be used to evaluate the constraints in our problem, most spatial and temporal locations are of no interest (those not containing supply wells) and do not appear as constraints in the formulation. If steady flow is assumed—often a reasonable approximation—the flow equation can be treated using a linear response function and represented by a

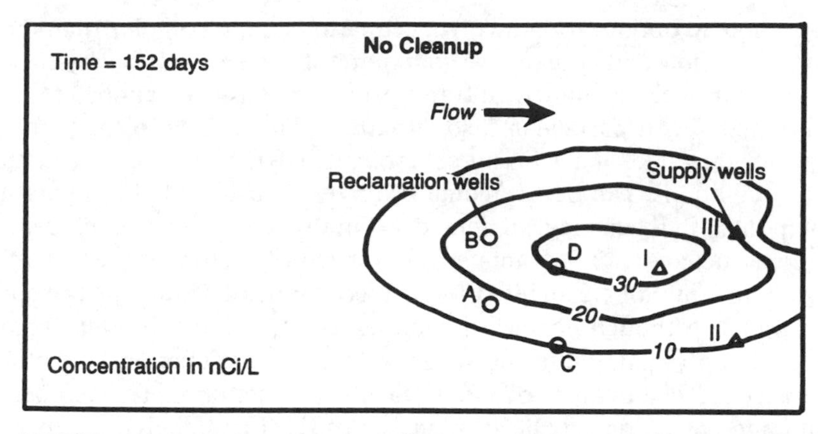

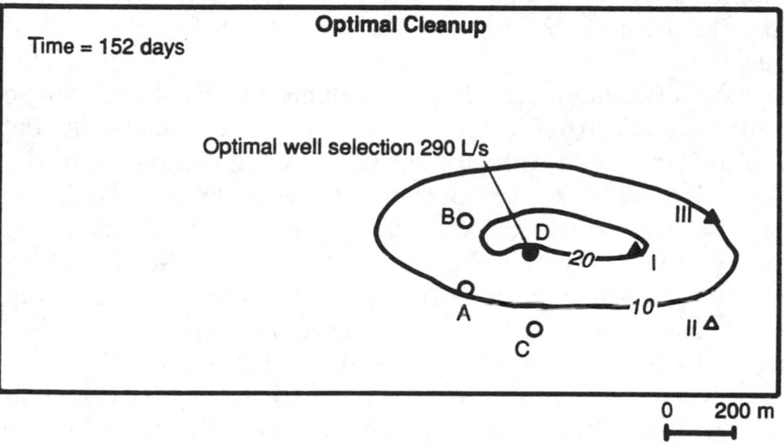

Figure B–3. Maps showing contours of simulated tritium plume after 152 days of travel for the cases of no cleanup and optimal cleanup.

small number of dense linear constraints to represent the hydraulic restrictions.

There are two approaches that have been used in dealing with combined simulation-nonlinear optimization problems, such as the one above. Both utilize the simulation model as a subroutine for function evaluations, $f(\mathbf{q}) = c\,(x,y,t)$. The first approach employs a numerical approximation of the Jacobian (Gorelick et al., 1984; Ahlfeld et al., 1986a, 1986b). The simulation model, SUTRA, served as a subroutine that simply provides concentrations over time at the control locations (supply wells), given a set of pumping/injection decisions. The subroutine is called repeatedly to construct the Jacobian using a difference approximation. The approach is quite flexible and general; it is directly applicable to optimization involving equations more complex than those for simple flow and nonreactive contaminant transport. It is very easy to implement once the numerical transport model has been applied to, and calibrated for, a particular field situation. The simulation model is a separate, testable, and efficient component. Difficulties, such as numerical oscillations, can be overcome prior to inclusion of the simulation component in the management model. Finally, the simulation model can be constructed to take advantage of the special structure of the system of algebraic equations generated by discretization of flow and transport equations. Although no special routines are required to obtain the numerical approximation of the Jacobian, numerous function evaluations are needed. The number of evaluations is a function of the number of pumping-/injection-rate decision variables. Unfortunately, this cost restricts the number of decision variables to a few tens for many applications.

A more efficient approach is to evaluate the Jacobian analytically. Here the partial differential equations are solved numerically, but the Jacobian is evaluated without numerical approximation. Ahlfeld et al. (1988a) discuss two ways of obtaining these analytic sensitivity coefficients. Both ways were derived (but not implemented) for advective-dispersive transport simulation alone by Samper and Neuman (1986). In the first approach, known as the continuous adjoint state equations method, one begins with the original partial differential equation and takes derivatives with respect to the decision variables. The resulting partial differential equation is multiplied by a function known as the adjoint and then integrated after manipulation. The numerical solution of the adjoint differential operator yields the sensitivity coefficients (see Cacuci et al., 1980; or Sykes et al., 1985). The second approach, which yields discrete adjoint state equations, begins with the numerical (e.g., finite element) approximations of flow and transport equations. These algebraic equations are differentiated analytically with respect to the decision variables. When transmissivity is constant over time (confined flow), the analytic Jacobian of the numerical approximation of the flow equation is trivial because this is a linear system. The analytic Jacobian for the transport equation is cumbersome to compute. The spatial discretization of the transport equation gives coefficients that are functions of velocity, nodal concentrations, and pumping-rate decision variables.

The analytic Jacobian was incorporated into a contaminant remediation model by Ahlfeld et al. (1988a). Wanakule et al. (1986) applied the method to an aquifer management model involving the nonlinear form of the flow equation that arises for unconfined flow, where transmissivity is a function of time and pumping. Lasdon et al. (1986) developed a management model based on a nonlinear partial differential equation similar to the flow equation applied to petroleum reservoir production. Ahlfeld et al. (1988b) applied their model, employing the analytically derived Jacobian, to the Woburn aquifer contamination problem in Massachusetts. Their problem contained 835 finite element nodes, with 12 three-month time steps, and 40 potential well sites. There were 80 pumping/injection decision variables. The computational advantages of using the analytic versus the numerical Jacobian were clearly demonstrated. The management model, which coupled their finite element simulator with MINOS, used analytic derivatives and was 12 to 25 times faster than when using the numerical Jacobian, on the basis of this and other field scale problems. A total of 461 simulation runs and 112 sensitivity evaluations were required. With the numerical Jacobian, far more simulation runs are required. For example, in the four-decision-variable problem, 128 simulations were required—48 function evaluations plus 80 for numerical evaluation of the Jacobian using MINOS. Similar problems using the numerical Jacobian with NPSOL require a total of about 40 function evaluations.

E. Remediation Design—Stochastic Formulation

The previous section described a combined simulation-management model for a problem that assumes complete certainty of prediction using the contaminant transport simulation model. This problem does not exist in the real world. The question is then: Does optimization make sense given our limited understanding about the real systems of interest? The answer is clearly no, unless the uncertainty is small, or unless we can categorize the uncertainty and incorporate reliability into the contaminant remediation design formulation.

Perhaps the simplest formulation involving reliability, r, is

$$\text{Minimize } Z = \sum_{i=1}^{n} q_i$$

Subject to

$$\text{Prob}\{c_j \leq c^*_j\} \geq r \quad \text{for contaminant constraint locations and times } j$$

$$q \geq 0 \quad \text{for all pumping wells}$$

The problem now requires that the probability of the concentrations exceeding the water quality standard at the supply wells must not be greater than a specified reliability, r, of perhaps 90 percent. The simulated concentrations are now defined probablistically because the parameters in our simulation model are treated as estimates rather than known constants. Therefore, they carry with them uncertainty that consequently generates uncertainty in the simulated concentrations.

Evaluating the uncertainty in concentrations predicted by a simulation-management model is a difficult matter. Selection of an approach to the problem is based upon the manner in which model uncertainty is perceived. If the simulation model parameters are viewed as "uniform" or "zonally uniform," but unknown and uncertain, one can estimate their values and uncertainties and subsequently introduce the effects of this uncertainty into the management model. It will be shown below that such an approach lends itself to formulation as a large-scale, nonlinear, chance-constrained (Charnes and Cooper, 1962) optimization problem. If, on the other hand, one considers the uncertainty in predicted concentrations to be the result of uncertain spatial variability of hydraulic conductivity, then Monte-Carlo-simulation approaches involving a sequence of conductivity realizations (maps) are possible. Here a map is a spatially distributed set of spatially correlated random hydraulic-conductivity values. It is worth stressing that mapping continuous fields of aquifer properties is not possible with current technology and therefore their heterogeneous character can only be quantitatively described using spatial statistics (see Journel and Huijbregts, 1978; Vanmarcke, 1983).

1. Parameter Uncertainty

Recent research has addressed the problem of groundwater quality management under model parameter uncertainty. A three-staged approach was adopted by Wagner and Gorelick (1987). This approach was applied to the Gloucester landfill groundwater contamination problem by Gailey and Gorelick (1993). The first stage takes concentration and hydraulic head data and solves an inverse problem to identify the parameters needed for flow and transport equations. During this stage, the parameter covariance matrix is also estimated. In the second stage, a first-order first and second moment analysis is performed to translate estimated parameter uncertainty into estimated uncertainty in hydraulic heads and concentrations.The third stage solves the above formulation as a nonlinear chance-constrained programming problem. Here both the expected value and stochastic terms appearing in the chance constraints are nonlinear functions. In this formulation, probabilistically defined concentrations are now a complex nonlinear function of both the decision variables, $\mathbf{q}$, and the uncertain simulation model parameters.

This simulation-regression-management modelling procedure begins with the most readily available data in field situations, which are the measured hydraulic heads and concentrations. In the inverse problem, the flow and transport equations are used to minimize a function of the differences between *observed* data and *simulated* heads and concentrations. Among the unknown parameters are the hydraulic conductivity, the effective porosity, and the longitudinal and transverse dispersivities. These are now scalar quantities. Estimating their values requires the union of simulation and weighted nonlinear regression. Solution to the inverse problem gives the parameter estimates and the covariance of the parameter estimates (Donaldson and Tryon, 1983),

Including the effects of parameter uncertainty in the management model is the next step. A first-order Taylor series expansion of concentration about the expected value of the aquifer parameters is used. As shown by Wagner and Gorelick (1987), the first-order approximation of the expected value of concentration indicates that the expected values are equal to the simulated concentrations given using the mean parameters (approximated as the parameter estimates from the simulation-regression model).

The covariance of concentrations as a function of parameter uncertainty provides a measure of the predicted uncertainty in concentrations when the model parameters are uncertain. The approximate covariance of concentrations is a function of the covariance of the parameter estimates given by the simulation-regression procedure and the sensitivity of concentrations to changes in the parameters.

The remaining step is to evaluate the probabilistic water-quality constraint in the optimization formulation. This constraint can be transformed into a deterministic equivalent if the concentrations, as a function of the uncertain parameters, are normally distributed. This deterministic equivalent, known as a **chance constraint**, consists of two terms. The first term is the expected value component and the second term is the stochastic component. Both are nonlinear functions of the pumping rates (the decision variables). The expected value term is nonlinear because concentration is a nonlinear function of the decision variables, $\mathbf{q}$, as discussed in the deterministic case. The stochastic term is nonlinear because the Jacobian, the change in concentration with respect to parameter changes, is itself a nonlinear function of pumping, $\mathbf{q}$. To handle this nonlinearity, the stochastic term is evaluated repeatedly using the first-order approximation.

It is worth noting that if the reliability level, r, is 50 percent, then the standard normal cumulative distribution corresponding to this value is zero. The stochastic term drops out, and we are left with the deterministic formulation of the previous section. If the parameter estimates are used in the deterministic formulation, the water quality constraints have a 50–50 chance of being violated. For this reason, that formulation is called **risk neutral**. Formulations in which the reliability of meeting the constraints is greater than 50 percent are known as **risk averse**.

Wagner and Gorelick (1987) demonstrated the above approach for remediation under parameter uncertainty for both steady-state and transient contaminant transport. They demonstrated how to evaluate the cost of restoration as a function of the reliability of cleanup using sensitivity analysis. Owing to the extreme nonlinearity of the problem, computational requirements were high. For a five-decision-variable case involving a simulation model with 336 finite elements nodes and 80 time steps, 20 hours of central processing unit (CPU) time on a 5 MIPS minicomputer were used. For a case involving only steady-state transport (equivalent in size to a problem with just one time step) but with 20 decision variables, 8 hours of CPU time were required.

2. Spatially Variable Aquifer Properties

As mentioned earlier, aquifer properties, such as hydraulic conductivity, are spatially variable and complete maps of their values are never known. Sedimentary and erosional processes are the primary agents that create hydraulic conductivity variations. These processes operate in complex ways and lead to hydraulic properties of the media with some degree of spatial connectivity or persistence. In the absence of maps of their values, one can say they are correlated over space. If one sees a small area of high conductivity, it is probable that some zone around this small area will contain similar values.

Consideration of spatial variability of hydraulic conductivity leads to a description of flow and transport in aquifers using stochastic partial differential equations. Hydraulic conductivity is treated as a random variable with some spatial structure. It is virtually impossible to define and work with a different probability distribution function at each point in space and with a multidimensional joint distribution function. A simpler model treats the hydraulic conductivity as a second-order stationary spatial process. Here the expected value of the quadratic variation between two conductivity values, separated by a distance, Δx, does not depend on the location of the values but only on their separation distance and, in some cases, their directional dependence.

Under the stationarity assumption and considering the hydraulic conductivity variation to be lognormally distributed, the spatial correlation can be defined using a variety of functions. Typically, spatial correlation

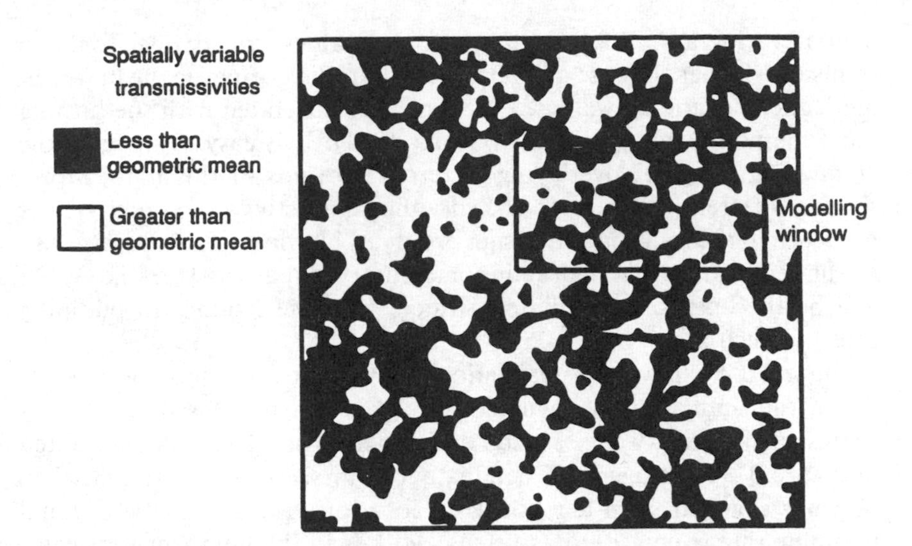

Figure B–4. Spatially variable transmissivities (hydraulic conductivity times saturated thickness) generated using an exponential correlation structure.

models are defined by an exponential or spherical relation known as a variogram. More complex models exist but are used less often (for details see texts such as Journel and Huijbregts, 1978; or de Marsily, 1986). These models contain two important statistical parameters, the variance of log-conductivity and the correlation scale. The variance describes the natural variability of the log-conductivity field, and the correlation scale describes the length over which two conductivity values remain correlated.

Methods aimed at estimating the statistical parameters of spatially variable hydraulic conductivity are discussed by Hoeksema and Kitanidis (1984) and Kitanidis (1987). The "geostatistical approach" uses local, sparse, hydraulic conductivity data as well as hydraulic head data that are related to hydraulic conductivity through the flow equation. Estimation of the covariance functions employs techniques such as maximum likelihood, restricted maximum likelihood, co-kriging, and minimum-variance unbiased quadratic estimation and will not be discussed here.

Assuming that one can estimate the spatial statistics associated with hydraulic conductivity variation, it is a straightforward matter to generate realizations (maps) of conductivity values. It is noteworthy that these realizations can be conditioned so that each map honors certain local head or conductivity data at locations where measurements exist. Figure B–4 represents a large map whose spatial structure is given by an exponential variogram. For presentation, the values are mapped only in terms of whether they are greater or less than the geometric mean (arithmetic mean of the log values). The map was generated with the turning band method (Mantoglou and Wilson, 1981). It is easy to imagine the "modelling window" overlaying different portions of this map, representing different realizations of hydraulic conductivity (ergodicity). Let us return to the remediation design problem, but now let us operate with the understanding that hydraulic conductivity is spatially variable. What will be the optimal remediation strategy (well selection and pumping rates) in such a case?

Figure B–5 shows one realization (map) of hydraulic conductivity. When the simulation-management model is run for this map, as if it represented reality, we see several differences from the results presented in Figure B–3 (homogeneous conductivity). First, the optimal well selection has changed from location D to location A. Second, the optimal pumping rate is now 199 L/s versus 290 L/s in the homogeneous case. Third, we see that the plume follows a different migration path. This simple demonstration points to one possible Monte Carlo formulation for optimal design under spatial variability. We can repeat the above exercise with numerous realizations of hydraulic conductivity. For each realization, the simulation-management model is run. Each time, the optimal well location and pumping rate(s) are recorded. The most probable well sites and pumping rates could then be selected. If the realizations were unbiased, we would expect the optimal location(s) to be the same as those determined in the homogeneous case (Figure B–2). We could even restrict the potential well locations to those given by the optimal solution for the homogeneous problem.

The Monte Carlo formulation gives, in some sense, a reliable design solution, but it does not necessarily produce a robust design. That is, one desires a remediation that will work for all sample realizations of the hydraulic conductivity distribution. Using the above Monte Carlo approach, for any given realization the particular well selection and rate determination is not guaranteed to successfully clean up the aquifer if some other conductivity realization is tried.

The solution should provide optimal well locations and rates such that cleanup will be effective for each and every map. This notion of robustness was explored by Gorelick (1987) for contaminant capture involving only velocity controls (both magnitude and direction of groundwater flow) for a mixed-integer programming formulation involving solution to the flow equation. The formulation involves the simultaneous solution of many constraint sets, f_1 through f_n; however, each simulation is based upon a different conductivity realization. Ig-

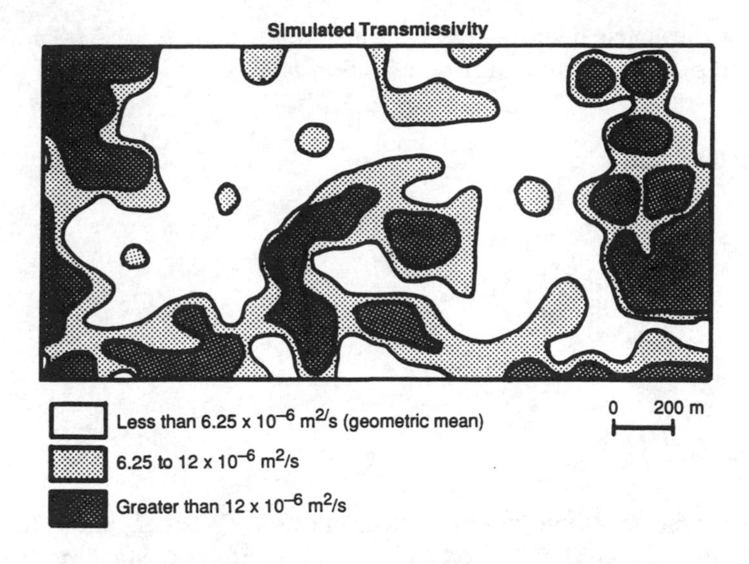

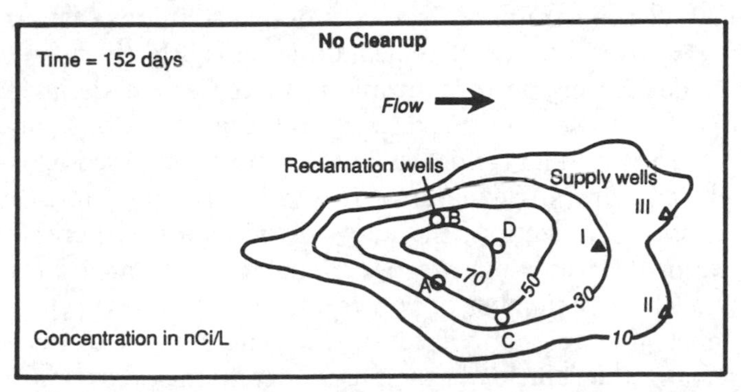

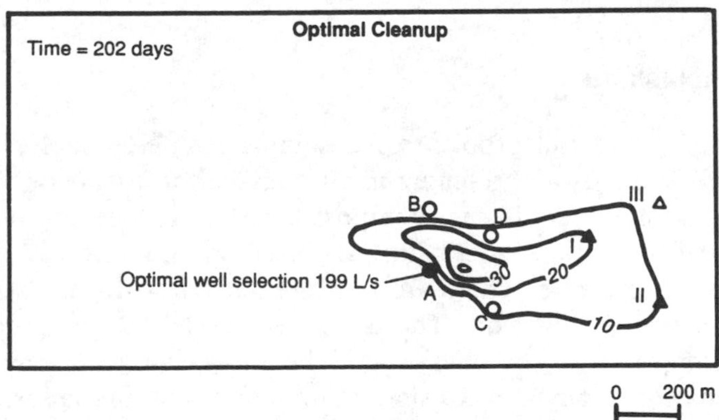

Figure B–5. Simulated transmissivities and corresponding simulated tritium plumes for the cases of no cleanup and optimal cleanup.

noring the linear hydraulic constraints, for simplicity, the formulation for the case of contaminant remediation is

$$\text{Minimize } Z = \sum_{i=1}^{n} q_i$$

Subject to

$$\begin{aligned} f_1\,(\mathbf{q}) &\le c^* \\ f_2\,(\mathbf{q}) &\le c^* \\ \cdot \quad & \quad \cdot \\ \cdot \quad & \quad \cdot \\ \cdot \quad & \quad \cdot \\ f_n\,(\mathbf{q}) &\le c^* \\ \mathbf{q} &\ge 0 \end{aligned}$$

Experiments employing this formulation (Wagner and Gorelick, 1989) have shown the method to be quite interesting. Stacking between 10 and 30 such constraint sets provides design solutions that are 92 to 99 percent effective in terms of remediation. This effectiveness was determined by conducting postoptimization Monte Carlo experiments. Using the optimal well selection and rates, 1,000 realizations of hydraulic conductivity served as data for deterministic simulation based on solution to the flow and transport equations. It is worth noting that compared to deterministic optimization, the above formulation requires additional computational effort roughly proportional to the number, n, of f_k simulation models. The number of constraints grows in direct proportion to the number of simulated realizations, n. The number of decision variables remains the same as in the purely deterministic optimization.

F. Conclusions

Although the full problem of designing a safe, reliable, and cost-effective groundwater contaminant remediation strategy when our models are uncertain is nearly impossible to solve, the combination of numerical simulation, uncertainty analysis, and nonlinear optimization holds much promise. There are few alternatives with the exception of trial-and-error simulation. The combined model can incorporate our best descriptions of the physics and chemistry that govern aquifer system behavior. Compared to simulation alone, nothing is lost. What is gained is that the joint models force the engineer to formulate the real problem of concern and to include, or consciously exclude, all design considerations and constraints. The joint approach forces one to face the issue of risk versus cost in water resources management. Typically, the matter of uncertainty in design studies involving simulation is treated by employing sensitivity analyses or bracketing studies. However, in nonlinear management problems such techniques should not be treated using post-design sensitivity studies. The ultimate formulations, from the outset, should include all quantitative aspects of what we know and what we do not know in the simulation-design process.

References for Appendix B

Ahlfeld, D. P., J. M. Mulvey, G. F. Pinder, and E. F. Wood, "Contaminated Groundwater Remediation Design Using Simulation, Optimization, and Sensitivity Theory 1, Model Development," *Water Resources Research*, Vol. 24, No. 3, pp. 431–442 1988a.

______, "Contaminated Groundwater Remediation Design Using Simulation, Optimization, and Sensitivity Theory 2, Analysis of a Field Site," *Water Resources Research*, Vol. 24, No. 3, pp. 443–452, 1988b.

Ahlfeld, D. P., J. M. Mulvey, and G. F. Pinder, "Designing Optimal Strategies for Contaminated Groundwater Remediation," *Advances in Water Resources*, Vol. 9, No. 2, pp. 77–84, 1986a.

______, "Combining Physical Containment With Optimal Withdrawal for Contaminated Groundwater Remediation," *in Proceedings of the VI International Conference on Finite Elements in Water Resources*, Springer-Verlag, New York, NY, pp. 205–214, 1986b.

Bakr, A. A., "Effects of Spatial Variations of Hydraulic Conductivity on Groundwater Flow," Ph.D. Thesis, New Mexico Institute of Mining and Technology, Socorro, NM, 1976 .

Cacuci, D. G., C. F. Weber, E. M. Oblow, and J. H. Marable, "Sensitivity Theory for General Systems of Nonlinear Equations," *Nuclear Science Engineering*, Vol. 75, pp. 88–110, 1980.

Charnes, A., and W. W. Cooper, "Chance Constraints and Normal Deviates," *Journal American Statistical Association,* Vol. 57, pp. 134–148, 1962.

Dagan, G., "Statistical Theory of Groundwater Flow and Transport: Pore to Laboratory, Laboratory to Formation, and Formation to Regional Scale," *Water Resources Research*, Vol. 22, No. 9, pp. 120s–134s, 1986.

de Marsily, G., *Quantitative Hydrogeology*, Academic Press, London, 1986.

Donaldson, J. R., and P. V. Tryon, *Nonlinear Least Squares Regression*

Using STARPAC: The Standards Time Series and Regression Package, Technical Note 1068-2, National Bureau of Standards, Boulder, CO, 1983.

Gailey, R. M. and S. M. Gorelick, "Design of Optimal Reliable Plume Capture Schemes: Application to the Gloucester Landfill Ground-Water Contamination Problem," *Ground Water*, Vol. 31, No. 1, pp. 107–114, 1993.

Gelhar, L. W., "Stochastic Subsurface Hydrology From Theory to Applications," *Water Resources Research*, Vol. 22, No. 9, pp. 135s–145s, 1986.

Gill, P. E., W. Murray, M. A. Saunders, and M. H. Wright, *Users Guide for NPSOL (version 4.0): A FORTRAN Package for Nonlinear Programming*, Report SOL 86-2, Department of Operations Research, Stanford University, Stanford, CA, 1986.

Gorelick, S. M., "Sensitivity Analysis of Optimal Ground Water Contaminant Capture Curves: Spatial Variability and Robust Solutions," *in Proceedings of the Solving Ground Water Problems with Models Conference,* Denver, CO, pp. 133–146, 1987.

Gorelick, S. M., C. I. Voss, P. E. Gill, W. Murray, M. A. Saunders, and M. H. Wright, "Aquifer Reclamation Design: The Use of Contaminant Transport Simulation Combined With Nonlinear Programming," *Water Resources Research,* Vol. 20, No. 4, pp. 415–427, 1984.

Hoeksema, R. J., and P. K. Kitanidis, "An Application of the Geostatistical Approach to the Inverse Problem in Two-Dimensional Groundwater Modeling," *Water Resources Research*, Vol. 20, No. 7, pp. 1003–1020, 1984.

Journel, A. G., and Ch. J. Huijbregts, *Mining Geostatistics,* Academic Press, London, 1978.

Kitanidis, P. K., "Parametric Estimation of Covariances of Regionalized Variables," *Water Resources Bulletin 23*, pp. 557–567, 1987.

Lasdon, L., P. E. Coffman, Jr., R. MacDonald, J. W. McFarland, and K. Sepehrnoori, "Optimal Hydrocarbon Reservoir Production Policies," *Operations Research*, Vol. 34, pp. 40–54, 1986.

Lefkoff, L. J., and S. M. Gorelick, "Design and Cost Analysis of Rapid Aquifer Restoration Systems Using Flow Simulation and Quadratic Programming," *Ground Water*,Vol. 24, pp. 777–790, 1986.

Mantoglou, A., and J. L. Wilson, *Simulation of Random Fields With the Turning Bands Method*, Massachusetts Institute of Technology, Department of Civil Engineering Report 264, Cambridge, MA, 1981.

Murtagh, B. A., and M. A. Saunders, *MINOS 5.0 User's Guide,* Report

SOL 83-20, Department of Operations Research, Stanford University, Stanford, CA, 1983.

Samper, F. J., and S. P. Neuman, "Adjoint State Equations for Advective-Dispersive Transport," *in Proceedings of VI International Conference on Finite Elements in Water Resources*, Springer-Verlag, New York, NY, pp. 423–428, 1986.

Sykes, J. F., J. L. Wilson, and R. W. Andrews, "Sensitivity Analysis for Steady-State Groundwater Flow Using Adjoint Operators," *Water Resources Research,* Vol. 21, No. 3, pp. 359–371, 1985.

Vanmarke, E., *Random Fields*, Massachusetts Institute of Technology Press, Cambridge, MA, 1983.

Voss, C. I., *SUTRA: A Finite Element Simulation Model for Saturated-Unsaturated, Fluid Density-Dependent Groundwater Flow With Energy Transport or Chemically-Reactive Single-Species Solute Transport,* United States Geological Survey Water Resources Investigation Report 84-4369, 1984.

Wagner, B. J., and S. M. Gorelick, "Optimal Groundwater Quality Management Under Parameter Uncertainty," *Water Resources Research,* Vol. 23, No. 7, pp. 1162–1174, 1987.

______, "Reliable Aquifer Remediation in the Presence of Spatially Variable Hydraulic Conductivity: From Data to Design," *Water Resources Research*, Vol. 25, No. 10, pp. 2211-2215, 1989.

Wanakule, N., L. W. Mays, and L. S. Lasdon, "Optimal Management of Large-Scale Aquifer: Methodology and Applications," *Water Resources Research,* Vol. 22, No. 4, pp. 447–466, 1986.

APPENDIX C

AQMAN and MINOS Data Files and Output

The use of AQMAN and MINOS for aquifer remediation design using a capture and containment system has been discussed in Chapter V of this report. In this appendix we provide an example set of data files along with the output the two codes produce. The example is one of simple hydraulic gradient control using three wells. The problem formulation was discussed in Chapter V.F.1. The formulation and the tableau that AQMAN calculates are shown in Tables C–1 and C–2. It should be noted that AQMAN does not print out the tableau in the form shown in Table C–2. Most of this information is contained in the MPS file that is shown in Table C–3. Constraints 9 through 12 in Table C–2 are pumping constraints that must be added manually to the MPS file produced by AQMAN. The modified MPS file is shown in Table C–4.

The input data needed to produce the MPS file are discussed in the AQMAN documentation (Lefkoff and Gorelick, 1987). There are three required data files.

1. The Trescott model input data file for two-dimensional confined groundwater flow. This data set is shown in Table C–5. The finite difference model grid and boundary conditions used in the example problem are shown in Figure C–1.
2. The AQMAN unit 13 data file that defines the gradient control pairs

Table C–1. Formulation for Three-Well Gradient Control Problem Without Integer Restrictions.

Mathematical Formulation	Explanation
Minimize $Z = Q_1 + Q_2 + Q_3$	Minimize total pumping in L/s
Subject to	
$H_{out} - H_{in} \geq 0$ for each control pair	Gradient control
$Q_1 + Q_2 + Q_3 \leq 8.0$	Limit on total pumping
$0 \leq Q_1 \leq 3.0$	Limit on pumping at well 1
$0 \leq Q_2 \leq 3.0$	Limit on pumping at well 2
$0 \leq Q_3 \leq 3.0$	Limit on pumping at well 3

Table C–2. Tableau for Three-Well Gradient Control Problem

Index	Variable Names			Type	RHS	Row Label	Explanation
	Q_1	Q_2	Q_3				
Portion of the tableau that AQMAN will place in the MPS file							
Z	0.001	0.001	0.001	=	0	OBJ	Objective
1	–0.015876	–0.078634	–0.007687	≤	–0.10	DIF01001	Gradient Pair 1
2	–0.030226	–0.030173	–0.006204	≤	–0.10	DIF01002	Gradient Pair 2
3	–0.078713	–0.015849	–0.005280	≤	–0.10	DIF01003	Gradient Pair 3
4	–0.046852	–0.006888	–0.004029	≤	–0.10	DIF01004	Gradient Pair 4
5	–0.007687	–0.078634	–0.015876	≤	–0.10	DIF01005	Gradient Pair 5
6	–0.006204	–0.030173	–0.030226	≤	–0.10	DIF01006	Gradient Pair 6
7	–0.005280	–0.015850	–0.078713	≤	–0.10	DIF01007	Gradient Pair 7
8	–0.004029	–0.006889	–0.046852	≤	–0.10	DIF01008	Gradient Pair 8
The constraints below must be added to the MPS file produced by AQMAN							
9	1.0	1.0	1.0	≤	8.0	TOTALQ	Total Pumping
10	1.0			≤	3.0	BOUNDS	Limit on Q
11		1.0		≤	3.0	BOUNDS	Limit on Q
12			1.0	≤	3.0	BOUNDS	Limit on Q

Table C–3. AQMAN MPS File

```
NAME          GRADCON3
ROWS
 L  DIF01001
 L  DIF01002
 L  DIF01003
 L  DIF01004
 L  DIF01005
 L  DIF01006
 L  DIF01007
 L  DIF01008
 N  OBJ
COLUMNS
    Q01001    OBJ        0.10000e-02
    Q01001    DIF01001  -0.15876e-01   DIF01002  -0.30226e-01
    Q01001    DIF01003  -0.78713e-01   DIF01004  -0.46852e-01
    Q01001    DIF01005  -0.76870e-02   DIF01006  -0.62035e-02
    Q01001    DIF01007  -0.52804e-02   DIF01008  -0.40293e-02
    Q01002    OBJ        0.10000e-02
    Q01002    DIF01001  -0.78634e-01   DIF01002  -0.30173e-01
    Q01002    DIF01003  -0.15849e-01   DIF01004  -0.68888e-02
    Q01002    DIF01005  -0.78634e-01   DIF01006  -0.30173e-01
    Q01002    DIF01007  -0.15850e-01   DIF01008  -0.68889e-02
    Q01003    OBJ        0.10000e-02
    Q01003    DIF01001  -0.76870e-02   DIF01002  -0.62035e-02
    Q01003    DIF01003  -0.52804e-02   DIF01004  -0.40293e-02
    Q01003    DIF01005  -0.15876e-01   DIF01006  -0.30226e-01
    Q01003    DIF01007  -0.78713e-01   DIF01008  -0.46852e-01
RHS
    RHS       DIF01001  -0.10000e+00   DIF01002  -0.10000e+00
    RHS       DIF01003  -0.10000e+00   DIF01004  -0.10000e+00
    RHS       DIF01005  -0.10000e+00   DIF01006  -0.10000e+00
    RHS       DIF01007  -0.10000e+00   DIF01008  -0.10000e+00
ENDATA
```

and the type of constraint (≤, >, or =). This data set is shown in Table C–6.

3. The AQMAN unit 14 data set that gives the locations of the potential remediation wells and the control locations. This data set is shown in Table C–7.

AQMAN produces an output file on unit 17 that gives important information about the simulation-optimization problem. It reports information, such as the control locations, the imposed limits on head, the potential pumping or injection well locations, the control pair locations, and the limits on the gradients for the control pairs. This output file is shown in Table C–8.

MINOS requires the SPECS file. This file indicates the size and nature of the problem to be solved using linear programming. The SPECS file appears in Table C–9. The full output from MINOS is shown in Table C–10.

Table C–4. AQMAN Modified MPS File

```
NAME          GRADCON3
ROWS
 L  DIF01001
 L  DIF01002
 L  DIF01003
 L  DIF01004
 L  DIF01005
 L  DIF01006
 L  DIF01007
 L  DIF01008
 L  TOTALQ
 N  OBJ
COLUMNS
    Q01001    OBJ        0.10000e-02
    Q01001    DIF01001  -0.15876e-01   DIF01002  -0.30226e-01
    Q01001    DIF01003  -0.78713e-01   DIF01004  -0.46852e-01
    Q01001    DIF01005  -0.76870e-02   DIF01006  -0.62035e-02
    Q01001    DIF01007  -0.52804e-02   DIF01008  -0.40293e-02
    Q01001    TOTALQ     1.0
    Q01002    OBJ        0.10000e-02
    Q01002    DIF01001  -0.78634e-01   DIF01002  -0.30173e-01
    Q01002    DIF01003  -0.15849e-01   DIF01004  -0.68888e-02
    Q01002    DIF01005  -0.78634e-01   DIF01006  -0.30173e-01
    Q01002    DIF01007  -0.15850e-01   DIF01008  -0.68889e-02
    Q01002    TOTALQ     1.0
    Q01003    OBJ        0.10000e-02
    Q01003    DIF01001  -0.76870e-02   DIF01002  -0.62035e-02
    Q01003    DIF01003  -0.52804e-02   DIF01004  -0.40293e-02
    Q01003    DIF01005  -0.15876e-01   DIF01006  -0.30226e-01
    Q01003    DIF01007  -0.78713e-01   DIF01008  -0.46852e-01
    Q01003    TOTALQ     1.0
RHS
    RHS       DIF01001  -0.10000e+00   DIF01002  -0.10000e+00
    RHS       DIF01003  -0.10000e+00   DIF01004  -0.10000e+00
    RHS       DIF01005  -0.10000e+00   DIF01006  -0.10000e+00
    RHS       DIF01007  -0.10000e+00   DIF01008  -0.10000e+00
    RHS       TOTALQ     8.0
BOUNDS
 UP BOUND     Q01001     3.0
 UP BOUND     Q01002     3.0
 UP BOUND     Q01003     3.0
ENDATA
```

Table C–5. Trescott Model Input Data File.

```
*******************   GRADCON3     *********************************************
*****
                        SIP        PUNC             NUME HEAD
      61        61        0         200

       1       500  .0000001  .0000001         0         0         0        10
    1.57         1         1



       1         1         1
 0.000000  0.000000  0.000000  0.000000  0.000000  0.000000  0.000000  0.000000
 0.000000  0.000000  0.000000  0.000000  0.000000  0.000000  0.000000  0.000000
 0.000000  0.000000  0.000000  0.000000  0.000000  0.000000  0.000000  0.000000
 0.000000  0.000000  0.000000  0.000000  0.000000  0.000000  0.000000  0.000000
 0.000000  0.000000  0.000000  0.000000  0.000000  0.000000  0.000000  0.000000
 0.000000  0.000000  0.000000  0.000000  0.000000  0.000000  0.000000  0.000000
 0.000000  0.000000  0.000000  0.000000  0.000000  0.000000  0.000000  0.000000
 0.000000  0.000000  0.000000  0.000000  0.000000
  0.         90.0000   90.0000   90.0000   90.0000   90.0000   90.0000   90.0000
  90.0000   90.0000   90.0000   90.0000   90.0000   90.0000   90.0000   90.0000
  90.0000   90.0000   90.0000   90.0000   90.0000   90.0000   90.0000   90.0000
  90.0000   90.0000   90.0000   90.0000   90.0000   90.0000   90.0000   90.0000
  90.0000   90.0000   90.0000   90.0000   90.0000   90.0000   90.0000   90.0000
  90.0000   90.0000   90.0000   90.0000   90.0000   90.0000   90.0000   90.0000
  90.0000   90.0000   90.0000   90.0000   90.0000   90.0000   90.0000   90.0000
  90.0000   90.0000   90.0000   90.0000   0.
                                      .
                                      .
                                      .
        For the steady state case, arbitrary values are entered here
        for the next 57 lines of the 61 x 61 matrix.
                                      .
                                      .
                                      .
  0.         81.9900   81.9900   81.9900   81.9900   81.9900   81.9900   81.9900
  81.9900   81.9900   81.9900   81.9900   81.9900   81.9900   81.9900   81.9900
  81.9900   81.9900   81.9900   81.9900   81.9900   81.9900   81.9900   81.9900
  81.9900   81.9900   81.9900   81.9900   81.9900   81.9900   81.9900   81.9900
  81.9900   81.9900   81.9900   81.9900   81.9900   81.9900   81.9900   81.9900
  81.9900   81.9900   81.9900   81.9900   81.9900   81.9900   81.9900   81.9900
  81.9900   81.9900   81.9900   81.9900   81.9900   81.9900   81.9900   81.9900
  81.9900   81.9900   81.9900   81.9900   0.
 0.000000  0.000000  0.000000  0.000000  0.000000  0.000000  0.000000  0.000000
 0.000000  0.000000  0.000000  0.000000  0.000000  0.000000  0.000000  0.000000
 0.000000  0.000000  0.000000  0.000000  0.000000  0.000000  0.000000  0.000000
 0.000000  0.000000  0.000000  0.000000  0.000000  0.000000  0.000000  0.000000
 0.000000  0.000000  0.000000  0.000000  0.000000  0.000000  0.000000  0.000000
 0.000000  0.000000  0.000000  0.000000  0.000000  0.000000  0.000000  0.000000
 0.000000  0.000000  0.000000  0.000000  0.000000  0.000000  0.000000  0.000000
 0.000000  0.000000  0.000000  0.000000  0.000000
      -1         1         1
0.0 0.0 0.0 0.0 0.0 0.0 0.0 0.0 0.0 0.0 0.0 0.0 0.0 0.0 0.0 0.0 0.0 0.0 0.0 0.0
0.0 0.0 0.0 0.0 0.0 0.0 0.0 0.0 0.0 0.0 0.0 0.0 0.0 0.0 0.0 0.0 0.0 0.0 0.0 0.0
0.0 0.0 0.0 0.0 0.0 0.0 0.0 0.0 0.0 0.0 0.0 0.0 0.0 0.0 0.0 0.0 0.0 0.0 0.0 0.0
0.0
0.0 1.0 1.0 1.0 1.0 1.0 1.0 1.0 1.0 1.0 1.0 1.0 1.0 1.0 1.0 1.0 1.0 1.0 1.0 1.0
1.0 1.0 1.0 1.0 1.0 1.0 1.0 1.0 1.0 1.0 1.0 1.0 1.0 1.0 1.0 1.0 1.0 1.0 1.0 1.0
1.0 1.0 1.0 1.0 1.0 1.0 1.0 1.0 1.0 1.0 1.0 1.0 1.0 1.0 1.0 1.0 1.0 1.0 1.0 1.0
0.0
0.0 0.0 0.0 0.0 0.0 0.0 0.0 0.0 0.0 0.0 0.0 0.0 0.0 0.0 0.0 0.0 0.0 0.0 0.0 0.0
0.0 0.0 0.0 0.0 0.0 0.0 0.0 0.0 0.0 0.0 0.0 0.0 0.0 0.0 0.0 0.0 0.0 0.0 0.0 0.0
0.0 0.0 0.0 0.0 0.0 0.0 0.0 0.0 0.0 0.0 0.0 0.0 0.0 0.0 0.0 0.0 0.0 0.0 0.0 0.0
0.0
                                      .
                                      .
                                      .
          For the next 55 lines of the 61 x 61 matrix, zeros are
          entered here.
                                      .
                                      .
                                      .
0.0 0.0 0.0 0.0 0.0 0.0 0.0 0.0 0.0 0.0 0.0 0.0 0.0 0.0 0.0 0.0 0.0 0.0 0.0 0.0
0.0 0.0 0.0 0.0 0.0 0.0 0.0 0.0 0.0 0.0 0.0 0.0 0.0 0.0 0.0 0.0 0.0 0.0 0.0 0.0
0.0 0.0 0.0 0.0 0.0 0.0 0.0 0.0 0.0 0.0 0.0 0.0 0.0 0.0 0.0 0.0 0.0 0.0 0.0 0.0
0.0
0.0 1.0 1.0 1.0 1.0 1.0 1.0 1.0 1.0 1.0 1.0 1.0 1.0 1.0 1.0 1.0 1.0 1.0 1.0 1.0
1.0 1.0 1.0 1.0 1.0 1.0 1.0 1.0 1.0 1.0 1.0 1.0 1.0 1.0 1.0 1.0 1.0 1.0 1.0 1.0
1.0 1.0 1.0 1.0 1.0 1.0 1.0 1.0 1.0 1.0 1.0 1.0 1.0 1.0 1.0 1.0 1.0 1.0 1.0 1.0
0.0
```

Table C–5. Trescott Model Input Data File (continued).

```
0.0 0.0 0.0 0.0 0.0 0.0 0.0 0.0 0.0 0.0 0.0 0.0 0.0 0.0 0.0 0.0 0.0 0.0 0.0 0.0
0.0 0.0 0.0 0.0 0.0 0.0 0.0 0.0 0.0 0.0 0.0 0.0 0.0 0.0 0.0 0.0 0.0 0.0 0.0 0.0
0.0 0.0 0.0 0.0 0.0 0.0 0.0 0.0 0.0 0.0 0.0 0.0 0.0 0.0 0.0 0.0 0.0 0.0 0.0 0.0
0.0
   0.0005         0         0
        1         1         1
      109        73        49        33        22        15        10        10
       10        10        10        10        10        10        10        10
       10        10        10        10        10        10        10        10
       10        10        10        10        10        10        10        10
       10        10        10        10        10        10        10        10
       10        10        10        10        10        10        10        10
       10        10        10        10        10        10        10        15
       22        33        49        73       109
        1         1         1
      109        73        49        33        22        15        10        10
       10        10        10        10        10        10        10        10
       10        10        10        10        10        10        10        10
       10        10        10        10        10        10        10        10
       10        10        10        10        10        10        10        10
       10        10        10        10        10        10        10        10
       10        10        10        10        10        10        10        15
       22        33        49        73       109
        1         0         3        30       100       1.5       0.5
       40        23
       40        31
       40        39
```

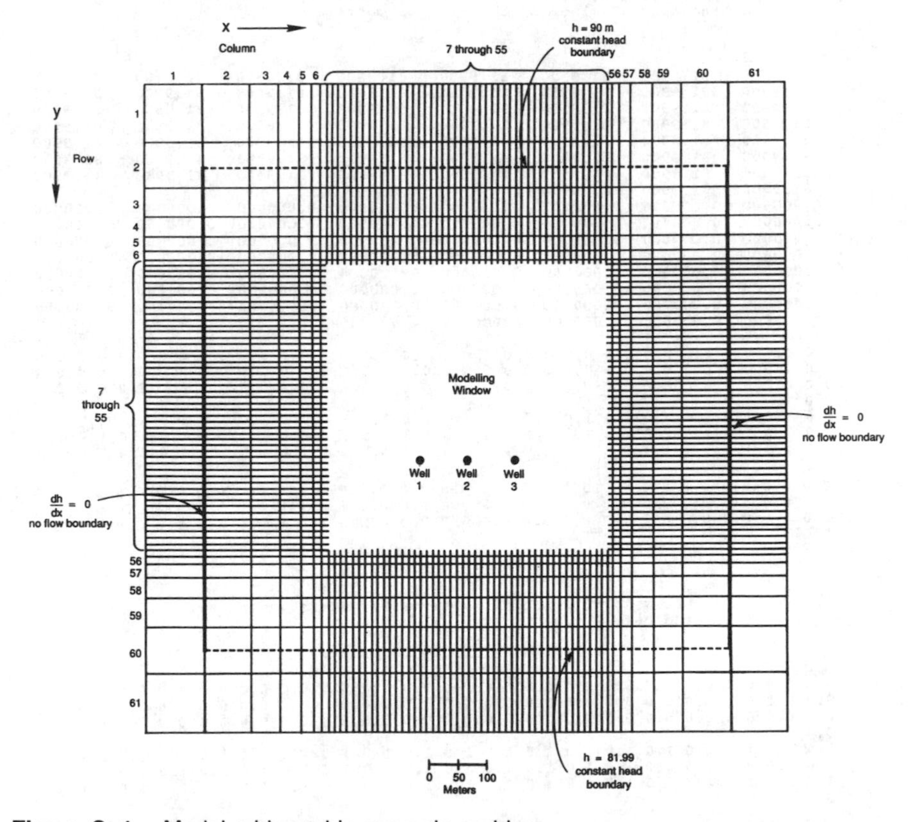

Figure C–1. Model grid used in example problem.

Table C–6. AQMAN Unit 13 Data File

```
    42          29          41          29          0
    42          27          41          27          0
    42          25          41          25          0
    42          20          41          20          0
    42          33          41          33          0
    42          35          41          35          0
    42          37          41          37          0
    42          42          41          42          0
       10.0          0.00      G
       10.0          0.00      G
       10.0          0.00      G
       10.0          0.00      G
       10.0          0.00      G
       10.0          0.00      G
       10.0          0.00      G
       10.0          0.00      G
```

Table C–7. AQMAN Unit 14 Data File

```
LINE  3 Well Gradient Control
         3         16          1         1.5          8
      30.0
       0.5
         0
        40         23          0                  -0.001
        40         31          0                  -0.001
        40         39          0                  -0.001
        41         20                     1
        41         25                     1
        41         27                     1
        41         29                     1
        41         33                     1
        41         35                     1
        41         37                     1
        41         42                     1
        42         20                     1
        42         25                     1
        42         27                     1
        42         29                     1
        42         33                     1
        42         35                     1
        42         37                     1
        42         42                     1
      0.00
      0.00
      0.00
      0.00
      0.00
      0.00
      0.00
      0.00
      0.00
      0.00
      0.00
      0.00
      0.00
      0.00
      0.00
      0.00
```

Table C–8. AQMAN Unit 17 Output File

CONTROL LOCATIONS AND UNSTRESSED HEADS

	Period	Location #	I-Loc.	J-Loc.	Unmanaged Head	KEYWL
1	1	1	41	20	84.9950000	0
2	1	2	41	25	84.9950000	0
3	1	3	41	27	84.9950000	0
4	1	4	41	29	84.9950000	0
5	1	5	41	33	84.9950000	0
6	1	6	41	35	84.9950000	0
7	1	7	41	37	84.9950000	0
8	1	8	41	42	84.9950000	0
9	1	9	42	20	84.8950000	0
10	1	10	42	25	84.8950000	0
11	1	11	42	27	84.8950000	0
12	1	12	42	29	84.8950000	0
13	1	13	42	33	84.8950000	0
14	1	14	42	35	84.8950000	0
15	1	15	42	37	84.8950000	0
16	1	16	42	42	84.8950000	0

USER IMPOSED LIMITS ON HEAD AT CONTROL LOCATIONS

Period	Loc. #	I-Location	J-Location	Limit	Type	KEYGRD	KDEFHD
1	1	41	20	0.		1	0
1	2	41	25	0.		1	0
1	3	41	27	0.		1	0
1	4	41	29	0.		1	0
1	5	41	33	0.		1	0
1	6	41	35	0.		1	0
1	7	41	37	0.		1	0
1	8	41	42	0.		1	0
1	9	42	20	0.		1	0
1	10	42	25	0.		1	0
1	11	42	27	0.		1	0
1	12	42	29	0.		1	0
1	13	42	33	0.		1	0
1	14	42	35	0.		1	0
1	15	42	37	0.		1	0
1	16	42	42	0.		1	0

WELL LOCATIONS AND TYPE

Period	Loc. #	I-Loc.	J-Loc.	KEYQ	Fixed or Unit Rate
1	1	40	23	0	-.1000e-02
1	2	40	31	0	-.1000e-02
1	3	40	39	0	-.1000e-02

CONTROL PAIR LOCATIONS AND DEFINITIONS

Pair #	1st I-Loc.	1st J-Loc.	2nd I-Loc.	2nd J-Loc.	KDEFGR
1	42	29	41	29	0
2	42	27	41	27	0
3	42	25	41	25	0
4	42	20	41	20	0
5	42	33	41	33	0
6	42	35	41	35	0
7	42	37	41	37	0
8	42	42	41	42	0

USER IMPOSED LIMITS ON HEAD DIFFRENCE AT CONTROL PAIRS

Period	Pair #	Conversion Factor	Difference Limit	Type
1	1	10.0000	0.	G
1	2	10.0000	0.	G
1	3	10.0000	0.	G
1	4	10.0000	0.	G
1	5	10.0000	0.	G
1	6	10.0000	0.	G
1	7	10.0000	0.	G
1	8	10.0000	0.	G

Table C–9. MINOS Specs File

```
BEGIN  SPECS

   MINIMIZE

   OBJECTIVE                          OBJ

   ROWS                               400
   COLUMNS                            600
   ELEMENTS                          4000

   BOUNDS                           BOUND

   ITERATIONS                         500

   MPS FILE                            10

   SOLUTION                           YES
   LOG FREQUENCY                        1
```

Table C–10. MINOS Output File

```
OPTIONS file
------------
     BEGIN  SPECS

        MINIMIZE

        ROWS                          400
        COLUMNS                       600
        ELEMENTS                     4000

        ITERATIONS                    500

        MPS FILE                       10

        SOLUTION                      YES
        LOG FREQUENCY                   1

     END

 Reasonable WORKSPACE limits are          0 ...   25269
 Actual     WORKSPACE limits are          0 ...  300000 ...   300000   words of  Z.
1
 MPS file
 --------
      1    NAME           GRADCON3
      2    ROWS
     13    COLUMNS
     32    RHS
     38    BOUNDS
     42    ENDATA


 Names selected
 --------------
 OBJECTIVE       OBJ        (MIN)       1
 RHS             RHS                    9
 RANGES                                 0
 BOUNDS          BOUND                  3

 PARTIAL PRICE section size (A)             0
```

Table C–10. MINOS Output File (continued)

```
PARTIAL PRICE section size (I)              1

Nonzeros allowed for in LU factors     199840


Matrix statistics
-----------------
                 Total      Normal        Free       Fixed     Bounded
Rows                10           9           1           0           0
Columns              3           0           0           0           3

No. of matrix elements          30    Density     100.000
Biggest and smallest coeffs        1.0000e+00  4.0293e-03
(excluding free rows and RHS)
1

Initial basis
-------------

No basis file supplied


Scaling
-------
                Min elem     Max elem        Max col ratio

After  0        4.03e-03     1.00e+00              248.18
After  1        2.59e-01     3.86e+00               14.91
After  2        2.59e-01     3.86e+00               14.91
After  3        2.59e-01     3.86e+00               14.91

Norm of fixed columns and slacks         2.53e-01
(before and after row scaling)           1.53e+01

CRASH option    1
Free rows       1    Free columns       0    Slacks       9    Triangular columns       0    Remainder       0


Iterations
----------

   Itn  Nopt  Ninf  Sinf,Objective      LU
```

Table C–10. MINOS Output File (continued)

```
       1     1     8  3.16227766e+00      10
       2     2     6  1.51885166e+00      10
       3     1     2  1.80558799e-01      10

 Itn       3 -- Feasible solution.  Objective =    6.372752245e-03

       4     1     0  6.37275224e-03      19
       5     1     0  5.69365318e-03      28
       6     1     0  5.09783268e-03      29
       7     1     0  4.75129399e-03      37
       8     1     0  4.75129386e-03      38
1

 EXIT -- OPTIMAL SOLUTION FOUND

 No. of iterations                  8      Objective value      4.7512938554e-03

 No. of degenerate steps            1      Percentage                      12.50

 Norm of X                  1.334e+00      Norm of PI                  1.000e+00

 Norm of X   (unscaled)     2.915e+00      Norm of PI  (unscaled)      1.000e+00
1
 NAME              GRADCON3                Objective value     4.7512938554e-03

 Status            OPTIMAL SOLN            Iteration      8      Superbasics     0

 OBJECTIVE         OBJ          (MIN)
 RHS               RHS
 RANGES
 BOUNDS            BOUND

 SECTION 1 - ROWS

  NUMBER   ...ROW.. STATE   ...ACTIVITY...   SLACK ACTIVITY  ..LOWER LIMIT.  ..UPPER LIMIT. .DUAL ACTIVITY    ..I

       4  DIF01001    BS          -0.13121          0.03121           NONE        -0.10000       0.                1
       5  DIF01002    BS          -0.10000          0.00000           NONE        -0.10000       0.                2
       6  DIF01003    BS          -0.17012          0.07012           NONE        -0.10000       0.                3
       7  DIF01004    UL          -0.10000         -0.                NONE        -0.10000      -0.01741           4
       8  DIF01005    BS          -0.13121          0.03121           NONE        -0.10000       0.                5
       9  DIF01006    UL          -0.10000         -0.                NONE        -0.10000      -0.02889           6
      10  DIF01007    BS          -0.17012          0.07012           NONE        -0.10000       0.                7
      11  DIF01008    UL          -0.10000         -0.                NONE        -0.10000      -0.00121           8
      12  TOTALQ      BS           4.75129          3.24871           NONE         8.00000       0.                9
      13  OBJ         BS           0.00475         -0.00475           NONE            NONE      -1.00000          10

1
 SECTION 2 - COLUMNS

  NUMBER  .COLUMN. STATE   ...ACTIVITY...   .OBJ GRADIENT.  ..LOWER LIMIT.  ..UPPER LIMIT. REDUCED GRADNT    M+J

       1  Q01001     BS          1.81301          0.00100      0.                3.00000      -0.00000          11
       2  Q01002     BS          1.12528          0.00100      0.                3.00000      -0.00000          12
       3  Q01003     BS          1.81300          0.00100      0.                3.00000      -0.00000          13
 ENDRUN
```

Reference for Appendix C

Lekoff, L. J., and S. M. Gorelick, "Design and Cost Analysis of Rapid Aquifer Restoration Systems Using Flow Simulation and Quadratic Programming," *Ground Water*, Vol. 24, pp. 777–790, 1987.

APPENDIX D

Water-Quality Standards Under the Safe Drinking Water Act

The process of determining whether a contaminated aquifer requires remediation is influenced by numerous factors, including (1) the regulatory status of the contaminated site—CERCLA, RCRA, or other; (2) the hydrogeologic setting; and (3) the social, economic, and environmental importance of the contaminated aquifer. The process of determining contamination levels at which a remediation effort can be terminated is influenced by the same factors. While there is no universal set of water-quality standards that can be automatically applied to these decision-making processes for any type of site, the water-quality standards promulgated under the Safe Drinking Water Act (SDWA) do receive widespread use for defining cleanup action levels and remediation target concentrations.

The SDWA was enacted in 1974 and amended in 1986. Regulations for the SDWA are codified in the Code of Federal Regulations, 40 CFR Parts 141, 142, and 143. The Act itself is found in 42 U.S.C. § 300f–300j-11. The 1974 version is Public Law No. 93–523, 88 Stat. 1661 (1974); the 1986 amendments are Public Law No. 99–339, 100 Stat. 666 (1986). The purpose of the Act is to protect the quality of drinking water supplied by public water systems that include facilities or equipment to collect and treat water. Accordingly, the water-quality regulations promulgated under the Act are intended to apply to drinking water at the point of exit from the treatment plant. However, when SDWA regulations are used for the practice of developing aquifer remediation systems, water-quality standards are often applied to contaminant concentrations in groundwater that is still in an aquifer.

Two types of water-quality regulations have been developed under the SDWA. The National Primary Drinking Water Regulations specify maximum contaminant levels (MCLs) for contaminants that are thought to have a negative impact on human health. If it is not "economically or technically feasible" to measure the concentration of a contaminant, a required treatment technique may be specified in place of an MCL. For water systems in which such a contaminant is known to exist, the treatment system must include the required treatment technique. As of this writing, treatment techniques are specified for two organic compounds used in water treatment corrosion control and for four microbiological contaminants. MCLs and "required treatment techniques" are enforceable standards.

Also published under the SDWA are maximum contaminant level goals (MCLGs) that are nonenforceable health goals. MCLGs are intended to "define the maximum contaminant level at which no known or anticipated adverse" health effects occur in humans (40 CFR Part 141, § 141.2). Under some circumstances, MCLGs may be used to define enforceable standards for remediation systems.

The second class of water-quality standards developed under the SDWA are the National Secondary Drinking Water Regulations. These regulations specify secondary maximum contaminant levels (SMCLs) for contaminants that affect the aesthetic qualities of drinking water. SMCLs do not play a significant role in the development of standards for groundwater remedial action systems.

The National Primary Drinking Water Regulations are often used in the development of remedial action standards for groundwater contamination problems at CERCLA sites. Under CERCLA, MCLs are often adopted as the applicable or relevant and appropriate requirements (ARARs) needed to protect an aquifer that is a current or potential source of drinking water. Potential water supplies are determined by classifying aquifers as Class I, II, or III (U.S. EPA, 1988). Any Class I or II aquifer is considered a potential source of drinking water and is likely to be subject to MCLs applied as enforceable standards. Since all CERCLA cleanup levels are determined on the basis of health-risk assessments, MCLGs may also be applied to Class I or II aquifers as enforceable standards when a contamination problem poses unusual risks (U.S. EPA, 1988).

Groundwater is considered Class III if it contains more than 10,000 parts per million (ppm) of total dissolved solids (TDS) (U.S. EPA, 1988). MCLs may be applied to Class III aquifers when contamination in the Class III aquifer threatens a Class I or II aquifer through some hydraulic connection. In general, however, MCLs are not used as enforceable standards for Class III groundwater.

For RCRA sites, the maximum permissible contaminant levels for a contaminant may be defined as the background or upgradient concentration, the SDWA MCL, or a RCRA alternate concentration limit (ACL). The reader is directed to 40 CFR Parts 240 through 280 for additional information on RCRA regulations.

As of February 1991, final or proposed MCLs have been set for 89 contaminants. This list is supposed to be updated with standards for 25 new contaminants every 3 years after 1991 (U.S. EPA, 1990). Table D–1 provides a list of MCLs and MCLGs. New MCLs and MCLGs and revisions to current standards are published in the *Federal Register*. Additional information on water-quality regulations under the SDWA may be obtained by calling the Safe Drinking Water Hotline at (800) 426–4791 or by writing to

U.S. Environmental Protection Agency
Drinking Water Standards Division
401 M Street, S.W. (WH–550D)
Washington, D.C. 20460
(202) 382–7575

Table D–1. National Primary Drinking Water Regulations for Organic and Inorganic Contaminants as of January 30, 1991

Contaminant	MCL (mg/L)	MCLG (mg/L)	Status[a]
Inorganics			
Antimony	0.010. = /0.005[b]	0.003	P
Arsenic	0.05	0	F
Asbestos	7 million fibers per liter (larger than 10 μm)	7 million fibers per liter (larger than 10 μm)	F
Barium	2	2	P
Beryllium	0	1	P
Cadmium	0.005	0.005	F
Chromium	0.1	0.1	F
Copper	1.3	1.3	P
Cyanide	0.2	0.2	P
Fluoride	4	4	F
Lead (current)	0.05	0	F
Lead (proposed)	0.005	0	P
Mercury	0.002	0.002	F
Nickel	0.1	0.1	P
Nitrate (as N)	10	10	F
Nitrite (as N)	1	1	F
Selenium	0.05	0.05	F
Silver	Deleted and reproposed as SMCL		
Sulfate	400	400	P
Thallium	0.002/0.001[b]	0.0005	P
Total Nitrate and Nitrite	10	10	F
Volatile Organics			
Benzene	0.005	0	F
Carbon Tetrachloride	0.005	0	F
1,2-Dichloroethane	0.005	0	F

Table D–1. National Primary Drinking Water Regulations for Organic and Inorganic Contaminants as of January 30, 1991 (continued)

Contaminant	MCL (mg/L)	MCLG (mg/L)	Status[a]
Volatile Organics (continued)			
1,1-Dichloroethylene	0.007	0.007	F
cis-1-2,Dichloroethylene	0.07	0.07	F
o-Dichlorobenzene	0.6	0.6	F
para-Dichlorobenzene	0.075	0.075	F
trans-1,2-Dichloroethylene	0.1	0.1	F
Dichloromethane (methylene chloride)	0.005	0	P
1,2-Dichloropropane	0.005	0	F
Ethylbenzene	0.7	0.7	F
Monochlorobenzene	0.1	0.1	F
Styrene	0.1	0.1	F
Tetrachloroethylene	0.005	0	F
Toluene	1.0	1.0	F
1,1,1-Trichloroethane	0.2	0.2	F
1,1,2-Trichloroethane	0.005	0	P
Trichloroethylene	0.005	0	F
Vinyl Chloride	0.002	0	F
Xylenes (total)	10.0	10.0	F
Other Organics			
Acrylamide	TT[c]	0	F
Alachlor	0.002	0	F
Aldicarb	0.001	0.003	P
Aldicarb Sulfane	0.002	0.003	P
Aldicarb Sulfoxide	0.001	0.003	P
Atrazine	0.003	0.003	F
Carbofuran	0.04	0.04	F
Chlordane	0.002	0	F
2,4-D	0.07	0.07	F
Delapon	0.2	0.2	P
1,2-Dibromo-3-Chloropropane (DBCP)	0.0002	0	F
Di(ethylexyl)adipate	0.5	0.5	P
Di(ethylexyl)phthalate	0.004	0	P
Dinoseb	0.007	0.007	P
Diquat	0.02	0.02	P
Endothall	0.1	0.1	P
Endrin	0.002	0.002	P
Epichlorohydrin	TT[c]	0	F
Ethylene dibromide	0.00005	0	F
Glyphosate	0.7	0.7	P
Heptachlor	0.0004	0	F
Heptachlor epoxide	0.0002	0	F
Hexachlorobenzene	0.001	0	P
Hexachlorocyclopentadiene	0.05	0.05	P
Lindane	0.0002	0.0002	F
Methoxychlor	0.04	0.04	F
Oxamyl (Vydate)	0.2	0.2	P

Table D–1. National Primary Drinking Water Regulations for Organic and Inorganic Contaminants as of January 30, 1991 (continued)

Contaminant	MCL (mg/L)	MCLG (mg/L)	Status[a]
Other Organics (continued)			
PAH[Benzo(a)pyrene]	0.0002	0	P
PAH[Benz(a)anthracene]	0.0001	0	T
PAH[Benzo(b)fluoranthene]	0.0002	0	T
PAH[Benzo(k)fluoranthene]	0.0002	0	T
PAH[Dibenz(a,h)anthracene]	0.0003	0	T
PAH[Indeno(1,2,3-c,d)pyrene]	0.0004	0	T
Polychlorinated biphenyls (PCBs) (as decachlorobiphenyl)	0.003	0	F
Pentachlorophenol	0.001	0	P
Picloram	0.5	0.5	P
Simazine	0.001	0.001	P
Toxaphene	0.003	0	F
Trihalomethanes (total)	0.1	NA[d]	F
1,2,4-Trichlorobenzene	0.009	0.009	P
2,3,7,8-TCDD (Dioxin)	5×10^{-8}	0	P
2,4,5-TP (Silvex)	0.05	0.05	F
Microbiological			
Total Coliforms	Varies[e]	0	F
Giardia lamblia	TT[f]	0	F
Legionella	TT[f]	0	F
Viruses	TT[f]	0	F
Physical			
Turbidity	1.0 NTU[g]	NA[d]	F
Radionuclides			
Gross alpha	15 pCi/L	NA[d]	F
Beta paticle and photon activity	4 millirems per year	NA[d]	F
Radium-226 and radium-228	5 pCi/L (combined)	NA[d]	F

[a]P = proposed; F = final.
[b]Practical quantitation limit.
[c]Required treatment technique (TT) for acrylamide and epichlorohydrin is Polymer Addition Practices.
[d]NA = not available.
[e]No more than 5 percent of samples may be positive. If fewer than 40 samples collected per month, no more than 1 may be positive.
[f]Treatment technique requires disinfection that must be 99.9 percent effective.
[g]NTU = National Turbidity Units.

References for Appendix D

U.S. Environmental Protection Agency, *Guidance on Remedial Actions for Contaminated Ground Water at Superfund Sites*, EPA/540/

G–88/003, OSWER Directive 9283.1–2, Office of Emergency and Remedial Response, Washington, D.C., 1988.

———, *Fact Sheet, Drinking Water Regulations under the Safe Drinking Water Act,* Criteria and Standards Division, Office of Drinking Water, Washington, D.C., 1990.

Index